THE FRONTIERS COLLECTION

THE FRONTIERS COLLECTION

Series Editors:
A.C. Elitzur   M. Schlosshauer   M.P. Silverman   J. Tuszynski   R. Vaas   H.D. Zeh

The books in this collection are devoted to challenging and open problems at the forefront of modern science, including related philosophical debates. In contrast to typical research monographs, however, they strive to present their topics in a manner accessible also to scientifically literate non-specialists wishing to gain insight into the deeper implications and fascinating questions involved. Taken as a whole, the series reflects the need for a fundamental and interdisciplinary approach to modern science. Furthermore, it is intended to encourage active scientists in all areas to ponder over important and perhaps controversial issues beyond their own speciality. Extending from quantum physics and relativity to entropy, consciousness and complex systems – the Frontiers Collection will inspire readers to push back the frontiers of their own knowledge.

Other Recent Titles

Weak Links
Stabilizers of Complex Systems from Proteins to Social Networks
By P. Csermely

The Biological Evolution of Religious Mind and Behaviour
Edited by E. Voland and W. Schiefenhövel

Particle Metaphysics
A Critical Account of Subatomic Reality
By B. Falkenburg

The Physical Basis of the Direction of Time
By H.D. Zeh

Mindful Universe
Quantum Mechanics and the Participating Observer
By H. Stapp

Decoherence and the Quantum-To-Classical Transition
By M. Schlosshauer

The Nonlinear Universe
Chaos, Emergence, Life
By A. Scott

Symmetry Rules
How Science and Nature Are Founded on Symmetry
By J. Rosen

Quantum Superposition
Counterintuitive Consequences of Coherence, Entanglement, and Interference
By M.P. Silverman

Gregg Jaeger

ENTANGLEMENT, INFORMATION, AND THE INTERPRETATION OF QUANTUM MECHANICS

Prof. Gregg Jaeger
Boston University
College of General Studies
Quantum Imaging Lab.
871 Commonwealth Ave.
Boston MA 02215-2421
USA
e-mail: jaeger@bu.edu

Series Editors:

Avshalom C. Elitzur
Bar-Ilan University, Unit of Interdisciplinary Studies, 52900 Ramat-Gan, Israel
email: avshalom.elitzur@weizmann.ac.il

Maximilian A. Schlosshauer
Niels Bohr Institute, Blegdamsvej 17, 2100 Copenhagen, Denmark
email: schlosshauer@nbi.dk

Mark P. Silverman
Trinity College, Dept. Physics, Hartford CT 06106, USA
email: mark.silverman@trincoll.edu

Jack A. Tuszynski
University of Alberta, Dept. Physics, Edmonton AB T6G 1Z2, Canada
email: jtus@phys.ualberta.ca

Rüdiger Vaas
Posener Str. 85, 74321 Bietigheim-Bissingen, Germany
email: Ruediger.Vaas@t-online.de

H. Dieter Zeh
Gaiberger Straße 38, 69151 Waldhilsbach, Germany
email: zeh@uni-heidelberg.de

ISBN 978-3-642-10070-3   e-ISBN 978-3-540-92128-8

DOI 10.1007/978-3-540-92128-8

Frontiers Collection ISSN 1612-3018

Cover design: KuenkelLopka GmbH, Heidelberg

Printed on acid-free paper

9 8 7 6 5 4 3 2 1

springer.com

To my wife, Savina, and our daughters, Alia and Talas

Preface

Entanglement was initially thought by some to be an oddity restricted to the realm of thought experiments. However, Bell's inequality delimiting local behavior and the experimental demonstration of its violation more than 25 years ago made it entirely clear that non-local properties of pure quantum states are more than an intellectual curiosity. Entanglement and non-locality are now understood to figure prominently in the microphysical world, a realm into which technology is rapidly hurtling. Information theory is also increasingly recognized by physicists and philosophers as intimately related to the foundations of mechanics. The clearest indicator of this relationship is that between quantum information and entanglement. To some degree, a deep relationship between information and mechanics in the quantum context was already there to be seen upon the introduction by Max Born and Wolfgang Pauli of the idea that the essence of pure quantum states lies in their provision of probabilities regarding the behavior of quantum systems, via what has come to be known as the *Born rule*. The significance of the relationship between mechanics and information became even clearer with Leo Szilard's analysis of James Clerk Maxwell's infamous demon thought experiment.

Here, in addition to examining both entanglement and quantum information and their relationship, I endeavor to critically assess the influence of the study of these subjects on the interpretation of quantum theory. The deepest implications of quantum phenomena remain controversial in large part because there remains a need to more adequately interpret quantum theory itself. For example, physicists and philosophers hold a variety of increasingly subtle and radically differing interpretations of the quantum state, ranging from (i) that it is merely a representation of the knowledge of an agent regarding the world or (ii) that it merely links the preparation of systems and the registration of later measurement results without being of ontological significance, to (iii) that it directly describes a continually growing number of real universes that jointly constitute a unique 'multiverse' of incredible size or (iv) that it is the only truly real entity that can be associated with a physical system.

Because both physical and philosophical approaches to foundational problems of physics are involved in this, it is important to address the negative impressions to which the differences of methodology between the larger scientific and philosophical communities have given rise. In that regard, one can do no better than to recall the following comment of Michael Redhead in his 1993 Tarner Lectures at Cambridge. "One must admit that many physicists would dismiss the sort of questions that philosophers of physics tackle as irrelevant to what they see themselves as doing... Either these metaphysical questions arise, they would say, as a result of philosophers involving themselves with the technicalities of theoretical physics, which they, the philosophers never really understand, or it is the physicists themselves who in some cases get sidetracked and ensnared by the temptation to indulge in the subtle sophistry of the philosophers posing unanswerable questions, a subject where there is no discernable progress on premises from which an argument could be launched, where every conceivable position has been argued for by some group of philosophers and equally refuted by another group... It may not come as a surprise to learn that philosophers generally regard physicists as naive people, who do physics in an uncritical way, rather like a child riding a bicycle, quite innocent of the subtleties of rigid-body dynamics!" [372].

Readers are requested to remove any caricatures from their minds, should they have previously entertained them. It behooves one to consider physics and philosophy as constructively coming together wherever both are deployed seriously and properly because our subjects necessitate the addressing of questions both philosophical and scientific. Physicists typically approach problems within a clearly defined mathematical framework, whereas philosophers typically emphasize logical and conceptual rigor and may be more flexible in their use of formalism in their attempts to surmount fundamental difficulties. Each approach has its strengths and weaknesses. Specialists in the foundations of physics tend to be concerned less with the department in which the office of a colleague is located than about whether he or she has presented a clear analysis of a problem. The work considered in this book provides evidence as to why this is so. I believe that work is a sufficient basis for the rejection of the forms of chauvinism to which Redhead refers, particularly on the part of some physicists toward philosophy.

The greatest minds behind the quantum theory worked actively with tools from both areas and used them to engage each other in valuable discussions and to define lines of research that have played an important role in our understanding of the physical world. Albert Einstein was correctly concerned that Niels Bohr was, in his subtle use of philosophically inspired concepts while building what has come to be known as the Copenhagen interpretation, perhaps *too* adept at reassuring physicists that quantum theory could be well founded by making use of ideas from outside physics proper. However, Bohr's interpretation achieved the status of an orthodoxy only *after* he was able to defend his interpretation with a remarkable degree of success against repeated logical *and* physical challenges from Einstein. Furthermore, Bohr's approach was likely so long sustained as such due to Werner Heisenberg's reworking of Bohr's approach into a helpful tool for using the quantum formalism. There should be no doubt that a useful basis for engaging the quantum world was achieved with the aid of the Copenhagen interpretation even though, from the metaphysical realist philosophical perspective traditionally assumed in physics, which itself is less easily combined with quantum mechanics, it is deficient. By contrast, more recently conceived interpretations of quantum mechanics, which may be of some practical benefit to physics in handling some newly considered situations, have yet to offer similar depth or comparable practical strength. Only Feynman's approach to quantum theory currently appears capable of supporting a new interpretation rivaling those of Dirac and von Neumann or of Bohr, Pauli, and Heisenberg.

To their credit, physicists proffering interpretations of quantum theory have often gone beyond the confines of physics when engaging fundamental problems. For example, Bohr, Heisenberg, and Pauli were engaged in modes of creative thinking physicists rarely consider. Similarly, Eugene Wigner seriously contemplated the possibility of a psychophysics. One does well to consider the methodology of the physicist–philosopher Abner Shimony, who has struggled with the deepest of foundational issues and has explained his own well informed use of philosophy in probing the foundations of quantum mechanics. "The language which we have employed for describing the conceptual innovations of quantum mechanics is quite philosophical. We have no apology for this language, because we consider it to be appropriate to the subject. We do not regard philosophy as an autonomous discipline, with a subject matter distinct from other disciplines, but rather as the general investigation of foundations questions and the general search for perspective. The change of framework in physics from classical to quantum mechanical is clearly a fundamental transformation of the conception of nature, and hence is a philosophical matter according to our usage of the term... a highly formal exposition of quantum mechanics is unclear concerning interpretation even though it is clear concerning structure, whereas the formal and philosophical expositions in combination may supplement each other and achieve a fuller clarification" ([407], p. 374).

Investigations of the foundational problems of quantum mechanics and the physics of computation have provided an important context for the emergence of quantum information science. The former two have also begun to benefit from the last, largely due to the importance of information and the possibility or impossibility of its transmission or transformation in different contexts. In my first book, *Quantum information: An overview*, I took pains to avoid engaging issues requiring substantial philosophical discussion or arguing for or against interpretations, as is appropriate in a technical overview. Here, by contrast, I take foundational issues head-on in order to elucidate the centrality of entanglement and information to quantum physics while discussing some of the same situations.

Because entanglement has long been identified as distinctive of quantum mechanics and has recently been shown to serve as an information theoretical resource, it is the primary subject of the opening chapter, which also includes a brief introduction to the mathematical formalism of quantum mechanics and an explication of fundamental concepts such as quantum interference and uncertainty in a manner emphasizing their foundational aspects and relation to information. The second chapter provides an overview of further mathematical formalism and analyses that have played an important role in clarifying the foundations of the theory. This includes a survey of quantum probability, quantum logic, some fundamental theorems of the foundation of quantum mechanics, the description and significance of quantum measurement, and important thought experiments conceived in the history of quantum theory, all of which set the stage for a careful examination of the interpretation of quantum mechanics. The third chapter critically examines the most prominent interpretations of standard quantum mechanics that have emerged in light of the results described in the first two chapters. This includes discussion of recent interpretations in which information is taken to play a dominant role. The ultimate focus of the book is the final chapter, which considers in detail quantum information and its relationship to quantum mechanics, returning to its relationship to entanglement. In addition to articulating that relationship, the final chapter includes critical assessments of the various claims regarding the nature of information and mechanics at the foundational level.

Engaging many issues in foundations of quantum mechanics at this advanced stage in the history of the subject involves formulations that some readers may find challenging. However, the reward of mastering them more than justifies the effort required.

Gregg Jaeger   Cambridge, MA, 2009

Acknowledgments

I thank Lucien Hardy and the Perimeter Institute for hosting me as a short-term visitor during the Spring of 2007, as well as Paul Busch, Joy Christian and others then at PI for discussions which served to stimulate my thinking in relation to several of the foundational issues addressed here, as well as Alisa Bokulich and Arkady Plotnitsky for challenging some of my previous ideas regarding the Copenhagen interpretation. I also thank Alexander Sergienko for his friendship and support. Finally, I thank Angela Lahee at Springer–Heidelberg for inviting and convincing me to contribute to the Frontiers Collection and for her editorial advice and John Stachel for reminding me of the significance of Feynman's process approach to quantum mechanics.

Contents

Preface vii

Acknowledgments xi

1 Superposition, Entanglement, and Limits of Local Causality   1
   1.1   Quantum Interference 2
   1.2   Quantum Indeterminacy and Uncertainty 6
   1.3   Quantum States and Entanglement 12
   1.4   Quantum Entanglement Measures 24
   1.5   Surprising Implications of Entanglement 27
   1.6   The EPR Program and Absence of Local Causality 32
   1.7   Problems with Hidden-Variables Models 37
   1.8   Bell's Theorem and Independence Conditions 40
   1.9   Conditions Contradicted by Quantum Mechanics 44
   1.10 Operations, Communication, and Entanglement 47

2 Quantum Measurement, Probability, and Logic 55
   2.1   Logic and Mechanics 61
   2.2   Probability and Quantum Mechanics 68
   2.3   The Completeness of Quantum Mechanics 71
   2.4   Problems with Measurement in Quantum Mechanics 74
   2.5   Elements of Quantum Measurement Theory 78
   2.6   Advances in Quantum Measurement Theory 86
   2.7   Schrödinger's Cat and Wigner's Friend 89

3 Interpretations of Quantum Mechanics 95
   3.1   Interpretation and Metaphysics 104
   3.2   The Basic Interpretation 117
   3.3   The Copenhagen Interpretation 124
   3.4   Orthodoxy and Explanation in Quantum Physics 136
   3.5   The Collapse-Free Approach 139

    3.6   The Naive Interpretation 165
    3.7   The Radical Bayesian Interpretation 170
    3.8   The Process Interpretation 179
    3.9   Interpretational Underdetermination 185

4   **Information and Quantum Mechanics**.....................189
    4.1   The Theory of Information 194
    4.2   The Quantum Theory of Information 200
    4.3   Entropy in Quantum Measurement Theory 209
    4.4   Quantum Communication and Its Limitations...............211
    4.5   Quantum Information Processing and Speedup 215
    4.6   Protocols and the Nature of Quantum Information 224
    4.7   Informational Interpretations of Quantum Mechanics........232
    4.8   Entanglement 'Thermodynamics' 245
    4.9   Information and Entanglement 248
    4.10 The Great Arc 255

A   **Appendix**..259
    A.1   Mathematical Elements 259
    A.2   The Standard Postulates 262
    A.3   The Dirac Notation 263
    A.4   The Classification of Entangled States 264
    A.5   Elements of Traditional and Quantum Logic 267
    A.6   C*-Algebras 269

References...271

Index..295

1

Superposition, Entanglement, and Limits of Local Causality

A distinctive characteristic of quantum theory is that a pure quantum state precisely characterizes the values of only half of the physical magnitudes of the system it describes: Canonically conjugate magnitudes are not jointly precisely specifiable as they are in classical mechanics. Classically, the space of particle joint position–momentum states, phase space, is a six-dimensional manifold of points; in quantum mechanics, states can be associated only with finite *areas*, for which the product of the variances of these quantities are less than half the quantum of action $\hbar$, in an analogous space. In the case of a fundamental system such as an electron, position or momentum may be precisely specified, but the precision of their *joint* specification is inherently limited, as expressed in the Heisenberg–Robertson uncertainty relation, which follows from the quantum superposition principle. Because the physical properties of signals constrain communication, the uncertainties of states used for signaling and correlations between transmitter and receiver being of paramount significance, quantum mechanics also endows quantum information with a character distinct from that of classical information.

The foundational aspects of both quantum mechanics and quantum information science are most evident in situations having no classical analogue, such as those where quantum entanglement is present. Indeed, entanglement and quantum information can be closely related. Furthermore, quantum entanglement is required for the general description of multi-particle systems. Erwin Schrödinger viewed entanglement, which involves extraordinarily strong correlations between subsystems in multi-partite systems such as those involved in communication, as most distinctive of quantum mechanics [394]. Nonetheless, the superposition principle remains the primary element of the theory, as Richard Feynman emphasized, in that both quantum uncertainty and entanglement are implied by it. Foundational implications of all three were explored by Schrödinger in 1935 in his famous "cat paper." Like entanglement already did then, quantum information now also serves as an important probe of the quantum world.

G. Jaeger, *Entanglement, Information, and the Interpretation*
of Quantum Mechanics, The Frontiers Collection,
DOI 10.1007/978-3-540-92128-8_1, © Springer-Verlag Berlin Heidelberg 2009

1.1 Quantum Interference

The conception of entanglement has evolved from Schrödinger's relatively simple initial one to a rather subtle one. In recent years, entanglement has also been empirically studied with increasing care using the highly developed technique of interferometry, which has long served as a key tool of experimental physics. Indeed, it is through interference that entanglement is most clearly manifested [12, 33, 254, 255, 422].

Quantum interference arises due to the coherent superposition of quantum states corresponding to distinct values of a physical magnitude. Such superpositions occur when there is indistinguishability in principle, by a precise measurement under fixed conditions, of alternative sequences of quantum states that originate with a common initial preparation and arrive at a specific later eigenstate. That is, interference can be observed whenever there exist several indistinguishable possibilities for the physical system in question to reach a detectable final state [168, 171]. The distinguishability of such alternatives plays a central role in the quantification of information as encoded in quantum signals. The communication of information using quantum systems is very naturally characterized within quantum mechanics, because the very use of quantum states involves considering two basic elements of signaling, namely, state preparation and state measurement.

Paul Dirac viewed quantum interference as the interference of a physical system with itself, as opposed to with other systems as occurs in classical physics [139]. Quantum interference is similar to but importantly different from the more familiar interference effect in classical physics wherein patterns with discernible regularities are observed, for example, on the surface of a body of water when the wakes of passing ships meet. A crucial difference is that in classical physics what are added may be non-statistical quantities, whereas in quantum physics what fundamentally are added are complex-valued state amplitudes, the squares of which provide only *probabilities* of physical events. Thus, the quantum wave-function does not directly describe any substance; there is no *thing* the parts of which *wave*.

A simple apparatus for observing quantum interference is the Young double-slit interferometer, illustrated in Figure 1.1, that often appears in textbooks on quantum mechanics, as in the Feynman Lectures [171]. In the basic double-slit experiment, many identically prepared systems such as electrons are directed precisely normally toward a double-slit diaphragm and, if not absorbed by it, continue on to an opaque screen that acts as an array of area detectors. Taking $a_1(x)$ to be the quantum amplitude corresponding to the passage through one slit of the diaphragm toward the spatial point x on the measurement screen which, like the diaphragm, is oriented perpendicularly to the direction of the initial beam, the probability density of later finding these systems at x upon measurement is found to be $p_1(x) = |a_1(x)|^2$. Letting $a_2(x)$ similarly be the amplitude corresponding to passage through the second slit toward the same position x, the probability density for arrival at x is

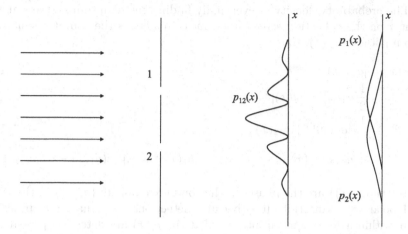

Fig. 1.1. The Young double-slit experiment. Every elementary system, such as an electron, can pass through slit 1 and/or slit 2 and be detected near a point, parameterized by x, and exhibit interference on an opaque detection screen. In a statistical measurement on an ensemble of many particles, that is, a collection of particles identically prepared, coming precisely from the left, there will be high values and low values of the detection probability $p_{12}(x)$. If instead only slit 1 or slit 2 were made available to incoming particles, two non-periodically modulated distributions, $p_1(x)$ and $p_2(x)$, respectively, would instead result. If the systems were prepared in an incoherent state before reaching the two slits, or equivalently were the environment before the slits decoherence-inducing, the detection probability instead would be $p_1(x) + p_2(x)$, the sum of the distributions at far right.

$p_2(x) = |a_2(x)|^2$. The amplitude for systems being found at x when *both* slits are passable, so that either slit might be entered on the way to the screen, is

$$a_{12}(x) = \frac{1}{\sqrt{2}}\big(a_1(x) + a_2(x)\big) \, , \qquad (1.1)$$

according to the superposition principle,[1] where the factor of $1/\sqrt{2}$ appearing here confers proper normalization on the total probability describing the ensemble. The *ensemble*, a concept pioneered by Ludwig Boltzmann and Willard Gibbs (*cf.* [479], p. 75; [496], p. 62) is a set of similar systems; in quantum mechanics, these systems can be considered similar on the basis of their physical history, that is, their preparation and, in particular, are understood as a Gibbs ensemble for which squared amplitudes provide the statistics.[2]

---

[1] A set of postulates of quantum mechanics, of which the superposition principle is the primary one, is provided in the appendix.

[2] For more detail, see Section 1.7.

The probability density of eventually finding the quantum systems at the point x on the collection screen upon measurement is the complex square of the amplitude $a_{12}(x)$, that is,

$$p_{12}(x) = |a_{12}(x)|^2 \qquad (1.2)$$

$$= \frac{1}{2}|a_1(x) + a_2(x)|^2$$

$$= \frac{1}{2}\Big[|a_1(x)|^2 + |a_2(x)|^2 \qquad (1.3)$$

$$+|a_1(x)a_2(x)|\big(\exp\big[i(\theta_2(x) - \theta_1(x))\big] + \exp\big[i(\theta_1(x) - \theta_2(x))\big]\big)\Big] ,$$

where the $\{\theta_i(x)\}$ are the phases of the complex numbers $\{a_i(x)\}$ in the standard polar representation. It is because detections, as a matter of practice, occur within a finite spatial interval that the $p_i(x)$ are detection-probability *densities* rather than simple probabilities. The correctness of this quantum-mechanical description has been extensively confirmed by observation, even when the systems are sent into this apparatus *one at a time*; such independency from intensity was first clearly observed in a related "feeble light" diffraction experiment by Geoffrey Ingram Taylor [81, 444] and is also exhibited in the interference of massive electrically neutral particles, such as neutrons (*cf.* [516, 517]). That and similar later results support the characterization of Dirac that quantum interference is the self-interference of individual systems.

When the above apparatus is instead modified so that only one slit in the diaphragm of Figure 1.1 is available at a time, with both being made available in the full course of the experiment on the entire ensemble of systems— assuming that diffraction from each slit is negligible—interference *vanishes*. The observed probability distribution in such a modified experiment is simply proportional to the right-hand side of Equation 1.4, which describes a detection pattern similar to that in the corresponding classical mechanical experiment. Only then is it appropriate to add the measurement counts of the two detected quantum subensembles, because the histories are fully distinguishable in principle—only one history of the particle is possible for each quantum system—precluding quantum interference; the difference between the interference patterns, shown in Figure 1.1, appearing in these two different experiments illustrates the difference between what is observed in quantum state superposition and what is observed in state mixing.

The important difference between this quantum mechanical experiment and the analogous one in which particles are described by *classical* mechanics is that in the quantum case the probability density is not additive:

$$p_{12}(x) \not\propto p_1(x) + p_2(x) . \qquad (1.4)$$

Again, this arises because, in quantum mechanics, complex probability amplitudes are added and then squared in the mathematical description of the experiment. The predicted probability of a detection event in an interval Δx about a point x corresponding to the resolution of the detector-screen is obtained by integrating the probability density $p_{12}(x)$ over the interval. Interference is reflected in the presence of a modulated "interference term," which is proportional to $\sqrt{p_1(x)p_2(x)}\cos\left(\theta_2(x) - \theta_1(x)\right)$, that arises from their being complex-valued, which implies a non-zero-valued empirical visibility, V. The visibility is defined as the difference of maximum and minimum detection-event rates of the pattern, which yield probabilities through the relative frequencies they provide, divided by their sum; in the absence of this term, $V = 0$. Notably, a loss of quantum interference can also occur due to quantum state decoherence, such as when one modifies the experiment by introducing strong phase noise between the double-slight diaphragm and the detection screen.

However, one must consider not only the differences arising in such a formal treatment of interference but also those arising from differences in theory *interpretation*. From the logical perspective, which is taken up in detail in Chapter 2, the propositions P_1, P_2 that a system detected in an interval Δx of the screen prior to striking the screen had passed through one or the other of the two slits, respectively, and the proposition $P_1 \lor P_2$ that either one *or* both may have been traversed are ultimately relevant. The conditional probability that the system arrived in Δx by first passing through *either* of the screens can also be used to describe interference; even in the case of very simple situations such as this, interpretative background assumptions, which are typically required for explaining experimental results, can change the significance of an experiment. Consider the result of giving up a standard interpretational assumption, for example, the eigenvalue-eigenstate link.

"Even interference phenomena, by themselves, say nothing about whether or not observables have determinate values in the absence of measurements. The usual story, in the case of a double-slit photon interference experiment, for example, is that you get the wrong distribution of hits on the screen behind the slits, if you assume that each individual photon goes through one or the other of the two slits, when the photon is in a quantum state that is a linear superposition of a state in which the photon goes through slit 1 and a state in which it goes through slit 2.... when the state of the photon is the linear superposition, the observable A, corresponding to localization in the plane of the slits to slit 1 or slit 2, has no determinate value... The interference pattern appears to be incompatible with the assumption that the photon either goes through slit 1 or through slit 2, exclusively... The loophole in the argument is the link between attributing a definite value to the observable A and attributing a specific quantum state to the photon. This is the orthodox (Dirac–von Neumann) interpretation principle; specifically, the eigenvalue to eigenstate link. If we reject this principle, then we can attribute a determinate value to the observable A, [for example]..." ([87], pp. 192-193)

It may be that Schrödinger viewed entanglement rather than superposition as fundamental because, during at least some periods of his career, he believed that the quantum state might actually directly describe a physical substance (*cf.* [49]), in which case the character of quantum superposition would be less distinctive. Unlike Schrödinger, Feynman believed that the simple double-slit experiment already involves the essence of quantum mechanics, "its only true mystery," which arises from the above-mentioned probability amplitudes the calculus of which forms the basis of his own approach to the theory [168]. Indeed, the characteristic behavior of entangled systems, like that of non-composite systems, is also understandable in terms of the interference of indistinguishable possible histories, which are histories of *composite* systems, reflected in state superpositions—see, for example, [249, 254]. Interference can also be 'lost,' that is, may be not seen when there is entanglement with additional quantum systems, referred to as *ancillae*, allowing particle paths to be distinguished [254], as in the case discussed in Section 1.5 below.

The empirical signature of entangled quantum systems, which must consist of at least two subsystems by definition, is the occurrence of particularly strong interference effects of "fourth order" (in amplitude) under the proper experimental conditions, that is, interference observed not in simple events, which exhibit "second-order" effects, but rather in *joint-detection* events as seen, for example, in the seminal Hanbury-Brown–Twiss experiment [205].

1.2 Quantum Indeterminacy and Uncertainty

One of the foundational issues first to be addressed in the history of quantum mechanics is the question of the conceptual origin of the *a priori* limitations on the simultaneous determination of values of pairs of quantities, as characterized by Heisenberg relations. These quantitative relations are involved in interference phenomena in that, for example, they reflect the difference of quantum from classical behavior in the double-slit experiment, which is essentially illustrated by the differing ensemble detection patterns of Figure 1.1 in the case in which both slits are always available.

The wavelike appearance of the probability density for detections of systems and, hence, of detection count rates in specific position intervals, which can be made as small as practicable without affecting the above analysis, has often been viewed as due to a specifically quantum mechanical ambiguity of form; the interferogram of the double-slit experiment is intuitive for wavelike systems, although it emerges as a pattern in collections of discrete *localized* detection events such as produced by particles. It has accordingly been described as the manifestation of a qualitative quantum 'wave–particle duality.' Although often intuitively helpful, this less precise characterization is of limited value, as argued in 1960 by Alfred Landé, one of its strongest critics.

"Nor does the word 'duality' explain anything; it is a purely descriptive term. Even worse, the road to a better understanding has been
blocked for thirty years by the creed that duality itself represents the
bottom of theoretical analysis, and that 'we must try to understand
that there is nothing further to be understood' in quantum theory."
([292], p. x)

Landé's criticism was directed toward the positions of Heisenberg and Pascual
Jordan, in particular. Jordan had, for example, in 1944 written the following.

"Experiments which let the wave side of light emerge clearly force
(through the action that is connected with every experiment) the
corpuscular nature of light back into the indeterminate and unobservable; other experiments, which force the corpuscular side of light
into prominence, leave undefined and indiscernible all the properties which usually betray to us the wave nature of light. With this
wonderful device of complementarity nature combines in one and
the same physical object properties and regularities that contradict
each other so that they could never exist directly at the same time."
([263], p. 132)

For Heisenberg,

"the particle picture and the wave picture are merely two different
aspects of one and the same physical reality," (cf. [256], pp. 68-69)

perhaps less susceptible to the sort of "action" to which the "mutually exclusive" or "contradictory" properties of Jordan might be.[3]

For Niels Bohr, with whom this idea is most often identified, neither the
wave nor particle description ever applies *directly* to quantum objects, which
for him were "entities to which no specific properties or conceptions are applicable" ([354], p. 12). Nonetheless, it has been argued by a number of Bohr
scholars that wave–particle duality was the inspiration for his conception of
complementarity, to which Jordan referred ([233], Section 2.4 and references
therein). Heisenberg, whose name is now attached to the uncertainty principle,
characterized complementarity as follows.

"By this term... Bohr intended to characterize the fact that the same
phenomenon can sometimes be described by very different, possibly
even contradictory pictures, which are complementary in the sense
that both pictures are necessary if the 'quantum' character of the
phenomenon shall be made visible." ([223])

Heisenberg viewed the uncertainty principle as a "law of nature" and essential
for the understanding of quantum mechanics, much as Bohr viewed complementarity ([256], p. 59). Shortly after its introduction, it was regarded as a
fundamental principle of the theory by several other influential physicists as

---

[3] A characterization of quantum behavior in the double-slit experiment by Heisenberg can be found in [219], pp. 51-52.

well. Most importantly, perhaps, Pauli understood it as so in his influential article in the *Handbuch der Physik* [338], as did Hermann Weyl in his important early book *Gruppentheorie und Quantenmechanik* [491].

Today, physicists not specializing in the foundations of physics remain largely unaware of the subtleties associated with the uncertainty principle, under which uncertainty relations are primary to the theory, and its foundational potential, often using the relations as merely rules of thumb. On the other hand, the principle has also often been misinterpreted and over-applied by those outside the community of physicists. As Hans Bethe has commented,

> "The uncertainty principle has profoundly misled the lay public: they believe that everything in quantum theory is fuzzy and uncertain. Exactly the reverse is true. Only quantum theory can explain why atoms exist at all[, for example]." ([47])

Outside of certain interpretations of the quantum formalism, the word *uncertainty* as used in quantum theory should not be understood in an exclusively epistemic sense, which the term itself most naturally suggests.[4] Furthermore, as Max Jammer noted,

> "The term used by Heisenberg in these considerations was *Umgenauigkeit* (inexactness, imprecision) or *Genauigkeit* (precision, degree of precision). In fact, in his classic paper these terms appear more than 30 times (apart from the adjective *genau*), whereas the term *Unbestimmtheit* (indeterminacy) appears only twice and *Unsicherheit* (uncertainty) only three times. Significantly, the last term, with one exception (p. 186), is used only in the Postscript, which was written under the influence of Bohr." ([256], p. 61)

Indeed, the uncertainty principle was introduced by Heisenberg as the restriction that "canonically conjugate quantities can be determined simultaneously only with a characteristic inaccuracy" [216, 217]. For example, there is a trade-off between the precision of specification of position and momentum in the quantum double-slit experiment, whether by preparation or measurement. This appears to have been the original subject of interest to Heisenberg. Moreover, recent work in quantum measurement theory has shown that advances can be made by considering simultaneous values of complementary physical magnitudes, as shown below in Chapter 2.

The uncertainty relations are most often expressed in terms of the dispersions of Hermitian operators for quantum states.[5] The *dispersion* of an operator A, given a quantum state ρ, is

$$\mathrm{Disp}_\rho A \equiv \langle (A - \langle A \rangle \mathbb{I})^2 \rangle_\rho \tag{1.5}$$

$$= \langle A^2 \rangle_\rho - \langle A \rangle_\rho^2 , \tag{1.6}$$

---

[4] Instead, *indeterminacy* and *indefiniteness* are here used in reference to properties.

[5] Both of these are discussed in greater detail in the following section.

where the brackets on the right hand sides indicates the expectation value. The square root of the dispersion is the 'uncertainty'

$$\Delta A \equiv \sqrt{\text{Disp}_\rho A} \tag{1.7}$$

of A in state ρ; often, no specific state is explicitly written but is clearly implied. The product of joint uncertainties for two non-commuting quantum observables A and B, following from the postulates of quantum mechanics, takes the form of the *Heisenberg–Robertson relation*, namely,

$$\langle(\Delta A)^2\rangle_\rho\langle(\Delta B)^2\rangle_\rho \geq \frac{1}{4}|\langle[A,B]\rangle_\rho|^2 \ . \tag{1.8}$$

In standard quantum mechanics, *observables* are specific sorts of operators, which are more precisely defined in the next section, representing physical magnitudes.[6] Two operators A and B *commute* if the commutator $[A,B] \equiv AB - BA = \mathbb{O}$, $\mathbb{O}$ being the zero operator; operators for canonically conjugate quantities do not commute, that is, provide a non-zero value to the right-hand side of Equation 1.8. This relation clearly expresses a mathematically complementarity: If the right-hand side is non-zero, the uncertainty of one quantity is reduced when that of the other is increased. One can find the range of likely values of individual observables in many situations with extreme precision; as Bethe pointed out, there also not necessarily is an unavoidable "fuzziness" of individual quantities. Rather, it is only that any observable that does not commute with one that is at any given moment precisely determined will be poorly specified at the time.

Heisenberg believed that a physical magnitude, such as position, depends essentially on the experimental circumstances involved in providing its specification (*cf.* [281], p. 25).

> "If one wants to clarify what is meant by 'position of an object,' for example, of an electron, he has to describe an experiment by which the 'position of an electron' can be measured; otherwise this term has no meaning at all." ([216]; as translated in [256], p. 58)

In order to illustrate this, he introduced a thought experiment involving its measurement for an electron using a gamma-ray microscope [216, 217, 375]. It is important here to distinguish the relevance of experiments to quantities' actual specification from their providing an *explanation* of the behavior of the two quantities in conjunction with the Heisenberg relations; this is particularly important because the supposed 'Heisenberg effect,' the disturbance of the observed electron, by a gamma ray in this case, possibly corresponding to

---

[6] The term *observable* may have its origins in Heisenberg's characterization of matrix mechanics, in which these commutators first arose, as "a quantum theoretical mechanics, analogous to the classical mechanics in which only relations between observable quantities appear" ([215], [332] p. 253).

Jordan's 'forcing' of one or the other of its "dual" characters, is sometimes considered a *cause* of the corresponding behavior.[7] Specifically, the gamma-ray microscope example was introduced in order to illustrate the fact that, in such situations, an increase of accuracy of position measurement by shortening the wavelength of the gamma ray results also in a corresponding increase of momentum transfer to the observed particle because "the inaccuracy of the measurement of the position can never be smaller than the wave length of the light" ([219], pp. 47-48).

Heisenberg argued that

> "...in the act of observation at least one light quantum of the γ-ray must have passed the microscope and must first have been deflected by the electron. Therefore, the electron has been pushed by the light quantum, it has changed its momentum and its velocity, and one can show that the uncertainty of this change is just big enough to guarantee the validity of the uncertainty relations." ([219], pp. 47-48)

Although he believed that quantum uncertainty is manifested in the act of observation, Heisenberg explicitly cautioned against considering momentum transfer the *cause* of uncertainty in this example, at the prompting of Bohr[8] who saw it as due to the *principle of complementarity* [98]; Bohr argued that

> "The reciprocal uncertainty which always affects the values of those quantities is...essentially an outcome of the limited accuracy with which changes in energy and momentum can be defined..." ([57], p. 63)

The perspective of Bohr in this regard became so influential that "the term 'principle of complementarity' came often to be erroneously understood as a *synonym* for the Heisenberg relations."[9] As Jammer has noted,

> "That complementarity and Heisenberg-indeterminacy are certainly not synonymous follows from the simple fact that the latter...is an immediate mathematical consequence of...the Dirac–Jordan transformation theory, whereas complementarity is an extraneous *interpretative* addition to it. In fact, the quantum mechanical formalism with the inclusion of the Heisenberg relations can be, and has been, interpreted in a logically consistent way without any recourse to complementarity." ([256], pp. 60-61)

Bohr's articulation of complementarity as an interpretative principle is discussed at length in Section 3.4.[10] Heisenberg also later used the term *uncer-*

---

[7] For a discussion of the failure of the disturbance theory of the uncertainty relations associated with such a use of the relation, see [371], Section 2.5.

[8] See Bohr as quoted in Section 3.4, and the analysis of [373], pp. 14-17.

[9] Vladimir Fock, cited in translation ([256], p. 60).

[10] For an extended discussion of "Heisenberg's microscope" and an analysis of Heisenberg's reasoning involving it, see [256], Section 3.2.

tainty more freely but in regard to pertinent considerations of classical statistical mechanics. For example, in the context of observation, he commented that

> "[It] is very important to realize that our object has to be in contact with the other part of the world, namely, the experimental arrangement. . . at the moment of observation. . . [This] introduces a new element of uncertainty, since the measuring device is necessarily described in terms of classical physics; such a description contains all the uncertainties concerning the microscopic structure of the device which we know from thermodynamics, and since the device is connected with the rest of the world, it contains in fact the uncertainties in the microscopic structure of the whole world." ([219], p. 53)

In discussion with Einstein during the Fifth Solvay Congress, Bohr provided an analysis along similar lines to Heisenberg's discussions of the microscope but in the context of the double-slit experiment ([256], pp. 127-129). Einstein argued that the transverse momentum transferred by particles when passing the double-slit diaphragm could be measured with arbitrary precision and that at the same time the position could be arbitrarily precisely measured by sufficiently reducing the slit width. However, Bohr pointed out that the transferred momentum was sufficiently uncontrollable in the apparatus that the relation would actually be obeyed. Einstein had to agree. However, Einstein then further suggested that, by introducing a slightly modified version of the apparatus of Figure 1.1 wherein the initial beam is created by placing a single-slit diaphragm before the double-slit diaphragm and then hanging the latter diaphragm *on a spring*, a violation could be demonstrated. However, Bohr noted that in that case the difference of the particle momentum for the two cases of passing through one slit and passing through the other, $\delta p = \omega p = h\omega/\lambda$, where ω is the angular frequency and λ is the wavelength of the momentum. Therefore, if the second slit is considered to be a *quantum* system, then the required measurement of the resulting diaphragm momentum to the accuracy δp requires that the diaphragm have a position uncertainty of $\delta x = h/\delta p = \lambda/\omega$ which, by basic interferometric principles, would destroy the interference pattern. This apparatus was much later analyzed in great detail by William Wootters and Wojciech Żurek in ultimate agreement with Bohr's claim [507].

Ultimately, three distinct types of Heisenberg relation have been articulated. The first two of these were considered by Heisenberg between 1927 and 1930 [96, 216, 217]. The mathematical relation for quantum *states*, which is considered above in regard to position and momentum, is

$$(\Delta X)_\rho (\Delta P)_\rho \geq \frac{\hbar}{2} , \qquad (1.9)$$

where, again, the dispersions are calculated for the same state, ρ [375]. The form for simultaneous *measurement accuracies* of these two quantities is

$$(\delta x)(\delta p) \geq \frac{\hbar}{2} \ . \tag{1.10}$$

Specifically, Heisenberg described the outcome of a joint measurement of position and momentum in terms of the uncertainties of position and momentum for Gaussian states [216]. The third sort is an *accuracy–disturbance trade-off* relation for measurement sequences, which was first explicitly discussed by Pauli in 1933,

$$\delta x Dp \geq \frac{\hbar}{2} \ , \tag{1.11}$$

where Dp indicates the disturbance of what was initially a momentum eigenstate resulting from a position measurement of accuracy δx [179, 338]. It is noteworthy that this last relation has been explicated by reference to quantum information-theoretic concepts and causally so, as follows.

"The tradeoff between acquiring information and creating a disturbance is related to quantum randomness. It is because the outcome of a measurement has a random element that we are unable to infer the initial state of the system from the measurement outcome. That acquiring information causes a disturbance is also connected with the no-cloning principle... If we could make a perfect copy of a quantum state, we could measure an observable of the copy without disturbing the original..."([362], p. 10)

One sees that the specific Heisenberg relation pertaining in a given situation is determined by whether values are provided by states, simultaneous measurement accuracies, or measurement sequences, and can be understood as arising in different ways under different interpretations of the quantum formalism; it does not simply correspond to wave–particle duality.

Before continuing the discussion of foundational questions of quantum mechanics, it is helpful to consider more explicitly the mathematical elements and notation of modern quantum mechanics and the basic theory of quantum entanglement with which they are typically bound up. This constitutes the bulk of the following two sections. Those familiar with these, or who wish on a first reading to avoid the detailed consideration of them, may wish to move immediately on to Section 1.5, where additional experimental scenarios are discussed.

1.3 Quantum States and Entanglement

In the standard Hilbert-space formalization of quantum mechanics, pure states are described by *state-vectors*, $|\psi\rangle$, forming a linear state space attributed to the system in question. This is the mathematical content of the superposition principle: in the absence of further restrictions, which are referred to as *superselection rules*, it renders all sums of states in this space physically valid. In particular, each quantum system is attributed a separable complete complex vector space, the *Hilbert space* $\mathcal{H}$, constituting its state space, analogous to that of classical mechanics, via the set of trace-class Hermitian linear operators ρ acting in it, the *statistical operators*, which include both pure states and mixtures thereof, together with an appropriate inner product.[11]

State-vectors correspond to the pure statistical operators, that is, the *projectors*, $P(|\psi\rangle)$, onto one-dimensional rays; each ray corresponds to the normalized "ket" vectors $|\psi\rangle \in \mathcal{H}$ that spans it. Any vector element of a ray can be obtained from any other by multiplying it by a complex number; in this way, the elements of the ray form an equivalence class that is identified with a single normalized quantum state-vector $|\psi\rangle$. Accordingly, the Hilbert space in question is a projective Hilbert space. Any state ρ can then be formally expressed as a mixture of these projectors, that is, as an element of the convex set of real-weight linear combinations of pure states. Given any linear operator, O, on $\mathcal{H}$, the equation $(O - o\mathbb{I})|\psi\rangle = \mathbf{0}$ of the eigenvalue problem, for non-zero vectors $|\psi\rangle \in \mathcal{H}$, can be considered and its solutions indexed. When this equation is satisfied, the vectors and corresponding scalars are referred to as *eigenvectors* and *eigenvalues*, respectively; the eigenvectors for the problem are typically labeled using eigenvalues o_i, that is, written $|o_i\rangle$.

The "bra" state-vectors $\langle\psi|$ are elements of the Hilbert space $\mathcal{H}^*$ that is dual to $\mathcal{H}$: The bra is best thought of as a map from $\mathcal{H}$ into the complex numbers $\mathbb{C}$, namely, $\langle\psi| : |\xi\rangle \mapsto \langle\psi|\xi\rangle = re^{i\phi}$, where $r, \phi \in \mathbb{R}$, because the probabilities associated with observing specific values of quantum mechanical magnitudes are independent of the phase ϕ: $\left|(e^{-i\phi_i}\langle\psi_i|)(e^{i\phi_j}|\xi_j\rangle)\right|^2 = |\langle\psi_i|\xi_j\rangle|^2$ for all $\langle\psi_i| \in \mathcal{H}^*, |\xi_j\rangle \in \mathcal{H}$ and all $\phi_k = \phi_j - \phi_i$.[12] In order to associate states with values designating physical quantities, one then uses Hermitian linear operators O on $\mathcal{H}$, rather than functions of phase-space points as is done in the case of classical mechanics, a fact which is intimately related to the remarks made in the opening of this chapter. The projectors are outer products such that $P(|\psi\rangle)|\xi\rangle = \langle\psi|\xi\rangle = re^{i\phi}|\psi\rangle$, for some value of the pair r, ϕ for each $|\psi\rangle$, that is, $P(|\psi\rangle) \equiv |\psi\rangle\langle\psi|$. The mixed states are mathematically distinct from the pure states; it is particularly important to keep clearly in mind that mixed states *cannot* be written as linear combinations of state-vectors but only of the corresponding statistical operators, as shown below.

---

[11] An operator is *Hermitian* if it is equal to its Hermitian conjugate $O = O^\dagger$, see Appendix. An operator is *trace class* if its modulus $|O| = \sqrt{O^\dagger O}$ is finite. For Hermitian operators, the set of eigenvalues $\{o_i\}$, called the *eigenvalue spectrum*, is real and a complete basis for the space is formed by its eigenstate-vectors. If these values all differ for a given quantum state, the operator is *maximal* for it (as is the corresponding physical magnitude, *cf.* Section 2.3), and the set of eigenvalues is *non-degenerate*; otherwise, the set of eigenvalues is called *degenerate*. In the case of systems described by continuous observables, there generally does not exist a eigenvector basis for the system, so that there are no eigenvalues to consider although there does remain a well defined spectrum.

[12] *Cf.* Equation 1.29 below.

The spatial part of the full quantum state of any elementary particle is infinite-dimensional; the corresponding state-vectors are referred to as *wave-functions*. It is often sufficient for the purposes of calculation to consider only part of the full states of particles such as photons, leaving aside consideration of the spatial part; one can explain a broad range of quantum phenomena equally well by attending only to the spin subspaces of particles or similar subspaces of the full system Hilbert space. Two-level systems, such as spin-$\frac{1}{2}$ particles, which are now often referred to as *qubits*—misleadingly so because they are physical objects rather than information units—have a state space representable by that of the 2×2 complex Hermitian trace-one matrices, $H(2)$. The general pure state in a two-dimensional complex Hilbert space representing such a system has the following convenient representation

$$|\psi(\theta, \phi)\rangle = \cos(\theta/2)|0\rangle + e^{i\phi} \sin(\theta/2)|1\rangle \doteq \begin{pmatrix} \cos(\theta/2) \\ e^{i\phi} \sin(\theta/2) \end{pmatrix} , \qquad (1.12)$$

where $\theta \in (0, \pi)$, $\phi \in [0, 2\pi)$.[13] Each state in this Hilbert space can be equivalently specified by three numbers, the components of a real vector, called the *Stokes vector*, in a ball in the Euclidean space of probabilities, the expectation values of operators corresponding to three orthogonal directions, to which these matrices are naturally mapped; in the Stokes parameterization, the general two-level system state lies in the Bloch ball (*cf.* Figure 4.2 of Chapter 4), the boundary of which is the Poincaré–Bloch sphere corresponding to the set of pure states, $P(|\psi(\theta, \phi)\rangle)$, again the interior (mixed) states being the normalized convex linear combinations of these pure states or, equivalently, a convex combination $\rho \doteq \{p_i, P(|\psi_i\rangle)\}$ of projectors weighted by the $\{p_i\}$.

The structure naturally generalizing the Bloch ball to more general cases is the *convex set* [318]. The state of a generic quantum system within a Hilbert subspace of d dimensions is the convex set of statistical operators of the form $\rho = w\rho_1 + (1 - w)\rho_2$, where the ρ_i are well defined statistical operators and $0 \leq w \leq 1$, which represents the *mixing* of the two states with weights w and $1 - w$, respectively.[14] The statistical operator ρ of a quantum system can be viewed as a linear normalized map from the observables $O^{(i)}$ of the system to their expectation values, the averages of values for the ensemble, $\langle O^{(i)} \rangle_\rho = \mathrm{tr}(\rho O^{(i)})$; this formula implements the fundamental *Born rule*. The pure and mixed state sets can then naturally be distinguished by the expectation value of the statistical operator itself, known as the *purity* $\mathcal{P}(\rho)$, which is identical to the trace of the square of the statistical operator

$$\mathcal{P}(\rho) = \mathrm{tr}\, \rho^2 , \qquad (1.13)$$

which is a positive real number less than one. When the statistical operator ρ describes physical magnitudes taking a finite number d of values, d is the

---

[13] When $\theta = 0$ and π, ϕ is taken to be zero by convention.

[14] Recall that in the literature of quantum mechanics, physical magnitudes are typically referred to as *observables*. Additional reasons for this choice of terminology are given in the next chapter.

dimension of the Hilbert space $\mathcal{H}$ attributed to the system and $\frac{1}{d} \leq \mathcal{P}(\rho) \leq 1$. The dimension of the operator space associated with the system is then $d^2 - 1$. The quantum state is *pure* if $\mathcal{P}(\rho) = 1$, that is, the system state cannot be any more precisely specified, and is an extremal point of the convex set. In the practical context, pure states are idealizations.

The purity of a quantum state is constant under transformations of the form $\rho \to U\rho\, U^\dagger$, where U is unitary, most importantly under the temporal evolution over the interval $t - t_0$, typically taken to be $U(t, t_0) = e^{-\frac{i}{\hbar}H(t-t_0)}$, where H is the Hamiltonian operator, in accordance with the Schrödinger equation. Unitary linear operators, U, are those for which $U^\dagger U = UU^\dagger = \mathbb{I}$, where "$\dagger$" denotes Hermitian conjugation. The temporal evolution need not, however, involve a time-independent Hamiltonian; temporal evolution in quantum mechanics need not be so simple. Given two statistical operators ρ_1 and ρ_2, the following are equivalent: (i) ρ_1 and ρ_2 are *unitarily equivalent*: $\rho_2 = U\rho_1 U^\dagger$, for some unitarity U; (ii) ρ_1 and ρ_2 have *identical eigenvalue spectra*; (iii) $\mathrm{tr}\rho_1^r = \mathrm{tr}\rho_2^r$ for all $r = 1, 2, \ldots, n$, where $n = \dim\rho_1 = \dim\rho_2$. A quantum state is pure if and only if the statistical operator ρ is *idempotent*, i.e., a projector: $P^2(|\psi\rangle) = (|\psi\rangle\langle\psi|)^2 = \langle\psi|\psi\rangle|\psi\rangle\langle\psi| = |\psi\rangle\langle\psi| = P(|\psi\rangle)$. A quantum state is mixed if it is not pure, that is, if $\mathcal{P}(\rho) < 1$.

There is a fundamental difference between the superposition of quantum states and the *mixing* of them, for example, by combining ensembles of quantum systems. To see how these two mathematical operations (state-vector addition and statistical operator addition) differ, one can, for example, consider the (normalized) sum in the simple two-dimensional case

$$|\nearrow\rangle = \frac{1}{\sqrt{2}}(|\uparrow\rangle + |\downarrow\rangle) \tag{1.14}$$

of two orthogonal pure state-vectors $|\uparrow\rangle$ and $|\downarrow\rangle$. The vector superposition in Equation 1.14 is a pure state: $\mathrm{tr}\big(P^2(|\nearrow\rangle)\big) = \mathrm{tr}\big(P(|\nearrow\rangle)\big) = 1$. The similar linear combination formed by subtraction rather than addition is written $|\searrow\rangle$. The corresponding two projectors are $P(|\nearrow\rangle) = |\nearrow\rangle\langle\nearrow|$, $P(|\searrow\rangle) = |\searrow\rangle\langle\searrow|$. By contrast to the case of state-vector addition, the normalized sum of a pair of projectors, for example, $P(|\uparrow\rangle)$ and $P(|\downarrow\rangle)$ corresponding to pure states $|\uparrow\rangle$ and $|\downarrow\rangle$, in that case

$$\rho_{\mathrm{mix}} = \frac{1}{2}\big(P(|\uparrow\rangle) + P(|\downarrow\rangle)\big) , \tag{1.15}$$

is *mixed*: $\mathrm{tr}(\rho_{\mathrm{mix}}^2) = \mathrm{tr}\big((\frac{1}{2})^2[P^2(|\uparrow\rangle) + P^2(|\downarrow\rangle)]\big) = \frac{1}{4}\mathrm{tr}(P(|\uparrow\rangle) + P(|\downarrow\rangle)) = \frac{1}{2}$. The state ρ_{mix} can also be written as a different such sum, namely,

$$\rho_{\mathrm{mix}} = \frac{1}{2}\big(P(|\nearrow\rangle) + P(|\searrow\rangle)\big) , \tag{1.16}$$

which illustrates the *non-uniqueness* of the decomposition of mixed states as convex combinations of pure states, which in general differ from eigenbasis to eigenbasis. Furthermore, the statistical operator corresponding to the normalized sum of $|\nearrow\rangle$ and $|\searrow\rangle$ is $P(|\uparrow\rangle) \neq \rho_{\text{mix}}$. The pure state $|\nearrow\rangle$ is the result of the quantum superposition of two state-vectors, whereas the fully mixed state, which can also be written $\rho_{\text{mix}} = \frac{1}{2}\mathbb{I}$, can be obtained by mixing two distinct pure ensembles and, therefore, cannot be represented as a projector. In general, given a finite set $\{P(|\psi_i\rangle)\}$ of projectors each corresponding to a distinct pure state $|\psi_i\rangle$ orthogonal to the others, any state ρ' that can be written

$$\rho' = \sum_i p_i P(|\psi_i\rangle) \,, \tag{1.17}$$

with $0 < p_i < 1$ and $\sum_i p_i = 1$, is a normalized mixed state. A typical mixed state has an *infinite* number of such decompositions.

There are several ways that quantum state mixtures have been interpreted.[15] Consider the more general mixture of two pure states

$$\rho = (1 - w)P(|\psi_1\rangle) + wP(|\psi_2\rangle) \,, \tag{1.18}$$

where $0 < w < 1$ is the relative weight of the two pure states; because any other mixed state can be obtained from one of this form by further mixing, it can be used to describe the three ways mixed states have been interpreted. The *statistical interpretation of mixtures* holds that such a mixture arises in nature exactly by the combination of *pure* states in the ratio $w : (1 - w)$. The *ignorance interpretation of mixtures* holds that mixed states arise as the result of the *ignorance* of cognitive agents, that is, observers about the 'actual' composition of collections of pure states in appropriate proportions. A third interpretation of mixed states in quantum mechanics is that they are primitive, that is, of the same fundamental character as pure quantum states, and only arise as *literal* mixtures in the case they are *operationally* so formed. Interpretations of mixed states often reflect specific interpretations of probability or quantum mechanics as a whole. Consider the following view of Heisenberg, an advocate of the Copenhagen interpretation of the theory, associating mixedness with ignorance.

> "The probability function combines objective and subjective elements. It contains statements about possibilities or better tendencies... [that] are completely objective, they do not depend on any observer; and it contains statements about our knowledge of the system, which of course are subjective in so far as they may be different for different observers. In ideal cases the subjective element... may be practically negligible... The physicists then speak of a 'pure case.'
> " ([219], p. 53)

Heisenberg argued that due to the classicality of measuring apparatus,

---

[15] For an extended discussion of *a priori* ways of approaching the question of the nature of quantum mixtures, see [177], pp. 78-83.

"After the [measurement] interaction has taken place, the probability
function contains the objective element of tendency and the subjective element of incomplete knowledge, even if it has been in a 'pure
case' before." ([219], p. 54)

When any quantum state ρ is considered in the matrix representation
provided by an Hermitian operator O, the state will be (at least partially)
coherent in the eigenbasis of the operator if and only if its matrix has non-
zero (typically complex-valued) off-diagonal elements, which, as a result, are
sometimes referred to as *coherences*. When a density matrix commutes with
the operator O, the state will be diagonal and hence not coherent with respect
to this operator; the fully mixed state $\rho_{\mathrm{mix}} = \frac{1}{d}\mathbb{I}$ in an d-dimensional complex Hilbert space commutes with *all* Hermitian operators and so contains *no*
coherence with respect to any observable, and so is sometimes referred to as
the *incoherent state*. Importantly, pure states typically evolve non-unitarily
from pure to mixed states when in non-trivial environments, a process referred
to as *decoherence* which is important, for example, for certain Collapse-Free
interpretations of quantum mechanics and, more practically, in quantum computing. Thus, the degree of coherence of a state can change with time and,
importantly, typically does so as a result of interactions with the system's
environment.

Entanglement is a pervasive quantum mechanical characteristic although
one that is manifested only in situations where *more than one* subsystem of
a mechanical system can be identified, so that each member of the ensemble
corresponding to its quantum state is collectively constituted by a collection
of parts; this is in contrast to generic quantum interference and quantum
uncertainty, which are often manifested by quantum mechanical ensembles,
and are relevant even when systems are not composed of parts, as well as to
state mixing which can occur in classical mechanics. Schrödinger introduced
the phrase *entangled state*, in German *verschränkter Zustand*, to designate
the non-separable pure states of quantum systems and considered it "the
characteristic trait of quantum mechanics" [394, 395]. Entanglement is associated with a sort of coherence that differs from the coherence seen in classical
physics even more so than the coherence seen in non-composite quantum systems; although classical waves are capable of producing interference patterns,
coherence in local classical mechanical systems is *incapable* of supporting the
extremely high-modulation interference phenomena observed when quantum
systems are in entangled pure states, as in situations wherein Bell inequalities,
which are discussed below, are violated.

Specific phenomena involving entanglement are directly related to the unusual joint probabilities of physical events and the behavior of systems conditioned by the behavior of others. Like single-system states, entangled states
are associated with interference that can be manifested when there is sufficiently strong state coherence within a composite system. Interference effects
can be manifested in systems in entangled states because, in their preparation,

there is indistinguishability in principle of alternatives for producing the same sets of *jointly occurring* measurement events on more than one subsystem. Recall that in the double-slit experiment systems hit the screen individually, producing a statistical pattern of detection with periodic modulation corresponding to interference. Similar but stronger modulation relative to classical behavior is possible in the joint detection events of several particles forming a compound system on a *pair* of detection screens [235, 254]; classical physics also allows for modulations of joint detections due to correlation, but classical coincidence probability distributions allow a maximum interference visibility value of less than $\frac{1}{\sqrt{2}}$, whereas an entangled quantum system can produce full interference visibility, that is, a visibility of $\frac{1}{1}$.

Schrödinger defined the *entangled pure states* as the pure quantum states $|\Psi\rangle$ of composite systems that cannot be represented in the form of simple tensor products of subsystem state-vectors, that is,

$$|\Psi\rangle \neq |\psi_1\rangle \otimes |\psi_2\rangle \otimes \cdots \otimes |\psi_n\rangle \;, \tag{1.19}$$

where $\otimes$ indicates the tensor product and the $|\psi_i\rangle$ are vectors providing the states of the subsystems, such as elementary particles [394, 395]. Those states of composite systems that can be represented as tensor products of subsystem states constitute the complement in the set of pure states, the *product states*.[16]

It is easy to determine whether any pure state of system consisting of only two subsystems is in an entangled state by making use of the Schmidt decomposition, which is always available for such systems [392], because any bipartite pure state $|\Psi\rangle$ can be written as a sum of bi-orthogonal terms: There always exists a way of writing $|\Psi\rangle$ in the form

$$|\Psi\rangle = \sum_i a_i |u_i\rangle \otimes |v_i\rangle \;, \tag{1.20}$$

with $a_i \in \mathbb{C}$, where the sets of vectors $\{|u_i\rangle\}$ and $\{|v_i\rangle\}$ consist of orthogonal unit vectors spanning the space of possible state-vectors for the system and the index i runs up to the *smaller* of the dimensions of the two subsystem Hilbert spaces. The probabilities that are the *squared magnitudes* of the Schmidt coefficients a_i are precisely those quantities unchanged by unitary operations performed locally on the individual subsystems. Any such vector-space basis $\{|u_i\rangle \otimes |v_i\rangle\}$ is referred to as a *Schmidt basis*. When the squared magnitudes of the coefficients a_i all differ, this decomposition is unique; one generally takes the amplitudes a_i to be real numbers, which is easily done by introducing any phases into the definitions of the $\{|u_i\rangle\}$ and $\{|v_i\rangle\}$. Thus, whether a pure state is entangled is clear: So long as there is *more than one* such amplitude of non-zero magnitude, $|\Psi\rangle$ is an entangled state, whereas when there exists

---

[16] In the Dirac notation used in this book, when notating quantum states associated with tensor product spaces, the symbol for the tensor product ("$\otimes$") is often omitted but implied.

only one such amplitude then the state is a product state, that is, is a non-entangled state. Nonetheless, it is important to recognize that the Schmidt decomposition is typically *not* available for multipartite systems, a point the significance of which is further discussed in Section 3.5.

A very simple but extremely historically and conceptually important example of an entangled state is the two-particle *spin-singlet* state

$$|\Psi^-\rangle = \frac{1}{\sqrt{2}}(|\uparrow\downarrow\rangle - |\downarrow\uparrow\rangle) \,, \tag{1.21}$$

which often appears in discussions of the foundations of quantum mechanics; in the literature of quantum information science specifically, where the spin eigenstates $|\uparrow\rangle$ and $|\downarrow\rangle$ are chosen to form the so-called *computational basis*, the standard basis vectors are written $\{|0\rangle, |1\rangle\}$ rather than $\{|\uparrow\rangle, |\downarrow\rangle\}$.

The Schmidt decomposition allows one not only to identify entangled states through the correlations they exhibit, but also allows one to quantify entanglement: For Hilbert spaces of countable dimension, the number of non-zero amplitudes a_i in the Schmidt decomposition of a quantum state, known as the *Schmidt number*, $\mathrm{Sch}(|\Psi\rangle)$, serves as a useful, although coarse quantifier of the *amount* of entanglement in a system. The *Schmidt measure* of entanglement of pure states

$$E_S(|\Psi\rangle) \equiv \log_2\left(\mathrm{Sch}(|\Psi\rangle)\right) \,, \tag{1.22}$$

is also often used for this purpose [154, 155]. For systems with more than two parts, a pure state is entangled if and only if, for at least one way of dividing the system into parts, the state describing such a split system is entangled as is, for example, the state $|\mathrm{GHZ}\rangle = \frac{1}{\sqrt{2}}(|\uparrow\uparrow\uparrow\rangle + |\downarrow\downarrow\downarrow\rangle)$ of three two-level systems, which yields a maximally entangled state $|\Phi^+\rangle$ for a bipartite splitting of the system into one spin and the remaining two together.

Historically, Albert Einstein, Boris Podolsky, and Nathan Rosen (EPR) were among the first to be deeply concerned by the implications of the strong correlations arising in systems in entangled quantum states. Indeed, they argued that quantum mechanics must be an *incomplete* theory, if it is to be understood as a fundamental local theory under a 'realist' interpretation, that is, if it assumes a local world and is essentially to describe that world, as opposed to describing only the knowledge of conscious agents.[17] This was very natural, because metaphysical realism had been widely assumed in physics. As Born put it in his 1950 Joule Memorial Lecture entitled *Physics and metaphysics*,

---

[17] The precise nature of the philosophical perspective this involves and those of various other interpretations of quantum mechanics which differ in this regard are discussed in detail in Chapter 3.

"The generation to which Einstein, Bohr and I belong was taught that there exists an objective physical world, which unfolds itself according to immutable laws independent of us... Einstein still believes that this should be the relation between the scientific observer and his subject." ([70])

By considering a two-particle system in the particular entangled quantum state

$$\Psi(x_1, x_2) = \int_{-\infty}^{\infty} \exp\left[\frac{i}{\hbar}(x_1 - x_2 + x_0)p\right] dp , \qquad (1.23)$$

where x_1 and x_2 are the particle positions, x_0 is a fixed distance and p is momentum [153], EPR presented a highly influential argument for the incompleteness of quantum mechanics, often called the *EPR paradox*, based on the assumption of several apparently natural broad conditions, discussed in Section 1.6 below.

John Stewart Bell further drew out the counter-intuitive implications of the presence of entangled states in quantum theory by delimiting the border between local classically explicable behavior and behavior that is not locally causal, with a theorem involving an inequality. This inequality must be obeyed by local (hidden-variables) theories that might be introduced in order to explain all correlations between two distant subsystems forming a compound system, such as one described by the above state when the particles are well separated. Schrödinger believed that such states of widely separated subsystems could not be realizable in practice [394]. However, it was subsequently found to be violated in essence by a broad range of quantum-mechanical systems, such as a pair of photons in the singlet state of Equation 1.21 [22]. Bell-type theorems are discussed further in Section 1.8, below. When asked to describe his theorem in "plain English," Bell said that

"It comes from an analysis of the consequences of the idea that there should be no action at a distance, under certain conditions that Einstein, Podolsky, and Rosen focussed attention on in 1935—conditions which lead to some very strange correlations as predicted by quantum mechanics." ([119], p. 45)

The two pure quantum states shown in Equations 1.21 and 1.23 are examples of pure state entanglement in bipartite systems. Although extremely illustrative, these examples are hardly representative of the majority of physical situations. Indeed, *pure* such states of two-particle systems are exceptional rather than typical in the world; typically, a system very soon interacts with a number of other systems so that, even if it were prepared in a pure state, it is typically described by a mixed state obtained from the greater system state by 'averaging over' the degrees of freedom associated with the systems with which it has interacted. In the last two decades, the definition of entanglement has been broadened beyond Schrödinger's original definition of Equation 1.19, particularly through the use of information theory. Entanglement, at least qualitatively, is readily extended to include such mixed states. A bipartite mixed state of a composite system of parts A and B is called *separable* when it can be given as convex combination of products of subsystem states:

$$\rho_{AB} = \sum_i p_i \rho_{Ai} \otimes \rho_{Bi}, \qquad (1.24)$$

where $p_i \in [0,1]$ and $\sum_i p_i = 1$, ρ_{Ai} and ρ_{Bi} being states on the respective subsystem Hilbert spaces, and the p_i being classical probabilities; entangled states of bipartite systems with components labeled A and B are typically denoted using "AB" in subscript, as in ρ_{AB}, or in superscript. The product states of the form $\rho_{AB} = \rho_A \otimes \rho_B$ correspond to situations in which the states ρ_A and ρ_B of the two subsystems are *entirely* uncorrelated. The entangled mixed states are precisely the inseparable states. However, it is not always possible to tell whether or not a given mixed state is separable. The problem of determining whether a given state of a composite system is entangled is known as the *separability problem*.

The separable mixed states 'contain' no entanglement, in the resource sense that has arisen since the arrival of quantum information science; they can be viewed as mixtures of product states and can be created by local operations and classical communication, which are discussed below, from pure product states. Separable states are those that can be jointly prepared by N spatially separated observers each preparing one local state $\rho_{A(i)}$ according to a shared set of instructions $\{p_i\}$ [348].[18] For example, in order to create a separable bipartite state, an agent in one localized region ("lab") needs merely to sample the probability distribution $\{p_i\}$ and share the corresponding measurement results with an agent in the other; the two agents can then create their own sets of suitable local states each in its separate location in (classical) correlation with the other. However, in the case of fully distributed composite systems, because not all entangled states can be converted into each other in this way, such transformations instead give rise to distinct classes of entangled states and different *sorts* of entanglement, as discussed in Section 1.10 below.

When there are correlations between observables of subsystems of systems in bipartite separable states, these can be fully accounted for locally in the above sort of way because the separate quantum subsystem states, even when located in spacelike-separated laboratories,[19] provide descriptions enabling such common-cause explanations of the joint properties of A and B. Such correlations are commonly referred to as *classical correlations*. Accordingly, the outcomes of local measurements on any separable state can be simulated by a local hidden-variables theory, that is, the behavior of systems described by such states can be accounted for using common-cause explanations.[20]

---

[18] The sense in which entanglement can be considered a resource is explained in Chapter 4.

[19] Laboratories are discussed formally in Section 1.9.

[20] Hidden-variables theories are discussed in Section 1.7.

The quantum states in which correlations between A and B can violate a Bell-type inequality are called *Bell correlated*, or *EPR correlated*. If a bipartite *pure* state is entangled, then it is Bell correlated with certainty, as was first pointed out by Sandu Popescu and Daniel Rohrlich [356] and by Nicolas Gisin in the early 1990s [192].[21] However, no simple logical relation between entanglement and Bell correlation holds for the *mixed* entangled states. For example, the Werner state

$$\rho_W = \frac{1}{2}\left(\frac{1}{4}\mathbb{I} \otimes \mathbb{I}\right) + \frac{1}{2}P(|\Psi^-\rangle) \tag{1.25}$$

is not Bell correlated although it is entangled, because ρ_W cannot be written a convex combination of product states.

Let us recall some of the basic elements of the mathematics of random processes, because these play an important role in quantum mechanics and the hidden-variables theories sometimes considered as possible alternatives to standard quantum theory. Expectation values are defined for random variables, such as the measurement outcomes in experiments on quantum systems; a *random variable* is a measurable deterministic function from a given sample space S, that is, the set of all possible outcomes of a given experiment, the subsets of which are known as *events* with those containing only a single element being the *elementary events*, to the real numbers; see Section 2.9. Given an elementary event ω, the value $X(\omega)$ is its *realization*. The *expectation value*, $E[Y(X)]$, of a function $Y(X)$ of a random variable X is given by a linear operator such that

$$E[Y(X)] = \sum_{i=1}^{n} Y(x_i)p(X = x_i) , \tag{1.26}$$

$$E[Y(X)] = \int_{-\infty}^{+\infty} Y(x)f(x)dx , \tag{1.27}$$

in the cases of discrete and continuous variables, respectively, where in the former $p(X = x_i)$ is the probability and in the latter $f(x)$ is the probability density function, both being collection of numbers between 0 and 1 summing to 1. A *stochastic process* is the generalization of deterministic temporal evolution by the consideration of a time-parameterized family of random variables and joint probability distributions. In particular, a stochastic process is a map $X : S \times T \to \mathbb{R}^N$ from each elementary event–time pair (ω, t) to the real numbers, providing a trajectory $t \mapsto X(\omega, t)$, where $X(\omega, t)$ is measurable for each value of t and N is the dimension of the generally vector-valued process.

The quantum mechanical expectation value of a physical magnitude represented by an Hermitian operator O of a system in a pure state given by a state-vector $|\psi\rangle$ is

---

[21] Note that not all such states are *Bell states*, that is, elements of the Bell basis as, say, $|\Psi^-\rangle$ and $|\Phi^+\rangle$ are.

$$\langle O \rangle_{|\psi\rangle} = \langle \psi | O | \psi \rangle = \sum_i o_i |\langle o_i | \psi \rangle|^2, \tag{1.28}$$

where $\{o_i\}$ is the set of eigenvalues comprising the *eigenvalue spectrum* of O. Expectation values thus take the form of *average values* for measurements on ensembles of quantum systems prepared in the same state under statistically ideal circumstances. A related mathematical theorem central to quantum mechanics is the *spectral theorem*: each Hermitian operator O can be written

$$O = \sum_i o_i P(|o_i\rangle) , \tag{1.29}$$

where $P(|o_i\rangle)$ is the projector onto the finite Hilbert subspace spanned by $|o_i\rangle$. This provides the *spectral decomposition* (or *eigenvalue expansion*) of the operator O.[22] When the state of the system is instead mixed, by necessity being described by a statistical operator ρ that is *not* a projector, the expectation value is

$$\langle O \rangle_\rho = \mathrm{tr}(\rho O) . \tag{1.30}$$

Consider now correlations specifically between pairs of subsystems, such as those underlying the observed interference patterns of joint measurements discussed above. In particular, consider a bipartite quantum system with the Hilbert space $\mathcal{H} = \mathcal{H}_1 \otimes \mathcal{H}_2$ in a possibly mixed state ρ. The pertinent observables $A^{(i)}$ in the two systems are uncorrelated between the two subsystems if one can write $\rho = \rho^{(1)} \otimes \rho^{(2)}$, where $\rho^{(i)} \in \mathcal{H}_i$ $(i = 1, 2)$. The expectation value of the product of the $A^{(1)}$ and $A^{(2)}$ on the subsystems can then be factored, that is,

$$\langle A^{(1)} \otimes A^{(2)} \rangle_\rho = \mathrm{tr}\big(\rho(A^{(1)} \otimes \mathbb{1})\big) \mathrm{tr}\big(\rho(\mathbb{1} \otimes A^{(2)})\big) \tag{1.31}$$

In this case, probability of outcomes of joint measurements of the $A^{(i)}$ is simply the product of the probabilities of outcomes of the measurements performed separately. When joint measurements are correlated, in that the subsystems are in the same state ρ_j $(j = 1, \dots, n)$ with probabilities p_j, the statistical operator is separable, having the form of Equation 1.24, and the expectation values of measurements of the physical magnitudes $A^{(i)}$ are instead non-trivially of the form

$$\sum_{j=1}^n p_j \mathrm{tr}\big(\rho_j^{(1)}(A^{(1)} \otimes \mathbb{1})\big) \mathrm{tr}\big(\rho_j^{(2)}(\mathbb{1} \otimes A^{(2)})\big) . \tag{1.32}$$

Any system with a density matrix non-trivially of the form shown in Equation 1.24 (with $n \geq 2$) is classically correlated, even if it can be created by mixing entangled states [489].

---

[22] This theorem does *not* hold for operators in *infinite-dimensional* Hilbert spaces, even when there does exist a countably infinite set of basis vectors, because there may not exist a countably infinite set of *eigenvectors* that form a basis. However, there do exist topologies on infinite-dimensional spaces for which the theorem in a generalized form (the *nuclear spectral theorem*) does hold, cf., e.g., [50].

The highest coincidence interference visibility obtainable in an experiment using *classically* uncorrelated states, including results predicted by local hidden-variables theory, is 0.5 [374]. By contrast, entangled states can attain visibilities of two-particle interference of up to 1.0. Bell-type inequalities, such as that appearing in Bell's theorem, which are also discussed below, can be shown to be violated once the visibility surpasses $1/\sqrt{2} \approx 0.71$.

1.4 Quantum Entanglement Measures

A practical entanglement measure quantifying mixed bipartite-state entanglement in a few important quantum system sizes is *negativity*. Negativity is defined in terms of the transpose of the density matrix representation of the statistical operator, as follows.

$$\mathcal{N}(\rho) = \frac{1}{2}\left(||\rho^{T_A}||_1 - 1\right) = \left|\sum_i \lambda_i\right|, \qquad (1.33)$$

where $|| \cdot ||_1$ is the trace-norm and i runs over the subset of negative eigenvalues of this density matrix; the operator ρ^{T_A} (or ρ^{T_B}) is positive if and only if the statistical operator ρ is separable, but only in the cases of 2×2, 2×3 dimensional systems [238] and systems in Gaussian states of infinite dimension [144]. The matrix elements of the partially transposed state are $\langle i_A j_B | \rho^{T_A} | k_A l_B \rangle \equiv \langle k_A j_B | \rho | i_A l_B \rangle$, stated above for transposition relative to subsystem A. When applied to an entangled state of appropriate dimension, such as a Bell state, the result of partial transposition is a matrix with at least one negative eigenvalue. The matrix property of *positivity of partial transpose* (PPT) is a necessary but not sufficient condition for separability when subsystems are of Hilbert spaces of dimension greater than two; for larger Hilbert spaces, there exist entangled states whose density matrices are *positive* under the partial transpose operation. The PPT-preserving class of quantum operations is that of bipartite quantum operations such that input states positive under partial transposition are mapped to states that are also positive under partial transposition.[23]

The negativity provides a criterion for determining state entanglement, known as the Peres–Horodečki (PH) criterion: When the statistical operator is separable, the matrix that results from the partial transposition operation is another statistical operator, whereas a state ρ is entangled if the partial transpose of the corresponding density matrix is negative, the result therefore being precluded from being well defined as a density matrix [348]. The PH criterion implies that both $\rho_A \otimes \mathbb{I} - \rho \geq \mathbb{O}$ and $\mathbb{I} \otimes \rho_B - \rho \geq \mathbb{O}$, where ρ_J ($J = A, B$) is the reduced state describing the individual subsystem J and $\mathbb{O}$

---

[23] See Section 1.10 for the relationship between the PPT-preserving operations and LOCC operations.

is the zero operator, which is the *reduction criterion* for entanglement;[24] the violation of the reduction criterion also implies separability of ρ in the case of two two-level systems and the case of one two-level system and one three-level system (*i.e.*, 'qutrit').

The reduced states are provided through the partial trace operation. Let $\{|u_i\rangle\}$ and $\{|v_j\rangle\}$ be bases for Hilbert spaces $\mathcal{H}_1$ and $\mathcal{H}_2$ of countable dimension, describing two subsystems 1 and 2, respectively, forming a composite system in state ρ. The set of vectors $\{|u_i\rangle|v_j\rangle\}$ ($i = 1, 2, \ldots$; $j = 1, 2, \ldots$) is then a basis for the Hilbert space of the total system, $\mathcal{H} = \mathcal{H}_1 \otimes \mathcal{H}_2$. Any operator O on $\mathcal{H}$, such as ρ, can be written in the form

$$O = \sum_{ij,kl} |u_i\rangle|v_j\rangle O_{ij,kl} \langle u_k|\langle v_l|, \tag{1.34}$$

where $O_{ij,kl}$ are the (scalar) matrix elements corresponding to O. Finding the partial trace is somewhat like finding the marginal distribution of a component of a two-dimensional random variable from the probability distribution of the latter in classical probability theory: The *partial trace* of O, with respect to the first subsystem, for example, is $\mathrm{tr}_1 O \equiv \sum_i \langle u_i|O|u_i\rangle$, which "averages over" the degrees of freedom of the second. In particular, the result of partial tracing the statistical operator ρ of the combined system over each of the subsystems individually is the pair of *reduced statistical operators* $\rho_1 = \mathrm{tr}_2\rho$, $\rho_2 = \mathrm{tr}_1\rho$, each describing the state of one subsystem, for example, in the case of the *dismissal* of the other subsystem. Importantly, the reduced statistical operator is the only statistical operator providing correct measurement statistics for subsystems [286]. Also, when the overall state ρ is an entangled pure state, the reduced states ρ_1 and ρ_2 describing the component systems are *mixed* rather than pure, which cannot occur for marginal distributions in classical mechanics, marking another difference between classical and quantum statistics.

There are currently two different, but relatable, general approaches to measuring quantum entanglement, one in terms of matrix operations and geometry, as in the case of the negativity, and one based on ideal limits of operations on quantum states, although there is no known *single* good entanglement measure applicable to all mixed states of systems with more than two subsystems. In the bipartite case, with the two subsystems labeled A and B, an instance of the former, the *concurrence*, is widely used in practical situations. For pure states of two spins, this quantity can be written $C(|\Psi_{AB}\rangle) = |\langle\Psi_{AB}|\tilde{\Psi}_{AB}\rangle|$, where $|\tilde{\Psi}_{AB}\rangle \equiv \sigma_2^{\otimes 2}|\Psi_{AB}^*\rangle$ which is referred to as the 'spin-flipped' state-vector [505]. The concurrence of a mixed two-qubit state, $C(\rho_{AB})$, can be expressed in terms of the minimum average pure-state concurrence $C(|\Psi_{AB}\rangle)$ where the required minimum is to be taken over all possible ways of decomposing the ensemble ρ_{AB} into a mixture of pure states $|\Psi_{AB}\rangle$,

---

[24] This reduction property is sufficient for the recoverability of entanglement by distillation, which is described in Section 1.10.

as in Equations 1.17-1.18. The concurrence of a general state two-spin state is then $C(\rho_{AB}) = \max\{0, \lambda_1 - \lambda_2 - \lambda_3 - \lambda_4\}$, where the λ_i are real square roots of eigenvalues of the matrix $\rho_{AB}\tilde{\rho}_{AB}$, ordered by decreasing size, and are non-negative. The standard 'operational' measure of entanglement in the quantum information literature, the *entanglement of formation*, $E_f(\rho_{AB})$, of a mixed state ρ_{AB} of a pair of two-level systems ('qubits') is defined in the large number limit of identical copies of the two-level system as the minimum number required to form the state ρ_{AB} by local operations and classical communication (LOCC) on this collection of copies. This quantity has the form of the binary entropy function, expressed in terms of the concurrence

$$E_f(\rho_{AB}) = h\left(\frac{1 + \sqrt{1 - C^2(\rho_{AB})}}{2}\right), \tag{1.35}$$

where $h(x) = -x \log_2 x - (1-x) \log_2(1-x)$, *cf.* Section 4.1 [505]. The square of the concurrence of a state is often referred to as its *tangle*.

For pure states, the entanglement of formation is the von Neumann entropy of the states of the individual components, equaling the number of Bell-state pairs convertible to subsystem states by LOCC again in the large number limit of converted bipartite systems. In fact, in the case of pure states, no classical communication is required in this limit [300]. Thus, the number of identical copies of the system available has an effect on the conversion properties of the state concerned [296]. This provides entanglement in units of "e-bits," that is, the minimum number of Bell states (*e.g.* the singlet of Equation 1.21) from which a state can be reached. Both the concurrence and the entropy-based measure are, in fact, consistent with a definition initially suggested by Abner Shimony [414]; that geometrical measure of the degree of entanglement identifies it as the distance of the state from the nearest factorable state, that is,

$$E_G(|\Psi\rangle) = \frac{1}{2} \min\big|\big|\ |\Psi\rangle - |\Xi\rangle\ \big|\big|^2, \tag{1.36}$$

where $|\Xi\rangle$ is a (normalized) product state in Hilbert space and the minimum is taken over the set of such normalized product states, provides the distance of the closest separable approximation [414].[25] This measure also relates directly to the definition of entanglement as non-factorability introduced by Schrödinger. Any monotonically increasing function of $E_G(|\Psi\rangle)$, giving the same ordering of normalized vectors $|\Psi\rangle$, serves as an equally good geometrical entanglement measure [504]. One can find the nearest separable state to a given state by solving the corresponding non-linear eigenvalue problem [486].

It is noteworthy that, like Schrödinger's definition, $E_G(|\Psi\rangle)$ is defined independently of explicit locality considerations. This is a particularly important attribute because the failure of local causality, for example, the use of

---

[25] The Hilbert-space angle $\phi \equiv \cos^{-1}(|\langle\Psi|\Xi\rangle|)$ is the natural distance between two state-vectors used here, and takes the state overlap to a distance function derivable from the Fubini–Study metric, a Riemannian metric on projective Hilbert-space [504].

the extent of Bell-inequality violation, has been found wanting as a measure of degree of entanglement, for reasons discussed below in Section 1.8. The other popular geometrical entanglement measure is the relative entropy 'distance' introduced by Vedral [466, 468], discussed in Section 4.5, is not a genuine metric in the mathematical sense, because it is asymmetrical in its arguments. Geometrical approaches have also been used in constructive, although imperfect attempts to provide an absolute multipartite-entanglement measure which would allow for a comprehensive classification of entangled states [18, 255, 407, 485]. This is perhaps not surprising, given that geometry and symmetry are naturally connected through the identification of invariant quantities and that symmetry has always played an important role in the investigation of quantum mechanics, most significantly in the case of Wigner's theorem, as in all parts of physics, *cf.* [500].[26]

In the case of systems with infinite degrees of freedom, the above geometrical definition also will not work, because there is always a separable state with an arbitrarily large amount of entanglement in a neighborhood surrounding any such entangled state [112, 156]. Nonetheless, geometrical methods have been used to make some progress in the classification of entangled states. Even for three parties, each in possession of a system described by a d-dimensional space, the combined pure state lies within in a d^3-dimensional Hilbert space and so depends on $2(d^3-1)$ real parameters, whereas the transformations used to unitarily transform, or equivalently to rotate this state, have only $3(d^2-1)$ independent real parameters. It is therefore often impossible to obtain the extraordinarily useful Schmidt decomposition for pure states.[27] Nonetheless, multipartite extensions of Schmidt decomposition can be found in special situations. For example, the construction of a generalized Schmidt decomposition may proceed in some three-spin systems [298, 347]. More generally, an N-partite pure state can be written in generalized Schmidt form if and only if *each* of its $N-1$ partite subsystem states is separable [446].[28]

1.5 Surprising Implications of Entanglement

Having recalled the formal elements of quantum mechanics and the basics of entanglement theory, let us return to the consideration of the simple double-slit experiment to see how entanglement bears on it in a larger context.

Recall that a set of alternative histories available to a system is required for its self-interference. For example, when only one of the two slits is available at a time in the double-slit experiment, no interference pattern appears on the

---

[26] Wigner's theorem can be stated as follows. Let $\mathcal{I} : u \to v$ be a length-preserving transformation (*isometry*) with respect to the Hilbert-space norm. Then $\mathcal{I}$ is either a unitary or anti-unitary transformation. For a proof, see [304], p. 101.

[27] As seen later in Chapter 3, this precludes one attempt straightforwardly to solve the quantum measurement problem.

[28] Multi-partite states are discussed in more detail in Section 1.10.

detection screen, as shown in Figure 1.1, because only one total history is possible from preparation to detection, whereas when two different histories are available an interference pattern can be seen. Indeed, the implementation of the final apparatus configuration can be delayed until *after* the system under measurement has entered the apparatus [492]. An interference-free detection pattern should result even if both slits are left available to incoming particles and one also detects particles after the two slits and this is so *before* the two alternatives (of having passed through slit 1 or slit 2) have become distinguishable from each other in any given run of the apparatus—say if substantially less than one particle is present in the apparatus during a configuration switching-time interval. Having alternative histories be distinguishable in no way rests on an assumption that the apparatus is a *classical* system; merely introducing a quantum system that *in principle* can be definitively verified as having entered a specific slit is sufficient to eliminate quantum interference.

A diverse set of experiments demonstrating such behavior has been carried out, *cf.* [227]. Wootters and Żurek successfully modeled the standard arrangement where path data is obtained by arranging for a momentum exchange between the photon and the initial slitted screen [507]. Scully, Englert and Walther (SEW) later introduced a different method, in an experiment in which laser-excited atoms forming a beam pass through an initial double-slit diaphragm with their possible paths continuing through two auxiliary microwave cavities that can be configured so as to allow the path information to be obtained before they exit another double-slit diaphragm, as shown in Figure 1.2 below [402]. This apparatus allows entanglement to arise between states corresponding to atomic-paths and those of cavity occupation. The interaction involved is too weak to lead to sufficient momentum transfer for path determination; momentum transfer is incapable of accounting for the destruction of the interference pattern that would typically be observed in a double-slit experiment. Interference can be restored by moving to an alternative configuration of the apparatus where an auxiliary system is introduced that possesses a suitable observable not commuting with the path-indicating operator that is precisely measurable by proper arrangement of the cavities. Because the data regarding paths that would be present is then no longer available, this phenomenon is called "quantum erasure" [400].

The SEW apparatus, allowing one to switch between the two pertinent configurations, is essentially a modification of the double-slit experiment of Figure 1.1. By switching between two configurations, data relating to one or the other non-commuting observable is erased. The enlarged apparatus allows alternation between the above two cases, with the option to make the choice of configuration at any time before the final screen is contacted, incorporating delayed choice. The auxiliary system can definitively indicate which of the two slits was entered by the primary system by exploiting the above-mentioned state entanglement [160, 402, 403]. However, the primary and auxiliary systems are allowed to interact in such a way that phenomena which *would have* occurred in one configuration are not exhibited in the other configuration. The incoming quantum ensemble is that of a beam of Rydberg atoms rather than of elementary particles, the laser is introduced as the first apparatus element and is oriented perpendicularly to the atom beam so as to allow its excitation, the auxiliary system consisting of a pair of micro-cavities is placed after it, and an additional double-slit diaphragm placed after the cavities, as shown in Figure 1.2. The two micro-cavities are each long enough that the atoms will de-excite with near certainty between their entrances and exits. Therefore, each will capture any radiation emitted from atoms entering it specifically, allowing the atoms of the beam to become entangled with the cavity pair before continuing on into the remainder of the system. The two cavities constituting the auxiliary system are positioned adjacent to each other but separated by a wall that is covered on each side by shutters which, when opened, allow captured radiation to be absorbed from either cavity without the discriminating from which cavity it came. A rapid switch of the shutter settings between open and closed positions allows the choice of configuration to be delayed until very near the time any atom strikes the screen.

To allow potential path data to be registered, the laser is made sufficiently powerful that, when turned on, it will excite every one of the beam atoms from its ground state to its excited state with near certainty. The atomic system can be considered prepared in the state $|\psi(r)\rangle|j\rangle = \frac{1}{\sqrt{2}}(|\psi_1(r)\rangle + |\psi_2(r)\rangle)|j\rangle$, where the position coordinate of the elementary particles of the standard experiment is replaced by that of the atomic center-of-mass position coordinate r and the atomic internal states are written $|j\rangle$, $j = 0, 1$, the ground and excited states, respectively. Without the laser on, all atoms are in the ground state $|0\rangle$. The atom beam is then in the pure product state $|\psi(r)\rangle|0\rangle$, so that its squared magnitude, the probability density of detected atoms at the screen position $r = R$, is

$$p(R) = \frac{1}{2}\Big[\big(||\psi_1(R)\rangle|^2 + ||\psi_2(R)\rangle|^2\big) + \big(\langle\psi_2(R)|\psi_1(R)\rangle + \langle\psi_1(R)|\psi_2(R)\rangle\big)\Big]\langle0|0\rangle$$

with $\langle0|0\rangle = 1$, that is, one finds the sort of interference pattern observed in the standard double-slit experiment when both slits are available. With the laser turned on and the shutters kept closed, with the atoms prepared in $|\psi(R)\rangle|1\rangle$, atomic radiation is deposited into one of the cavities and the state of the enlarged system must be considered, which is

$$|\Psi\rangle = \frac{1}{2}\big(|\psi_1\rangle|0\rangle|1_{C1}0_{C2}\rangle + |\psi_2\rangle|0\rangle|0_{C1}1_{C2}\rangle\big)$$
$$= \frac{1}{2}\big(|\psi_1\rangle|1_{C1}0_{C2}\rangle + |\psi_2\rangle|0_{C1}1_{C2}\rangle\big)|0\rangle \qquad (1.37)$$

where the subscripts $\{Ci\}$ indicate the cavity pair with eigenstates $|k_{C1}l_{C2}\rangle$, with $k = 0, 1$ indexing the occupation eigenvalue of cavity 1 feeding slit 1 and $l = 0, 1$ indexing that of cavity feeding slit 2. Thus, with the laser turned on and cavity shutters kept closed, the external atomic state and the occupation state of the two-cavity system become entangled, whereas the internal atomic

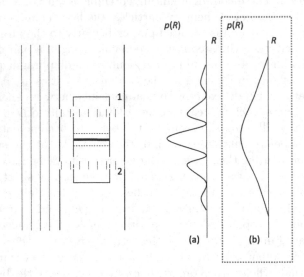

Fig. 1.2. Apparatus for quantum erasure. An enlargement of the standard double-slit apparatus of Figure 1.1 by the addition of intermediate microcavities with internal shutters (dark dashed lines) and a radiation absorber (thick solid line) introduced [402]. Excited atoms are input that de-excite with certainty within one of the cavities. (a) Atom detections when shutters are open; path data are unavailable because radiation is indiscriminately absorbed. (b) Atom detections when the radiation absorber is unreachable, with radiation is selectively contained in one cavity or the other; path data, which are incompatible with interference, is available. Opening the shutters, even after each atom has passed the double-slitted diaphragm, erases data associated with paths, which is irretrievable from the common radiation absorber, taking case (b) to case (a).

state factors out. The probability density for arrival of atoms at point R on the screen, shown in case (b) of Figure 1.2, is

$$p = \frac{1}{2}\Big[(||\psi_1\rangle|^2 + ||\psi_2\rangle|^2)$$
$$+ \langle\psi_1|\psi_2\rangle\langle 1_{C1}0_{C2}|0_{C1}1_{C2}\rangle + \langle\psi_1|\psi_2\rangle)\langle 0_{C1}1_{C2}|1_{C1}0_{C2}\rangle\Big]\langle 0|0\rangle \; ;$$

$\langle 1_{C1}0_{C2}|0_{C1}1_{C2}\rangle = 0$ and $\langle 0_{C1}1_{C2}|1_{C1}0_{C2}\rangle = 0$ imply that the terms including them are zero. The observed interference pattern of atoms striking the final screen is $p(R) = \frac{1}{2}||\psi_1(R)\rangle|^2 + \frac{1}{2}||\psi_2(R)\rangle|^2$, a probability sum corresponding to state mixture; the introduction of the cavities which selectively interact with passing atoms depending on their proximity to each slit allows for distinguishability of the paths of the atoms as long as their interior shutters are kept closed. This atomic detection pattern can be understood as that resulting from the enlargement of the system so as to include entangled subsystems.

Quantum path data encoded in this *de facto* two-cavity memory can readily be "erased" by switching instead to the configuration in which the internal shutters of the two cavities are opened, which allows the stored radiation to reach the photon absorber. In that case, because the radiation in the cavities from which path data might be retrievable is instead lost from them to the absorber, taking both cavity states to their ground states $|0_{C1}0_{C2}\rangle$,

$$|\Psi\rangle = \frac{1}{2}\left(|\psi_1\rangle|0\rangle|0_{C1}0_{C2}\rangle + |\psi_2\rangle|0\rangle|0_{C1}0_{C2}\rangle\right) = \frac{1}{2}\left(|\psi_1\rangle + |\psi_2\rangle\right)|0\rangle|0_{C1}0_{C2}\rangle .$$

The path data are, therefore, no longer encoded in them; interference reappears, as in case (a) of Figure 1.2, so that

$$p(R) = |\langle\Psi(R)|\Psi(R)\rangle| \tag{1.38}$$
$$= \frac{1}{2}\left[\left(||\psi_1(R)\rangle|^2 + ||\psi_2(R)\rangle|^2\right) + \left(\langle\psi_2(R)|\psi_1(R)\rangle + \langle\psi_1(R)|\psi_2(R)\rangle\right)\right] .$$

This situation can be viewed as an instance of the following general situation. Performing *different* measurements on individual d-dimensional subsystems of a bipartite system in a pure state $|\Psi\rangle_{AB}$, allows decompositions into mixtures of pure states of a subsystem ensemble to be operationally produced that differ but are described by the *same* statistical operator. To see this, consider two different decompositions of the same statistical state of the first system (atom) into d orthogonal pure states corresponding to possible measurement outcomes $\{i\}$:

$$\rho_A = \sum_i |a_i|^2 |\psi_i\rangle\langle\psi_i| , \tag{1.39}$$

$$\rho_A = \sum_i |a_i'|^2 |\psi_i'\rangle\langle\psi_i'| , \tag{1.40}$$

where $\{|\psi_i\rangle\}$, $\{|\psi_i'\rangle\}$ are differing bases for the subsystem Hilbert spaces. These two decompositions also have differing 'purifications' describing the total bipartite system, namely,

$$|\Psi\rangle_{AB} = \sum_i a_i |\psi_i\rangle|\chi_i\rangle , \tag{1.41}$$

$$|\Psi'\rangle_{AB} = \sum_i a_i' |\psi_i'\rangle|\chi_i'\rangle , \tag{1.42}$$

where $\{|\chi_i\rangle\}$ and $\{|\chi_i'\rangle\}$ are orthonormal bases for the Hilbert space of the second system (joint cavity); purifications are pure states of larger systems from which these states arise through the partial tracing out of degrees of freedom not associated with them.[29] With these expressions at hand, one

---

[29] *Cf.* Section 1.4.

sees that by differently measuring the second system, that is, in the different bases above, two different preparations are carried out providing ensembles described by the *very same* reduced statistical operator, namely, ρ_A.

The purifications $|\Psi\rangle$ and $|\Psi'\rangle$ are related to each other by a unitary transformation of the state of the second system acting only on the space of the second system, that is, of the form $\mathbb{I} \otimes U$. Thus, *both* of the two ensemble descriptions, that of Equation 1.41 and that of Equation 1.42, are obtainable by appropriate local actions on the second system alone. Likewise, one can consider different decompositions for the reduced state of the second system and find bases for the first system and purifications giving rise to that statistical operator by measurements of the first system [362].[30] This result illustrates quantum erasure in a more general context, in that it describes the effect of the 'choice' of measurement basis or, equivalently, of the choice of measurement apparatus on what is observed about a quantum system, A, when it is in contact with another, B, because knowledge obtained by a measurement of system B, when classically communicated to A, results in a change of the description of the subsystem at A. Any finite ensemble of bipartite quantum states can be remotely prepared by two agents in distant laboratories through local operations and classical communication using entanglement, although there is no question of this effect enabling superluminal signaling (*cf.* [404]).

Schrödinger called such a process *remote steering*, and believed that in practice such a procedure could *not* be carried out for well separated systems.

> "Attention has recently [in the EPR paper] been called to the obvious but very disconcerting fact that even though we restrict the disentangling measurements to *one* system, the representative obtained for the *other* system is by no means independent of the particular choice of observations which we select for that purpose and which by the way are *entirely* arbitrary. It is rather discomforting that the theory should allow a system to be steered or piloted into one or the other type of state at the experimenter's mercy in spite of his having no access to it." ([395])

Nonetheless, one can experimentally observe such surprising implications of the failure of local causality in systems prepared in entangled pure states.

1.6 The EPR Program and Absence of Local Causality

Nearly all those who made significant contributions to the construction and formalization of quantum mechanics, like Schrödinger, appreciated that the theory appeared to conflict with the traditional world view of working physicists. These investigators conceived and described to their colleagues a number

---

[30] This result, known as the *GHJW theorem*, was shown by Gisin, Hughston, Jozsa, and Wootters [191, 246], is similar to a result originally obtained by Schrödinger [272, 317]. It has also been extended by Cassinelli *et al.* [100].

of sorts of quantum physical situation in which the traditional assumptions evidently failed.

Let us begin by considering perhaps the most striking and boldest of these conceptions, the Einstein–Podolsky–Rosen scenario to which reference is made in the above comment by Schrödinger, which involves entanglement and the failure of local causality and inspired the later investigations of Bell and others that have come to be seen as constituting *experimental metaphysics* [408]. With it, EPR presented a forceful argument against the adequacy of the theory, under a number of assumptions within a particular realist world view. Their reasoning was designed to confront quantum mechanics with the heart of the traditional physical world view and to argue that quantum theory is inadequate by those lights by demonstrating a contradiction between the two sets of principles, one more mathematical and one more conceptual, on which real-world experiments would later bear. In particular, the argument of EPR was designed to demonstrate the inadequacy of the standard formulation of quantum mechanics for enabling a complete naturalistic description of the physical world. It provided a deeper method for the investigation of the physical and philosophical issues encountered in micro-physics than any before it. Their approach is now commonly referred to as the *EPR program*.

The physical situation presented by EPR shows that the lack of local causality inherent in quantum mechanics presents serious problems if one assumes quantum mechanics to be a complete fundamental theory, as it was then assumed to be, that is, one in need of no further physical elements and to be in accord with scientific realism.[31] Although some have disputed the fundamental character of non-relativistic quantum mechanics (*cf.* [187], p. 2), most if not all of its problems are inherited by its relativistic successors, which only introduce additional foundational questions. Einstein's personal view of the situation emphasized the importance of having physical theory locally and objectively describe the world.

"I just want to explain what I mean when I say that we should try to hold on to physical reality. We all of us have some idea of what the basic axioms of physics will turn out to be... whatever we regard as existing (real) should somehow be localized in time and space. That is, the real in part of space A should (in theory) somehow 'exist' independently of what is thought of as real in space B. When a system in physics extends over the parts of space A *and* B, then that which exists in B should somehow exist independently of that which exists in A. That which really exists in B should therefore not depend on what kind of measurement is carried out in part of space A; it should also be independent of whether or not any measurement at all is carried out in space A. If one adheres to this programme, one can hardly consider the quantum-theoretical description as a complete representation of the physically real." ([72], p. 164)

---

[31] *Cf.* Einstein's comment to Born quoted in Section 1.7.

This can be seen as Einstein's formulation of what later came to be known in the physics literature, albeit somewhat misleadingly from the philosophical point of view, as *local realism* because the EPR article stressed the importance of locality as an essential characteristic of objective physical objects.

Before the EPR paper, Einstein had presented a less compelling thought experiment directed toward this question, which was a simple situation not involving entangled states, in which the wave-function of a ball is restricted to two boxes, B_1 and B_2, which can be arbitrarily well separated in space [256]. This earlier thought experiment is sometimes referred to as the *Einstein box experiment* [326]. It should not, however, be confused with the better known experiment sometimes referred to as the *Einstein–Bohr box experiment*, which involved the determination of energy changes within the box by measurements of its weight, that is sometimes also said to present the *Alarm-clock paradox*.[32]

In Einstein's ball–box experiment, when a measurement is made to ascertain whether the ball is contained in one box, B_1, one is also able to determine whether the ball is contained in the other, B_2. Thus, Einstein argued, the wave-function provides an incomplete description of the ball's location; otherwise, the performance of the measurement on B_1 would immediately cause the ball to be present or absent in B_2 arbitrarily far away, at odds with relativity;[33] it would amount to there being a *non-local causal influence* at work. This early argument was unsuccessful, however, because the constraint of relativity that causal influences must propagate at speeds constrained by the speed of light is not relevant, because wave-function doesn't in general describe the density of a *substance*, despite the fact, for example, that Schrödinger had hoped such an interpretation might be viable; the example involves only a logical determination, not a causal influence, because there is no *physical contingency* involved. All that is required for logical consistency is that one assume that the ball must be in one and only one box at any given moment even when the wave-function is not entirely absent from either box [418]. Therefore, one is not driven to accept that a non-local causal influence is involved. For example, one could consistently hold that the quantum state is represented by different wave-functions in different frames of reference [175].

The EPR paper provides a much better argument than Einstein's original using the ball and box because it involves the consideration of two particles (in an entangled quantum state) which can be arranged to be spacelike separated from each other. Basic elements of a proper fundamental theory in the realist world view favored by Einstein were explicitly laid out in the EPR paper in the form of three conditions [153], even if not to the complete satisfaction of Einstein himself who, as it turned out, did not have as much control of its

---

[32] A related discussion involving the Stern-Gerlach apparatus can be found in [342], pp. 29-32.

[33] Related results pertaining to wave packet evolution, localization and causality are shown in [212–214].

final form as he would have liked.[34] The first is the *reality criterion*, presented as defining "physical reality" for the purposes of the scenario.

(1) "If, without in any way disturbing a system, we can predict
    with certainty (i.e., with probability equal to unity) the value
    of a physical quantity, then there exists an element of physical
    reality corresponding to this physical quantity."

This condition is, notably, also closely related to causality, in that, for example, indeterminism for some is the condition that "the state of a system at time t cannot in general be predicted with certainty given the history of its states priority to t" ([137], p. 19). The second is a *locality criterion*.

(2) "Since at the time of measurement the two systems no longer
    interact, no real change can take place in the second system
    in consequence of anything that may be done to the first
    system."

The third is a *completeness criterion*, the one which EPR ultimately argued quantum mechanics does *not* satisfy.[35]

(3) "Every element of the physical reality must have a counterpart
    in the physical theory."

The EPR argument, as presented, makes use of a quantum state, that of Equation 1.23, involving continuous variables. However, it is readily and perhaps better suited to physical systems involving discrete observables. In particular, without affecting their argument, one can apply it to the bivalent observables associated with a less controversial choice of state, namely, the spin-singlet state $|\Psi^-\rangle$ (*cf.* Eq. 1.21) as was done by David Bohm [54]. The original EPR state has been viewed as somewhat artificial in nature and as possessing mathematical peculiarities, not the least of which is that, despite the intentions of the authors, the perfect correlations of measurement outcomes at distant locations they take to be characteristic of the state when the particles are well separated, are not, strictly speaking, so. Moreover, the singlet state has the valuable property of remaining of the same anticorrelation-bearing form when re-expressed in any orthonormal eigenstate basis obtainable from the computational basis by rotating the eigenstates of either subsystem Hilbert space by an arbitrary non-zero angle ξ [54].

Using $|\Psi^-\rangle$, then, the EPR argument can be stated in terms of two propositions.[36]

---

[34] A reconstruction of the argument, a discussion of Einstein's original, solo incompleteness argument, and an analysis indicating reasons for Einstein's dissatisfaction are given in [124].

[35] Bernard d'Espagnat's discussion of the relation of often neglected question of measurability to these considerations is illuminating ([126], Section 7-1-1).

[36] See [407]. For a modern version of the EPR argument based on the logic of quantum conditionals, also see [415].

(I) If an agent can perform an operation that permits it to predict with certainty the outcome of a measurement without disturbing the measured spin, then the measurement has a definite outcome, whether this operation is *actually* performed or not.

(II) For a pair of spins in the state $|\Psi^-\rangle$, there is an operation that an agent can perform allowing the outcome of a measurement of one subsystem to be determined *without disturbing* the other spin.

By measuring the quantity corresponding to the projector $P(|\uparrow\rangle)$ for one spin, the value of the quantity corresponding to the projector $P(|\downarrow\rangle)$ onto the orthogonal state is also fully specified. By (II), one can similarly obtain the values of the same two observables of the second spin without influencing it, due to the perfect anti-correlation between spins in the composite system state $|\Psi^-\rangle$. By (I), the values of the second spin are definite. However, one could just as well have measured the values of the quantities corresponding to a *different* basis, such as the conjugate diagonal basis, those corresponding to $P(|\nearrow\rangle)$ and $P(|\searrow\rangle)$. But these other values must then be definite as well. Therefore, the value of the states of both systems for *all* values of ξ must be definite. The description of the system of particles by the quantum state $|\Psi^-\rangle$ is thus seen to be incomplete in a specific way.

However, the EPR sense of completeness is far from universally shared, and certainly not by non-realist followers of the Copenhagen school; it can be viewed as unnecessarily strong, due to the appearance of counterfactual events in (1) and (I), as was noted, for example, by Daniel Greenberger, Michael Horne, Shimony and Anton Zeilinger (GHSZ), who carefully assessed the EPR argument in light of contemporary interferometry, as follows.

"In the reality assumption the phrase 'can predict' occurs. The phrase is ambiguous, because it may be understood in the strong sense, that data are at hand for making the prediction, or in the weak sense, that a measurement could be made to provide data for the prediction. EPR assume the weak sense, and indeed unless they did so they could not argue that an element of physical reality exists for all components of spin, those which could have been measured as well as the one that actually was measured... The preference for one rather than the other of these two interpretations of the phrase is not merely a semantical matter, but is an indication of a philosophical commitment. Bohr believed that the concept of reality cannot be applied legitimately to a property unless there is an experimental arrangement for observing it, whereas Einstein regarded this view as anthropocentric and maintained that physical systems have intrinsic properties whether they are observed or not." ([194])

Despite the EPR argument, in 1964 Heisenberg, for example, continued to say that

"we have a consistent mathematical scheme [that] tells us everything
that can be observed. Nothing is in nature which cannot be described
by this scheme." (reported in [176])

The claim of EPR that the quantum state is an incomplete description
serves primarily as part of an interpretative program of quantum mechanics,
in that it suggests that individual systems have intrinsic properties about
which the quantum state provides only a statistical description through an
ensemble of what are in reality differing individual systems. A completed
state, containing information not present in the quantum state, would then
provide a straightforward explanation of the perfect correlations predicted
by quantum mechanics, which it does not obviously explain. Bell's theorem,
discussed below, later showed that such an approach is destined to failure,
because the empirically verifiable predictions of quantum mechanics for pairs
of two-level systems contradict the collection of assumptions of EPR.

1.7 Problems with Hidden-Variables Models

In the course of his rigorous mathematical formulation of quantum mechanics,
John von Neumann addressed the question of a hidden-variables alternative to
the approach to quantum mechanics that Dirac and others had been considering. For precision, let us from hereon refer to hidden-variables approaches,
theories, and models rather than hidden-variables interpretations.[37] Advocates of hidden-variables approaches to explaining the behavior of quantum
systems consider the quantum-mechanical specification of physical states to
be in some way incomplete, sometimes but not always along the lines argued
by EPR. In his formulation, von Neumann explicitly considered quantum mechanics to describe what he identified as a Gibbs ensemble,

> "an ensemble of very many (identical) mechanical systems, each of
> which may have an arbitrarily large number of degrees of freedom,
> and each of which is entirely separated from the others, and does not
> interact with any of them," ([477], Section V.2)

as opposed to a Maxwell–Boltzmann ensemble, wherein one considers

> "(interacting) components of a single, very complicated mechanical
> system with many (only imperfectly known) degrees of freedom..."
> ([477], Section V.2)

Von Neumann then provided a 'no-go' theorem against a class of hidden-
variables states potentially underlying the quantum ensembles, which is outlined below after the following brief description of various sorts of hidden
variables that can be considered in principle.

---

[37] The term *interpretation of quantum mechanics* is here reserved for ways of understanding the standard quantum formalism or subsets thereof, rather than modi-
fications or extensions of it that *add* to the theory, as argued in Chapter 3.

The first effectively hidden-variables model had been considered by Louis de Broglie in the mid-1920's [121, 122], before the introduction of New quantum mechanics or von Neumann's formulation of the theory. The early hidden-variables theory outlined by de Broglie was later more fully developed, in particular, by Bohm who took hidden variables up again in the early 1950's [52]. The possibility of extra 'hidden variables' (or 'hidden parameters') had also been considered by Born almost immediately after he introduced the Born rule which would, with their introduction, lose its fundamental status [67]; Born was initially inclined to view the wave-function as a "guiding field" for the behavior of traditional particles.

> "[T]he guiding field, represented by the scalar function ψ of the coordinates of all the particles involved and the time, propagates in accordance with Schrödinger's differential equation...only a probability for a certain path is found, determined by the value of the ψ function." ([71], pp. 207-208)

The probabilistic interpretation of the distribution of complex-squares of state-vector amplitudes is commonly called the *Born hypothesis* or *Born rule,* although Born had not been fully committed to this view; it was Pauli who clearly interpreted the modulus square of the wave-function as a general probability density, as opposed to providing probabilities for energies or angular momenta of stationary states [337].[38] Advocates of recent interpretations of quantum mechanics, such as those of Henry Krips ([281], p. 118) and Hugh Everett III [163–165], often seek to *derive* the Born rule rather than to hold it as an independent theoretical postulate.

The 'hidden' variables are those parameters not in the quantum state that would ostensibly complete the specification of the full set of physical magnitudes for the system; they are not truly hidden in the sense of being in principle *inaccessible.* Hidden-variables theories of quantum phenomena can be formalized by reference to a putative *complete state,* λ, which in such theories is taken to render the quantum-mechanically pure state $\rho = |\psi\rangle\langle\psi|$ as a statistical state in the ordinary sense.[39] The space of conjectured completed states is, by definition, constrained by statistical principles, at least by the need to preserve the functional subordination in the space of quantum observables and the preservation of the convex structure of the set of quantum states. Such theories are offered in order to explain the inability of an experimenter to prepare ensembles of quantum systems having zero dispersion.

---

[38] For historical discussions of this understanding of the state-vector, see [27], pp. 48-49, and [256], Section 2.4.

[39] Note that any ensemble approach to quantum statistics in which precise values are attributed to all dynamical variables at all times has been referred to as a "Gibbs ensemble" interpretation [496] to which Schrödinger refers [394], somewhat differently from von Neumann's use of the term quoted above [477], *cf.* Section 3.3 here.

The mathematical setting for the consideration of hidden variables theories therefore involves an observable $\mathcal{O}$ taking the value $\mathcal{O}_{|\psi\rangle}(\lambda)$ in a quantum-pure state $|\psi\rangle$ described by the map $\mathcal{O}_{|\psi\rangle} : \Lambda \to \mathbb{R}$, Λ being the domain of possible values of hidden-variables λ. Assuming a probability measure μ that can be used to characterize the degree of ignorance as to the value of λ, so that $\{\Lambda, \mu\}$ constitutes a standard probability space, one has a probability density function $\sigma_{|\psi\rangle}$ for each $|\psi\rangle$. The probability that the hidden variable lies in the interval $\lambda + d\lambda$ is given by $\sigma_{|\psi\rangle}(\lambda)d\lambda$; the expectation value of $\mathcal{O}$ is

$$\langle \mathcal{O} \rangle_{|\psi\rangle} = \int_{\lambda} \mathcal{O}_{|\psi\rangle}(\lambda)\sigma_{|\psi\rangle}(\lambda)d\lambda . \tag{1.43}$$

In this construction, the values of quantum observables are random variables over $\{\Lambda, \mu\}$. As Michael Redhead has pointed out, some hidden-variable treatments hold that the values of the quantum observables should not depend on quantum states but only on the hidden variables; one can incorporate the quantum state into the hidden variable, but this only renders the term *hidden variable* a greater misnomer (*cf.* [371], p. 47). Bell suggested the term *beable* as an alternative [24], which has the shortcoming of suggesting potentiality rather than actuality [270]. Shimony has offered the more appropriate term *existent* [409]. This suggests a yet more precise term, *existent magnitude*.

The simplest sort of hidden-variables model is that in which λ provides definite values to all physical magnitudes of a quantum system that correspond to the quantum pure states $P(|\psi\rangle)$. Such models, sometimes referred to as *non-contextual* hidden-variables theories, determine the value of a quantity obtained by measurement, regardless of which other quantities are simultaneously measured along with it, and specify the complete state of the overall system composed of the measured system together with the measurement apparatus. The *contextual* hidden-variables models, which were introduced by Bell [23], require not only λ but also other relevant parameters associated with the conditions of their measurement to assign each projector a definite value.[40] Other categories pertaining to hidden-variables theories are the following. *Algebraic contextuality* involves the specification of any other quantities that are measured jointly with the quantity of interest. *Environmental contextuality* involves there being some non-quantum-mechanical interaction between the system subject to measurement and its environment that occurs before measurement and that influences the value of the measured magnitudes. The *stochastic hidden-variables theories* require the hidden variables and experimental parameters to specify the probabilities of measurement outcomes corresponding to projectors. In the case of *non-local* hidden-variables theories, the action on a subsystem of a composite system may have an immediate effect on another, spacelike-separated system.

---

[40] Thus, for example, Stanley Gudder has considered the context to be a maximal Boolean subalgebra of the lattice of quantum Hilbert subspaces [200].

One can imagine a situation wherein the measurement of a given quantity attributed to a quantum-pure ensemble of systems gives different values even though all members of the ensemble have the same specification. Then, either there exist different subensembles distinguished by some hidden variable outside of the quantum description, or the measured dispersion of values arises directly from nature. In the former case, there must be as many subensembles as there are different results. The "no-go" theorem proved by von Neumann addresses hidden-variables theories through an analysis of dispersion-free states [477]; a state $|\psi\rangle$ is *dispersion-free* when the dispersion $\mathrm{Disp}_{|\psi\rangle}O$ of all physical magnitudes O is zero when the system is in it. This pertains, for example, to the Naive (realist) interpretation of quantum mechanics.[41] The theorem can be viewed as seeking to show that no dispersion-free descriptions exist that enable a hidden-variables description of quantum phenomena.

The "no-go" theorem states that no hidden-variables model exists that satisfies the following assumptions about operators in relation to physical properties the behaviors of which are to be described by the model.

(1) Any real linear combination of three or more Hermitian self-adjoint operators represents a measurable quantum magnitude.

(2) The corresponding linear combination of subsystem expectation values is the expectation value of that combination of operators.

However, von Neumann's result is less than definitive regarding the existence of a well defined hidden-variables theory. Although the second condition seems natural to impose on the dispersion-free states because it is satisfied by quantum-mechanical operators, there is no a *priori* support for this condition for the individual dispersion-free states, which are to be averaged over.

1.8 Bell's Theorem and Independence Conditions

Before Bell's investigations of the constraints imposed by local causality had begun, Einstein stressed the importance of locality for physics as follows.

> "Unless one makes this kind of assumption about the independence of the existence (the "being-thus") of objects which are far apart from one another in space—which stems in the first place from everyday thinking—physical thinking in the familiar sense would not be possible. It is also hard to see any way of formulating and testing the laws of physics unless one makes a clear distinction of this kind." ([72], p. 170)

Bell drew out the implications of such an assumption in the context of joint measurements, greatly clarifying the difference of the behavior of composite physical systems manifesting internal correlations that might be produced

---

[41] This interpretation is discussed in detail in Section 3.5

by hidden variables from the correlations predicted by quantum mechanics that they are incapable of simulating. In particular, he proved a now famous theorem involving an inequality that the correlations of local-causal theories must obey and that those of theories violating local causality need not.

Benefitting from the initial investigation of EPR by contemplating the implications of their premises for predicted joint measurement statistics, Bell first defined a specific sense of locality: A (hidden-variables) model describing a bipartite system is (*Bell-*)*local* if a definite probability is assigned to the event of there being a positive measurement outcome for every one of the bivalent physical magnitudes of each subsystem by the complete state of the joint system independently of measurements performed on the other subsystem, even when the subsystems are relatively spacelike separated. His theorem considers a putative complete state λ of the pair of particles that fully specifies *all* the "elements of physical reality" present in the pair at any given instant. As Shimony has tersely put it,

> "[This complete state] determines the results of measurements on the system, either by assigning a value to the measured quantity that is revealed by measurement regardless of the details of the measurement procedure, or by enabling the system to elicit a definite response whenever it is measured, but a response which may depend on the macroscopic features of the experimental arrangement or even on the complete state of the measured system together with that arrangement." ([417])

A specific inequality was shown to hold for local models based on such states. The term *Bell-type theorem* now refers to the collection of results having in common with Bell's original result the demonstration of the impossibility of common-cause explanations of all quantum mechanical predictions, that is, explanations of quantum correlations under the assumption of the Bell locality condition (*cf.* [244], Section 8.7).

The measured bivalent physical magnitude of each subsystem was taken by Bell for specificity to be the spin along a single direction as in Bohm's analysis. Bell then considered a probability measure, $\mu(\lambda)$, on the entire space Λ of parameters providing complete states λ. The expectation values, $E^{\mu(\Lambda)}$, of the bivalent quantities, as random variables, were therefore taken to be

$$E^{\mu(\Lambda)}(\hat{\mathbf{n}}_1, \hat{\mathbf{n}}_2) = \int_\Lambda A_\lambda(\hat{\mathbf{n}}_1) B_\lambda(\hat{\mathbf{n}}_2) d\mu(\lambda) , \qquad (1.44)$$

where $\lambda \in \Lambda$, and $A_\lambda(\hat{\mathbf{n}}_1)$ and $B_\lambda(\hat{\mathbf{n}}_2)$ are measurement outcomes along specific directions $\hat{\mathbf{n}}_1$ and $\hat{\mathbf{n}}_2$ on subsystems A and B, respectively. This leads straightforwardly to the *Bell inequality*,

$$\left| E^{\mu(\Lambda)}(a, b) - E^{\mu(\Lambda)}(a, c) \right| \leq 1 + E^{\mu(\Lambda)}(b, c) , \qquad (1.45)$$

where $\{a, b, c\}$ is any set of three angles specifying directions of measurement in planes normal to the line of particle propagation; see Figure 1.3 [22, 24].

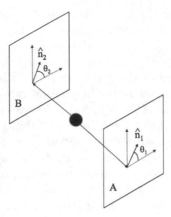

Fig. 1.3. Schematic of apparatus for a Bell-type inequality test using polarization interferometry. Two non-orthogonal states parameterized by angles θ_1 and θ_2, respectively, from a set of specific values are measured in planes normal to the axis of particle (counter-)propagation in two laboratories A and B.

As Jon Jarrett has explained, Bell's locality condition can be identified with the conjunction of two logically independent conditions [257], later descriptively renamed *parameter independence* (PI), which regards the choice of measurement in the distant laboratory, and measurement *outcome independence* (OI) by Shimony [407]. PI is the condition that the probability of a measurement outcome in one laboratory is independent of the *particular* measurement chosen to be made in the other laboratory, once λ is determined. OI is the condition that the probability of a measurement outcome in one laboratory is independent of the measurement outcome found in the other laboratory, although possibly dependent on the specific choice of measurement made in the other lab and dependent on λ. A background assumption is that the probability measure μ is independent of the specific choices of measurements in the two laboratories ([419], p. 118).

Subsequent inequalities were obtained based on weaker assumptions that still suffice for distinguishing Bell local from Bell non-local correlations. For example, the *Clauser–Horne* (CH) *inequality* is the straightforward algebraic result that probabilities constrained by Bell's locality condition obey the relation

$$-1 \leq p_{13} + p_{14} + p_{23} - p_{24} - p_1 - p_3 \leq 0 \tag{1.46}$$

as well as all inequalities resulting from permutations of the above indices, where p_1 and p_3 are the probabilities that the first particle is found along the *first* of the four directions $\{a, b, c, d\}$ and the second particle is found along the *third* direction; p_{ij} stands for the joint probability of finding the first particle along the direction i and the second particle along direction j, $1 \leq i, j \leq 4$. These particles correspond to quantum two-level systems. No special restrictions are placed on the complete-state space Λ or on the probability distribution used in the derivation of the CH result, for example. Indeed, the CH inequality follows from the elementary algebra of probabilities which, as such, lie in the interval $[0, 1]$. Nonetheless, the proofs of the above results do continue to assume that the experimental arrangement prescribes a probability distribution over state specifications that provides the above probabilities as expectation values over complete states.

In practical experimental situations, it is generally impossible to have control of the putative complete state λ of such a composite quantum system, greatly restricting the susceptibility of the early Bell-type inequalities to empirical testing. For this reason, John Clauser, Horne, Shimony, and Richard Holt (CHSH) modified the assumptions on the form of measured quantities so as to allow for meaningful Bell locality tests in any experimental arrangement sufficiently similar to that shown in Figure 1.3. The practically testable inequality CHSH obtained, now known as the *CHSH inequality*, can be written simply as

$$|S| \leq 2 , \tag{1.47}$$

for $S \equiv E(\theta_1, \theta_2) + E(\theta_1', \theta_2) + E(\theta_1, \theta_2') - E(\theta_1', \theta_2')$, where the Es are expectation values of the products of measurement outcomes given parameter values θ_i and θ_i' of the two different directions $\hat{n}_i$ for the same laboratory i relative to a reference direction as shown [107]. The correlation coefficients contributing to S can be expressed in terms of normalized experimental detection rates. The CHSH inequality is the Bell-type inequality to which reference is now most often made in the physics literature.

The first experiment (later) considered to have demonstrated non-classical behavior in the sense of violating the restrictions of local causality was that reported in 1950 by Wu and Shaknov [509], for two-level spin systems in a singlet state [54]. A maximum violation of this inequality by a factor of $\sqrt{2}$ beyond its bound can be achieved, for example, for a system in the fully entangled state $|\Phi^+\rangle = \frac{1}{\sqrt{2}}(|\uparrow\uparrow\rangle + |\downarrow\downarrow\rangle)$, where $|\uparrow\rangle$ indicates, for example, photon polarization oriented along one of the orthogonal axes of the plane indicated in Figure 1.3 and $|\downarrow\rangle$ indicates polarization oriented along the other, by performing measurements with $\theta_1 = \frac{\pi}{4}$, $\theta_1' = 0$, $\theta_2 = \frac{\pi}{8}$, and $\theta_2' = \frac{3\pi}{8}$; these angles represent steps of $\frac{\pi}{8}$ radians, where the two angles in each lab, that is, on each side of the apparatus differ by $\frac{\pi}{4}$ radians [417]. Since its introduction, the extent of the empirical value of $|S|$ beyond 2 has served experimentalists as a figure of merit for sources of non-separable quantum states.

Violations of Bell-type inequalities by microscopic systems have been demonstrated many times since in a number of contexts; results systematically show the violation of Bell-locality bounds in accordance with quantum-mechanical predictions [352]. Bell-type inequalities for pairs of systems of arbitrarily high countable dimension have also been found, showing that it is not only for two-level system pairs that correlations are constrained by local causality [114]. Because they rely on fundamental properties of probability, the expressions bounding the probabilities and expectation values in these inequalities can be derived, for example, by enumerating all conceivable classical possibilities. These can be viewed as extreme points spanning the classical correlation polytopes, the faces of which are expressed by Bell-type inequalities. These and other polytopes are discussed in detail in Chapter 4.

Because Bell-type inequalities involve sums of (joint) probabilities and expectation values, in order to show the incompatibility of the predictions of quantum mechanics with these inequalities, the quantum counterparts and expectation values can simply be substituted for the probabilities and expectation values appearing in them. Even though Bell inequality violation was empirically demonstrated without a locality "loophole"—which would be due to a possible dependence of outcomes of detectors on state sources—by the experiments of Alain Aspect *et al.* [11, 12] as well as later stricter tests (*e.g.* [487]), other potential experimental loopholes remain. However, their closing is not expected to have an effect on the results obtained.

Perhaps the greatest shortcoming of the Bell-type inequalities is that their violation is limited in its value for testing state *entanglement*. Indeed, it is unknown whether one will find them violated for many non-separable mixed states. Some states *can* be transformed so as to violate a Bell-type inequality by appropriate operations,[42] which serve to "distill" correlations; states that can be made to violate a Bell inequality in this way are referred to as *distillable states*. Importantly, all entangled states of pairs of spin-1/2 particles are distillable [239]. However, there are higher-dimensional bipartite states that are incapable of being so distilled, which are referred to as *bound entangled*. In any event, violations of Bell-type theorems strongly point out the surprising nature of quantum mechanics, and have motivated physicists and philosophers to seek alternative interpretations of the theory, which are discussed in Chapter 3.

1.9 Conditions Contradicted by Quantum Mechanics

In the case of perfectly correlated states of composite systems, the outcome of a measurement on one system can be predicted with certainty given the outcomes of appropriate measurements on other subsystems. Remarkably, a decade after Aspect's tests of the CHSH inequality, it was shown by Greenberger, Horne, Shimony, and Zeilinger (GHSZ) that the premises of the Einstein–Podolsky–Rosen paper become *inconsistent* when applied to systems possessing three or more subsystems, even for the cases involving such perfect correlations [194].

The GHSZ demonstration shows that the incompatibility of the EPR assumptions with quantum mechanics is stronger than that indicated by the

---

[42] In particular, by CLOCC operations, see Section 1.10.

violation of the Bell and CHSH inequalities, in that in the case of a pair of two-level systems there is no internal contradiction at the level of perfect correlations. Indeed, Bell produced an explicit model for the case of a pair of spin-1/2 particles demonstrating the consistency of the EPR conditions with the perfect correlations predicted by quantum mechanics [23]. Furthermore, the contradiction between quantum mechanical predictions and the Bell and CHSH inequalities are expressions violated only by *statistical* predictions of quantum mechanics, rather than by individual events.

In the lead up to the exceptionally clear exposition of GHSZ, Greenberger, Horne, and Zeilinger (GHZ) demonstrated the inconsistency in a new way in systems consisting of three or more correlated spin-1/2 particles [195]. Because this showed that the incompatibility of quantum mechanics with the EPR assumptions arises at the level of perfect correlations rather than statistical predictions and did not require the use of an inequality, these results are often referred to as "Bell's theorem without inequalities." For example, the correlations predicted for the outcomes of measurements of systems in the state

$$\frac{1}{\sqrt{2}}(|\uparrow\uparrow\uparrow\rangle - |\downarrow\downarrow\downarrow\rangle) \tag{1.48}$$

were shown to contradict the EPR assumptions as follows [44, 194, 195]. The state of Equation 1.48, now referred to as the *Greenberger–Horne–Zeilinger* (GHZ) *state*, is an eigenvector of all the operators $\sigma_x \otimes \sigma_y \otimes \sigma_y$, $\sigma_y \otimes \sigma_x \otimes \sigma_y$, $\sigma_y \otimes \sigma_y \otimes \sigma_x$, with corresponding eigenvalue $+1$ and of the operator $\sigma_x \otimes \sigma_x \otimes \sigma_x$ with corresponding eigenvalue -1. Measurement of the observables σ_x or σ_y on any two of the three two-level systems involved allows the outcome for a corresponding third measurement to be inferred. The EPR assumptions would then allow one to assign definite values to the local quantities $\sigma_x^{(i)}$ and $\sigma_y^{(i)}$, described by a function taking $\sigma_x^{(i)}$ and $\sigma_y^{(i)}$ each to the set $\{-1, +1\}$, where the superscript indicates the subsystem in question. There is, therefore, an immediate contradiction between the EPR premises and the perfect correlations predicted by quantum mechanics.

Prior to the GHZ paper, Paul Heywood and Redhead had also pointed out a contradiction between the conjunction of two conditions related to Bell locality with quantum mechanics, by considering a pair of three-level spin subsystems in a state such that the compound system they jointly form has zero spin in all directions, namely, $|\Psi\rangle = \frac{1}{\sqrt{3}}(|+\rangle|-\rangle + |-\rangle|+\rangle - |0\rangle|0\rangle)$, where $+, -, 0$ indicate the corresponding spin-projection eigenvalues [228]. In this context, they introduced the notion of a *locally maximal* magnitude for a bipartite system and two quantities $A^{(i)}$, respectively corresponding to the operators $A^{(1)} \otimes \mathbb{I}$ and $\mathbb{I} \otimes A^{(2)}$, where $A^{(i)}$ is maximal for the subsystem $i = 1, 2$. Such magnitudes are non-maximal for the composite system.

The two conditions introduced by Heywood and Redhead are *ontological locality* (OLOC) and *environmental locality* (ELOC). The first is the requirement that if physical magnitudes P and P′ are locally maximal for the first subsystem and Q and Q′ are so for the second, then the magnitude P⊗𝕀 is a unique magnitude for the composite system, as is 𝕀⊗Q. The second condition is the requirement that local magnitudes cannot be changed by changing a spacelike separated piece of measuring apparatus. More specifically, then, OLOC is the condition that

"If $\mathcal{H}_1$ and $\mathcal{H}_2$ are the Hilbert spaces for two spatially separated systems and $(A \otimes \mathbb{I})$ is a locally maximal operator, then $[A \otimes \mathbb{I}]^{|\phi\rangle}_{\{X\}} = [A \otimes \mathbb{I}]^{|\phi\rangle}_{\{Y\}}$ for any state $|\phi\rangle$ of the joint system where X and Y are both maximal operators on $\mathcal{H}_1 \otimes \mathcal{H}_2$ and $[X, Y] \neq 0$." ([371])

(The notation $[O]^{|\phi\rangle}_{\{P\}}$ means that the value of magnitude O in the quantum state $|\psi\rangle$ depends only on an equivalence class of one-to-one functions generated by maximal magnitude P.) This condition is designed to ensure that locally maximal magnitudes are not 'split' by contextuality of an ontological nature in relation to the specification of differing maximal magnitudes for the composite system. Precisely stated, the other condition, ELOC is that

"If S_1 and S_2 are two spatially separated systems, Q is [a magnitude] for S_1 and X and Y maximal [magnitudes] for the joint system $(S_1 + S_2)$, then if the difference between an apparatus set to measure X and one set to measure Y is only in the setting of that part of it located at S_2, $[Q \otimes \mathbb{I}]^{|\phi\rangle}_{\{X\}}(X) = [Q \otimes \mathbb{I}]^{|\phi\rangle}_{\{X\}}(Y)$." ([371])

This second condition is designed to ensure that the value possessed by a local magnitude cannot be changed to altering the arrangement of a remote element of the total measuring apparatus. Heywood and Redhead then pointed out that ELOC has no implication for local causality except in the presence of a condition like OLOC. They then successfully argued that a theory wherein physical magnitudes take only values that occur with non-zero probability comes into contradiction with the conjunction of the two conditions of 'ontological and environmental locality.'

As shown later in Chapter 4, the various conditions so far described here can be linked to information-theoretic situations and precisely related to entangled quantum states. In order to set the stage for this later connection, it is necessary to consider modern characterizations of several of the physical situations previously considered. These involve, in particular, formalizations of local quantum operations subsuming the sort first considered by EPR, Bell, and others engaged in "experimental metaphysics." Like the quantum-logical investigation of quantum mechanics discussed in the following chapter, this is carried out through further moves of a fairly technical nature, which are now briefly summarized, rounding out this largely introductory chapter.

1.10 Operations, Communication, and Entanglement

The lack of local causality associated with quantum entanglement in bipartite pure states is exhibited in systems with components located in spacelike-separated regions. In the language of quantum information science, such distinct regions are referred to as *laboratories* and the physical systems located in them are attributed distinct Hilbert spaces, so that the corresponding compound systems are describable by states in the Hilbert space that is formed by taking the tensor product of the Hilbert spaces of the components; each laboratory is taken to contain an agent physically described as a system that is capable of performing quantum operations on subsystems within its laboratory and that has the potential to communicate with agents in other laboratories. Communication and its relation to quantum entanglement is discussed in this section assuming at most *classical* communication between agents.[43] Investigations of this sort have led to an understanding of quantum entanglement as an information-processing resource, which has been seen to have a significance beyond its historical one related mainly to Bell inequality violation.

The fundamental task of communication between agents is the communication of information with the greatest possible accuracy. A classical information source can be defined via a sequence of probability distributions over sets of symbols produced in a number of emissions by a transmitter into a communication channel leading to a receiver.[44] In order to represent realistic physical situations, in addition to taking into account the ability of agents to communicate information among themselves on the basis of which, for example, quantum signal states can be locally manipulated by them, one must take into account the ability of natural environments to induce transformations of quantum states. The latter, despite being quantum processes, are not directly describable by the standard unitary evolution appropriate for closed systems when considering limited regions of the universe. Natural physical environments typically dephase or in other ways decohere subsystem states, with implications for their ability to efficiently perform communication and information processing tasks, and in many cases influence state non-locality and entanglement (*cf., e.g.*, [7]).

When a quantum system is in contact with an environment, it is an open rather than a closed system. Fortunately, both situations can often be readily described by the class of completely positive trace-preserving (CPTP) linear transformations, $\rho \to \mathcal{E}(\rho)$, often called *operations*, taking statistical operators

---

[43] More general situations, where channels for the communication of quantum information are also sometimes considered to be available, as discussed in Chapter 4.

[44] This characterization is provided in full detail in Chapter 4.

to statistical operators, each described by a *superoperator*, $\mathcal{E}(\rho)$, satisfying the following conditions.[45]

(1) $\mathrm{tr}[\mathcal{E}(\rho)]$ is the *probability* that the transformation $\rho \to \mathcal{E}(\rho)$ occurs.

(2) $\mathcal{E}(\rho)$ is a linear convex map on statistical operators, that is,

$$\mathcal{E}\left(\sum_i p_i \rho_i\right) = \sum_i p_i \mathcal{E}(\rho_i), \qquad (1.49)$$

p_i being probabilities, so that $\mathcal{E}(\rho)$ extends uniquely to a linear map.

(3) $\mathcal{E}(\rho)$ is a completely positive (CP) map.

A linear map $L : \mathcal{L}(\mathcal{H}) \to \mathcal{L}(\mathcal{H})$, where $\mathcal{L}(\mathcal{H})$ is the space of bounded linear operators on $\mathcal{H}$, is said to be *positive* if $L(\mathcal{O}) \geq \mathbb{O}$ for all $\mathcal{O} \geq \mathbb{O}$, that is, all $\mathcal{O} \in B(\mathcal{H})$ for which $\langle \psi | \mathcal{O} | \psi \rangle \geq 0$ for all $|\psi\rangle \in \mathcal{H}$, where a map $L : \rho \mapsto L(\rho)$ is linear if $L(\rho) = p_1 L(\rho_1) + p_2 L(\rho_2)$ for $\rho = p_1 \rho_1 + p_2 \rho_2$.[46] A positive L is *completely positive* (CP) if, in addition, any $\mathbb{I}_N \otimes L \in B(\mathbb{C}^N \otimes \mathcal{H})$ is positive, for all $N \in \mathbb{N}$. A map $\mathcal{E} : \rho \to E(\rho)$ satisfies the three above conditions if and only if it can be written

$$\mathcal{E}(\rho) = \sum_i K_i \rho K_i^\dagger, \qquad (1.50)$$

for some set $\{K_i\}$ of not necessarily Hermitian linear operators for which $\mathbb{I} - \sum_i K_i^\dagger K_i \geq \mathbb{O}$ [279], an *operator-sum* representation. The K_i are the *decomposition operators* (or *operation elements*); $\{K_i\}$ is the *operator decomposition* of $\mathcal{E}(\rho)$.[47]

Operations on composite quantum systems have been classified as follows. The class of local operations ("LO") is that of operations that are carried out on individual subsystems located within the laboratories of their corresponding agents, including unitary operations as well as measurements occurring with the prescribed quantum likelihoods. The operations of classical communication ("CC") are information transfer acts between agents in separate laboratories carried out via classical means, and may be in one or two directions. Local operations together with those of classical communication ("LOCC")

---

[45] In the quantum information science literature, this term has been used more loosely to apply in cases where, for example, operator traces are *not* preserved. For example, see the discussion of LOCC operations below.

[46] The relation $\geq$ is defined as follows. $A \geq B$ if $A - B \geq \mathbb{O}$, where $\mathbb{O}$ is the zero operator. It is an ordering on the set of self-adjoint bounded operators.

[47] The *trace preserving* (TP) property for $\mathcal{E}(\rho)$, $\mathrm{tr}(\mathcal{E}(\rho)) = \mathrm{tr}(\rho)$ is equivalent to $\sum_i K_i^\dagger K_i = \mathbb{I}$, which is a completeness relation. Any CPTP map can be viewed as the result of a unitary transformation acting in a larger Hilbert space containing $\mathcal{H}_1$: for a state $\rho \in \mathcal{H}_1$ in the larger space $\mathcal{H}_1 \otimes \mathcal{H}_2$, there is a state $\rho' \in \mathcal{H}_2$ such that $\rho = \mathrm{tr}_2(\rho \otimes \rho')$. Thus, any CPTP map can be given in the form $\mathcal{D}(\rho) = \mathrm{tr}_2(U(\rho \otimes \rho')U^\dagger)$, for some unitary operator, U, where ρ' describes, for example, the state of the environment [279].

are operations on quantum systems by agents acting locally who are also capable of classically communicating. The distinction between LOCC and LO is significant in that classical communication between agents allows the local operations of an agent to be conditioned on outcomes of measurements that might be carried out by other agents.

LOCC operations consist of combinations of local unitary operations, local measurement operations, and the addition or disposal of parts of the total system. Those operations that are trace preserving are referred to as *LOCC protocols*.[48] In the case of operations on a number of copies of a quantum system for any of these classes, the adjective "collective" is added and the above acronyms are given the prefix "C," for example, the CLOCC class is that of collective location operations and classical communication. In cases where transformations are not achievable deterministically, but rather only with some probability, they are considered *stochastic operations* and the adjective "stochastic" is added as well as the prefix "S," as in SLOCC.

When considering two-component systems described by statistical operators ρ_{AB}, the corresponding agents are customarily labeled A and B as done occasionally above, indicating the native subsystems or laboratories of these agents, considered physically, that is, as automata, unless otherwise indicated. The actions of two agents capable of communicating can be correlated in ways describable as global operations in both their laboratories that are not necessarily describable as direct products of local operations. Because violations of locality in the traditional sense are insufficient for the characterization of entanglement in an arbitrary situation, despite their value in the investigation of entangled quantum states, the distinction between the violation of local causality and quantum state entanglement should always be borne in mind.

Any quantum operation O_{AB} is implementable by a pair of parties via LOCC when it is separable [468], that is, when it can be written as a convex sum of local operations, $O_{AB} = \sum_i p_i A_i \otimes B_i$, which guarantees that individual operations are effectively carried out independently in the two laboratories with probabilities p_i, although the converse is not true, which accounts for the possible local influence of communicated classical information. The class of quantum states that can be prepared from a product state by LOCC is known as the *locally preparable* class. Although LOCC enables correlated mixed states to be created from previously uncorrelated states, it does not in itself enable the creation of *entangled* states.

Operations are, in large part, transformations of quantum states that can be described in the mathematical language of group theory, as shown further below. In addition to the conventional requirement that a measure of entanglement be non-negative and normalized in the sense that it be unity for the Bell states, a fundamental pair of monotonicity conditions has been put forth for any candidate, below indicated generically as $E_X(\rho)$, to be good a measure of entanglement in contemporary treatments. This defines the class of

---

[48] It is important to note that a LOCC operation is *not* necessarily a TP operation.

entanglement monotones, which are functionals that characterize the strength
of genuinely quantum correlations through the requirement that no state be
convertible by local operations and classical communication (LOCC) to a
state having a greater value of the monotone. In particular, a quantity $E_X(\rho)$
is called an entanglement monotone if it satisfies

$$E_X(\rho) \geq \sum_i p_i E_X(\rho_i) \,, \tag{1.51}$$

and

$$E_X\left(\sum_i p_i \rho_i\right) \leq \sum_i p_i E_X(\rho_i) \,, \tag{1.52}$$

for all local operations giving rise to states ρ_i with probabilities p_i, where
at the end of the LOCC operation i, classical information is available with
probability p_i and the state is ρ_i [469].

The first of the above two conditions, sometimes referred to as the *fundamental postulate* of entanglement theory, requires *monotonicity on the average
for each local operation*. The second condition requires $E_X(\rho)$ to be a convex
function that is monotonic *under mixing*, that is, the discarding of available
information, and provides mathematical convenience; it is sometimes relaxed.
The Schmidt measure E_S and negativity $\mathcal{N}$, discussed above in Section 1.4,
are examples of entanglement monotones for bipartite quantum systems.

Consider two sets of entanglement monotones, $E_l^\Psi = \sum_{i=1}^n |a_i|^2$ and
$E_l^\Phi = \sum_{i=1}^n |b_i|^2$, where $l = 1, \ldots, n$, respectively obtained from the Schmidt
decomposition of two bipartite states $|\Psi\rangle, |\Phi\rangle$ and having n components with
Schmidt coefficients a_i, b_i. The pure state $|\Psi\rangle$ can be transformed with cer-
tainty by local transformations to the pure state $|\Phi\rangle$ if and only if $E_l^\Psi \geq E_l^\Phi$
for all $l = 1, \ldots, n$ [471]. The following conditions are therefore now commonly
required of acceptable measures of bipartite entanglement E_X on all states
ρ_{AB} of a pair of systems.

(i) $E_X(\rho_{AB}) = 0$ if ρ_{AB} is separable.
(ii) $E_X(\rho_{AB})$ is invariant under *all* local unitary operations $U_A \otimes U_B$,
    that is, $E_X(\rho_{AB}) = E_X\big((U_A \otimes U_B)\rho_{AB}(U_A \otimes U_B)^\dagger\big)$.
(iii) $E_X(\rho_{AB})$ cannot be increased by *any* LOCC transformation,
    that is, $E_X(\rho_{AB}) \geq E_X\big(\Theta(\rho_{AB})\big)$, where $\Theta(\rho_{AB})$ is a CPTP map.

The necessity of the first condition is obvious: separable states, specified by
Equation 1.24, are by definition *not* entangled. Conditions (ii) and (iii) are
necessary for entanglement to be considered a global property of quantum
systems; they render impossible the creation or distribution of entanglement
via LOCC *alone*. The last two conditions accord with each other because
local unitary operations are CPTP maps that can be inverted by local unitary
operations.

The proper asymptotic behavior of entanglement monotones in the limit
of many state copies, which is used in the analysis of entanglement as an operational resource, requires that further conditions be imposed. For example,

the following. (iv) The entanglement of n copies of a state ρ_{AB} is n times the entanglement of one copy, $E_X(\rho_{AB}^{\otimes n}) = nE_X(\rho_{AB})$, in particular for the standard case of n Bell singlets, each of which are conventionally taken to have entanglement equal to unity, that is, are n e-bits. An operational procedure illustrating this relation is discussed below. With this fourth condition, known as *partial additivity*, the pure state entanglement of bipartite quantum systems is *uniquely* described by

$$E(|\Psi\rangle_{AB}) \equiv S(\rho_X) = -\text{tr}(\rho_X \log_2\rho_X) \,, \qquad (1.53)$$

the von Neumann entropy, where ρ_X $(X = A, B)$ is the (reduced) statistical operator of either one of the two subsystems of the compound system in state $|\Psi\rangle_{AB}$ [358]. Full additivity would require that $E(\rho \otimes \sigma) = E(\rho) + E(\sigma)$; because bound entanglement, discussed below, may be activated, this condition is often viewed as unwarranted. The two conditions are sometimes simply replaced by the condition that, for pure states, the measure reduce to this entropy.

The von Neumann entropy does have the property of being additive on pure states of the composite system: using this entropy of the subsystem reduced states as E_X, and labeling the individual particles of laboratories A and B in the n copies as A_i, B_i, the entanglement *additivity property* is

$$E\big(|\Psi\rangle_{A_1 B_1} \otimes |\Psi\rangle_{A_2 B_2} \otimes \cdots \otimes |\Psi\rangle_{A_n B_n}\big) = \sum_{i=1}^{n} E(|\Psi\rangle_{A_i B_i})$$

for all pure states $|\Psi\rangle_{AB}$. In the case of mixed states, additivity is desirable but must be explicitly imposed and one refers to the *additivity conjecture* for them.[49] For mixed states ρ_{AB} of a pair of two-level systems A and B, a good entanglement measure is the entanglement of formation, E_f, defined via the convex-roof construction as the minimum average marginal entropy of the single two-level-system reduced states for all possible decompositions of ρ_{AB} as a mixture of pure subensembles each described by a state $P(|\Psi_i\rangle_{AB})$, that is,

$$E_f(\rho_{AB}) = \min_{\{p_i, |\Psi_i\rangle\}} \sum_i p_i E(|\Psi_i\rangle_{AB}) \,, \qquad (1.54)$$

where $\{p_i, P(|\Psi_i\rangle)\}$ represents a decomposition of ρ_{AB}.[50] Note that, although it can be expressed directly in terms of the von Neumann entropy $S(\rho_A)$, the form provided here allows for explicit reference to the states of the pertinent pure sub-ensembles of two-level system pairs.

---

[49] Note, however, that a uniqueness theorem not assuming additivity has also been produced [469]. In the case of larger systems involving multiple parties, some of the above conditions must be slightly modified.

[50] This quantity is analogous to the total energy of thermodynamics, something discussed in detail in Chapter 4.

Again, it is possible to obtain pure entangled states that violate Bell inequalities beginning with mixed states that do *not* violate a Bell inequality by using entanglement distillation. Such entanglement distillation is accomplished using entanglement purification, that is, the local manipulation of a number of copies of a quantum state. Accordingly, the class of distillable states is that of the states n copies of which can be converted via LOCC into nr pure maximally entangled states, where $r > 0$. The pure states form two distinct classes, the preparable and the distillable states, corresponding to the product states and all other states, respectively. Distillation can mitigate the detrimental effect of quantum decoherence. In the quantum information processing context, entanglement has been viewed as a resource similar to energy that can take several interchangeable forms and can be distributed among quantum systems. Such problems were among the first considered in quantum information science, which has emerged naturally from the study of physical correlations in the study of quantum systems.

In order to find exactly how much of the resource of bipartite entanglement they share, two parties can concentrate Bell singlet states between them; they can distill, by CLOCC from a number, n, of copies of an initial bipartite pure (not necessarily *maximally*) entangled state $|\Phi\rangle_{AB}$, the greatest number $k < n$ of singlet states possible: $|\Phi\rangle_{AB}^{\otimes n} \to |\Psi^-\rangle_{AB}^{\otimes k}$, the resulting state clearly containing k e-bits of entanglement. Distillation can be carried out with an efficiency given by the von Neumann entropy $S(\rho)$, where ρ is the reduced statistical operator of a subsystem of AB [33]. This is a reversible process, in the sense that there is an asymptotic scheme in which the inverse conversion $|\Psi^-\rangle_{AB}^{\otimes k} \to |\Phi\rangle_{AB}^{\otimes n}$ can be performed, again via CLOCC, with *equal* efficiency. The monotonicity condition (iii) implies that no entanglement distillation scheme can perform better than such an asymptotic scheme. One thus sees that the original shared state contained k e-bits of entanglement, shared between the two parties. The limit $n \to \infty$ is associated with the use of a standard unit of entanglement to describe the transformation process associated with condition (iv); taking this limit provides one with a well-defined ratio characterizing the conversion process of a whole number of states to a whole number of states because the entanglement of formation may take any rational value.

Entanglement, like heat energy, cannot be increased by local operations on remote subsystems. The reversible transformations, consisting of only local operations that transform one entangled state into another, produce the analogue of the Carnot cycle. This highly suggestive analogy has stimulated an investigation into the depth of the similarities between quantum information theory and thermodynamics, something taken up in detail in Section 4.8. The Bell state entanglement resource allows *global* quantum operations to be performed with the aid of quantum teleportation, which is also discussed in detail in Chapter 4. The associated functional, the *entanglement of distillation*, $D(\rho_{AB})$, is defined as the *maximum fraction of singlets* that can be extracted, that is, distilled from n copies of ρ_{AB} by the CLOCC transformation $\rho_{AB}^{\otimes n} \to P(|\Psi^-\rangle)^{\otimes k}$ in the asymptotic limit as $n \to \infty$:

$$D(\rho_{AB}) = \limsup_{n \to \infty} \left(\frac{k}{n}\right), \tag{1.55}$$

where k depends on n. This quantity can be viewed as analogous to thermodynamical free energy, and so is sometimes called the *free entanglement*. It expresses, for example, the utility of a given entangled mixed state for performing quantum teleportation.

According to condition (iii), it must be the case that $D(\rho_{AB}) \leq E_f(\rho_{AB})$, which reflects the irreversibility character of state mixing. For *pure* states of a pair of qubits, both $D(\rho_{AB})$ and $E_f(\rho_{AB})$ are equal to the entropy $S(\rho_A)$ of the reduced statistical operator $\rho_A = \mathrm{tr}_B(\rho_{AB})$, that is, $E_f\big(P(|\Psi\rangle_{AB})\big) = D\big(P(|\Psi\rangle_{AB})\big) = S(\rho_A)$, where $|\Psi\rangle_{AB}$ is the pure state in question. In the manipulation of n entangled pairs of particles in state $|\Psi\rangle_{AB}$, the optimal probability of obtaining k singlets tends to 1 when $k < D(P(|\Psi\rangle_{AB}))$, in the infinite n limit; it is not possible to achieve the desired conversion for finite n.

For mixed states ρ_{AB}, it is also natural to consider the difference

$$B(\rho_{AB}) \equiv E_f(\rho_{AB}) - D(\rho_{AB}), \tag{1.56}$$

between the entanglement of formation and the entanglement of distillation, known as the *bound entanglement*. The bound entanglement is clearly non-negative: $B(\rho_{AB}) \geq 0$. Bound states are *not* distillable but the formation of a single copy of one requires entanglement; this can be viewed as arising from extreme state mixing. Note, however, that the existence of bound states does not preclude situations where forming a larger number of copies may require a *vanishingly small* amount of entanglement per copy; the states that do not violate the PH criterion form a known such class. Non-violation of this criterion is preserved under LOCC. It is currently an open problem whether one can determine whether an arbitrary mixed state is distillable. An algorithm exists for the problem of finding whether a mixed state can be (locally) prepared [143]—this is a computationally NP-hard problem, however.

The main direction of investigation of quantum entanglement in recent years has been through the consideration of entanglement as a resource for information processing, which is taken up in Chapter 4. This has been shown to bear on the central issues of the foundations of quantum mechanics. The connection has been made in several divergent ways, each of which depends to some extent on previous attempts to understand quantum measurement, quantum probability, and logic, which are taken up in the next chapter. These are themselves are tied up with the long quest to provide a fully satisfactory interpretation of quantum mechanics, which is taken up in Chapter 3.

Quantum Measurement, Probability, and Logic

In the centuries preceding the development of quantum mechanics, the conception of mechanical systems as objects existing independently of conscious agents and possessing physical magnitudes that can, in principle, be arbitrarily well specified was rarely questioned by physicists. In classical mechanics, that is, that of the tradition of Newton, Lagrange, and Hamilton, the full set of physical magnitudes describing each physical system is precisely determined at all times by a collection of six parameters, the *dynamical variables* of vector position $\mathbf{q}$ and vector momentum $\mathbf{p}$, together constituting the state $(\mathbf{q}, \mathbf{p})$, in accordance with Hamilton's partial differential equations for the Hamiltonian function $H(\mathbf{q}, \mathbf{p})$, and all imprecision of state specification is entirely due to the ignorance of agents as to this objective state.

In classical mechanics, the domain of the dynamical variables constitutes the *state space*, in which states evolve in time along precisely specifiable trajectories in such a way that each magnitude is at every time uniquely determined by the state at any earlier time, that is, there is strict *causality*. Bohm described the introduction of causality as a principle of physics as follows.

> "[O]ne begins to consider the possibility that in processes by which one thing comes out of others, the constancy of certain relationships is no coincidence. Rather, we interpret this constancy as signifying that such relationships are *necessary*, in the sense that they could not be otherwise, because they are inherent and essential aspects of what things are. The necessary relationships between objects, events, conditions, or other things at a given time and those at later times are then termed *causal laws*." ([53], pp. 1-2)

The state of a compound system is also taken to be fully described through those of its subsystems as specified by their dynamical variables, that is, their (generalized) positions and momenta subject to the appropriate form of the causal law of motion. This is the view of the world as a clockwork. In the event that the state of a classical system is incompletely known to a conscious agent, the agent can describe the system statistically through the behavior of a corresponding ensemble of possible copies of the system in states compatible with his knowledge, each copy possessing precise values for its physical magnitudes some or all of which are not precisely known by that agent. Such ensembles can be used by another agent with more complete knowledge of the states of the individual copies for communication with that agent by sending these more precisely specified members of the ensemble as signals the receipt of which provide information to the more ignorant agent when received.

Hans Reichenbach argued that inferences to determinism, such as captured by Bohm's simple reconstruction, are suspect, especially when they involve inferences from behavior at one length scale to determinism at another.

> "The idea of determinism, i.e., of strict causal laws governing the elementary phenomena of nature, was recognized as an extrapolation inferred from the causal regularities of the macrocosm. The validity of this extrapolation was questioned as soon as it turned out that macrocosmic regularity is equally compatible with irregularity in the microcosmic domain, since the law of great numbers will transform the probability character of the elementary phenomena into the practical certainty of statistical laws. Observations in the macrocosmic domain will never furnish any evidence for causality of atomic occurrences so long as only effects of great numbers of atomic particles are considered." ([373], p. 1)

Already early in the twentieth century, the deterministic conception of mechanics was brought into question not only in the realm of quantum theory but also in relation to the mechanics of classical systems having evolutions with extreme sensitivity to initial conditions for arbitrary initial conditions [384]. It is noteworthy that radioactive decay, observed as early as 1896, has often retrospectively been viewed as pivotal for the early acceptance of *indeterminism*, even though that appears not to have actually been the case. In particular, "van Brakel (1985) has surveyed the literature and come to the conclusion that 'before 1925 there is no publication in which the 'indeterministic' nature of radioactive decay is considered to be a remarkable aspect of the phenomenon'... On the other hand, he finds many publications claiming such a role, all written *after* 1928" ([479] p. 140, [459]). That is, there is a distinct possibility that the introduction of the New quantum mechanics of microscopic systems motivated a revisionist account of the initial significance of radioactivity for the increased acceptance of indeterminism.

The traditional pre-quantum mechanical picture of matter included causal relations between systems located in space and time and involved in processes under which they move between initial and final states in a *continuous* manner and may interact with one another. A classic statement of determinism,

following from causality, was made by Pierre Simon, Marquis de Laplace in
reference to the entire universe within space and time as a physical system:[1]

> "We ought then to regard the present state of the universe as the
> effect of its anterior state and as the cause of the one which is to
> follow. Given for one instant an intelligence which could compre-
> hend all the forces by which nature is animated and the respective
> situation of the beings who compose it—an intelligence sufficiently
> vast to submit these data to analysis—it would embrace in the same
> formula the movements of the greatest bodies of the universe and
> those of the lightest atom; for it, nothing would be uncertain, and
> the future, as well as the past, would be present to its eyes." ([123],
> pp. 3-4)

Laplace's characterization has two aspects, that of the first sentence, which
is metaphysical, and that of the second, which is epistemic in that it has to
do with what is in principle *calculable*. What is *in practice* calculable may
be significant as well, because, for example, the universe may have a finite
number of components that could be used by agents, such as ourselves, for
the performance of calculations, a point which is addressed later in Chapter
4.

For the evolution of a complex system, even a classical one, to be deterministic, a number of assumptions must be valid, including the assumption
that the system of interest is truly *closed*. Although a finite complete universe
is a closed system by definition, any part that human beings might actually
comprehend may never be closed, as reflected in the view of Émile Borel that
the classical description of gas "composed of molecules with positions and velocities which are rigorously determined at a given instant is. . . a *pure abstract
fiction*" because one is driven, for the purposes of practical physics, to consider the *external* forces acting on them as indeterminate [66]. Furthermore,
von Smoluchowski's 1918 model of radioactive decay based on sensitivity to
initial conditions suggested early on that causality and random phenomena
are not inherently incompatible.

> "As an explanation of the origins of the random variables observed in
> classical systems, he suggested what is known as sensitive dependence
> on initial conditions. In order to show that there is no contradiction
> between 'lawlike' causes and 'random' effects, von Smoluchowski con-
> structs a mechanical model reproducing precisely the exponential law
> of radioactive decay. To complicate matters further, he says that he
> 'of course does not believe radium atoms really possess such a struc-
> ture' [as that of a tiny planetary system]. Instead, radioactivity can
> be taken 'as the most complete type of 'randomness'." ([479], p. 140)

---

[1] Also see [244], Section 2.6. It is interesting to compare this with the perspective of
the Collapse-Free interpretation discussed in the following chapter, which assumes
the validity of a universal quantum state deterministically evolving.

The later popular idea that macroscopic behavior is essentially deterministic whereas microscopic behavior is essentially indeterministic can be viewed as the result of the long history of successes of classical physics in the macroscopic realm and the necessity of explaining later 'anomalous' phenomena in terms of the quantum mechanical behavior of microscopic systems, against a background of deeply embedded philosophical assumptions. As Dudley Shapere has pointed out,

> "Determinism was a guiding principle... because it was based on more general abstract or philosophical ideas, largely inherited from the Greeks (or before) about what an explanation ought to be. One could not have an explanation unless it explained every detail of experience and allowed the specific prediction or retrodiction of every detail of experience. That ideal of what an explanation ought to be and must be if the theory is to be explanatory is rejected in quantum mechanics." ([99], p. 148)

Although irreducible randomness, as opposed to limitations on practical descriptions free of stochastic elements, was increasingly accepted after the arrival of the New quantum theory, Schrödinger began his 1935 cat paper by reviewing the basic elements of the traditional approach, which he described as "an ideal of the exact description of nature," as a basis for criticizing the new state of affairs in physics vis-à-vis its conceptual foundations after the formalization of quantum mechanics had been essentially completed, although he strongly criticized "naive realist" interpretations of the theory [394].

Einstein was also gravely concerned about the status of quantum mechanics in relation to these long-standing ideals of natural philosophy. "That business about causality causes me a great deal of trouble... I would be very unhappy to renounce complete causality" ([72], p. 23). However, as Arthur Fine quotes Einstein saying in one of his letters to E. Zeisler, "For us causal connections only exist as features of the theoretical constructs" ([174], p. 87). Indeed, by 1950, Einstein had certainly come to see realism as a more crucial element of a proper physics than causality. "In the center of the problematic situation I see not so much the question of causality but the question of reality (in a physical sense)" (Quote from a 1950 letter to Jerome Rothstein, [174] p. 87). It is not that Einstein had a prejudice against probabilistic laws, but rather only against irreducibly probabilistic laws in a *fundamental* theory (*cf.*, [479], Chapter 4). Furthermore, as John Stachel has shown, "it was radical non-locality that most bothered Einstein, even more than its probabilistic element" ([429], p. 246, *cf.* [428]). Einstein sought *both* realism *and* locality in physical theory.

Few, if any, of the founders of quantum mechanics were entirely inflexible in the face of the new phenomena and the successes of quantum theory. Rather, they sought the best balance of fundamental principles that still allowed for the preservation of scientific naturalism. Thus, for example, Bohr's approach to quantum mechanics was an attempt to describe quantum phenomena within the broad outlines of traditional scientific methodology that emphasized exactly that the scope of some principles may be limited by that of others. This is most clear in his position that there is complementarity between the continuous space-time description of microscopic systems and their causal description. Related to but distinct from indeterminism, which is metaphysical and regards the relation between system magnitudes of differ-ent times, are *indeterminacy* (or *indefiniteness*), which is metaphysical and regards magnitudes at just one time, and *uncertainty*, which is epistemic and may relate to one time or several times.

> "Bohr uses the concept of 'complementarity' at several places in the interpretation of quantum theory. The knowledge of the position of the particle is complementary to the knowledge of its velocity or mo-mentum...; still we must know both for determining the behavior of the system. The space-time description of the atomic events is complementary to their deterministic description... [The change in the course of time of the probability function] is completely deter-mined by the quantum mechanical equation, but it does not allow a description in space and time. The observation, on the other hand, enforces the description in space and time but breaks the determined continuity of the probability function by changing our knowledge of the system." ([219], pp. 49-50)

His disciple Heisenberg, just quoted, was quite willing to abandon causality.

> "Since all experiments obey the quantum laws and, consequently, the indeterminacy relations, the incorrectness of the law of causality is a definitively established consequence of quantum mechanics itself." ([216], p. 197)

The differences of applicability of different sorts of 'Heisenberg relation' can lead to confusion in regard to the distinction between indeterminism and un-certainty. This relates to the fact that probabilities associated with quantum states ρ may have both an objective and a subjective aspect, with the bound-ary between them depending on the interpretation of the formalism assumed.

Eugene Wigner noted that in standard quantum mechanics there is "the possibility of an observation giving various possible results even on a system with a well defined and completely known state," where "the acausality of the theory manifests itself only at the observations undertaken" [502]. How-ever, the concept of *objective indefiniteness*, namely, that physical magnitudes inhere in an object without being simultaneously definite allows both inde-terminism and indefiniteness to obtain without essentially involving the mind in the description of measurement, as Wigner ultimately suggested. This idea has been explicated more recently by Shimony, who has articulated it in the following simple and striking context. "If... we concede that [a Bell state] Ψ is a complete description of the polarization state of the pair of photons [in-volved in demonstrating the violation of Bell's inequality], then we must accept the *indefiniteness* of the [relevant projections of] polarization of each. . . as an objective fact, not as a feature of the knowledge of one scientist or of all human beings collectively." Furthermore, "[w]e must also acknowledge *objective chance* and *objective probability*, since the outcome of the polarization analysis of each photon is a matter of probability" ([412], pp. 177-178). Shimony made the following suggestion for interpreting the quantum state.

> "It is convenient to use a term of Heisenberg to epitomize objective indefiniteness together with the objective determination of probabilities of the various possible outcomes; the polarizations of the photons are *potentialities*. The work initiated by Bell has the consequence of making virtually inescapable a philosophically radical interpretation of quantum mechanics: that there is a modality of existence of physical systems which is somehow intermediate between bare logical possibility and full actuality, namely, the mode of potentiality." ([412], pp. 177-178; also *cf.* [419], p. 108)

The interpretation of quantum theory, that is, the drawing out of the epistemic, metaphysical, and operational significance of the elements of the theory, especially of the state function $|\psi\rangle$, is necessary because the formalism, like that of any other physical theory is not, as is often suggested, self-interpreting.

In this chapter, various ways that physical magnitudes and measurements have been quantum mechanically characterized are considered. This will equip us with much of what we will need to appreciate more fully the import of interpreting quantum mechanics. After setting out pertinent elements of logic and probability theory, we consider fundamental results pointing out the above restrictions and explore some apparently paradoxical situations that arise in the application of the theory. For now, let us keep in mind the view of quantum mechanics that prevailed in 1935, as summarized by Schrödinger.

> "Continuing to expound the official teaching, let us turn to the already mentioned ψ-function. It is now the means for predicting probability of measurement results. In it is embodied the momentarily-attained sum of theoretically based future expectation, somewhat as laid down in a *catalog*. It is the relation- and determinacy-bridge between measurements and measurements, as in the classical theory the model and its state were. With this latter the ψ-function moreover has much in common. It is, in principle, determined by a finite number of suitably chosen measurements on the object, half as many as were required in the classical theory. Thus the catalog of expectations is initially compiled. From then on it changes with time, just as the state of the model of classical theory, in constrained and unique fashion ("causally")—the evolution of the ψ-function is governed by a partial differential equation (of first order in time and solved for $\partial\psi/\partial t$). This corresponds to the undisturbed motion of the model in classical theory. But this goes on only until one again carries out any measurement." ([394])

2.1 Logic and Mechanics

In characterizing a physical system at a given moment, one can both formally and operationally consider the set of *propositions*, that is, true/false questions about its magnitudes and their definiteness; one traditionally speaks of a system having or not having any specific value of any magnitude by virtue of that value lying or not lying in the corresponding Borel subset of the real numbers.

In the case of a classical mechanical system, there is a specific real spatial position vector of the particle and a specific real momentum vector, that is, a location in phase space, as long as this particle exists. For classical systems, complex propositions can be formed from the simplest ones via Boolean logical connectives; for any proposition regarding system properties, there is a subset of state space for which the proposition is true, any two propositions being physically equivalent if the same subset of state space attributes the same value to them, modulo sets of Lebesgue measure zero. One can always provide a characteristic function for all Borel subsets of the state space that provides the desired logical mapping to propositions. These functions are idempotent, that is, they are equal to their squares. However, when one attempts to extend this approach to the relation between states and properties to quantum mechanical systems, it fails, and does so in specific ways. In particular, the quantum state does not fully determine specific values of all their physical magnitudes: from the logical point of view, the distributive law fails.

The failure of the distributive law can be related to the term *observable* taking the place of the term *property* for the characterization of the physical magnitudes of quantum-mechanical systems. Recall that the quantum state of a system at a given time is capable of precisely specifying only a corresponding subset of the physical magnitudes, that is, the currently "observed" quantities and the others that are functions only of them, in the intervals between measurement events, this being so only as long as quantities other than these remain unmeasured, that is, "unobserved." As Dirac put it,

> "The expression that an observable 'has a particular value' for a particular state is permissible in quantum mechanics in the special case when a measurement of the observable is certain to lead to the particular value, so that the state is an eigenvalue of the observable... In the general case we cannot speak of an observable having a value for a particular state, but we can speak of its having an average value for the state. We can go further and speak of the probability of its having any specified value for the state, meaning the probability of this specified value being obtained when one makes a measurement of the observable." ([139], pp. 46-47)

These probabilities are the expectation values provided by the Born rule, although there is some debate as to propriety of this procedure in the context of traditional probability theory.

Bell argued that the use of the term *observable* can be misleading, much as Bethe argued that the term *uncertainty* can be.

"There is indeed much talk of 'observables' in quantum theory books. And from some popular presentations the general public could get the impression that the very existence of the cosmos depends on our being here to observe the observables. I do not know that this is wrong. I am inclined to hope that we are indeed that important. But I see no evidence that this is so in the success of contemporary quantum theory." ([24], p. 170)

Bell, Einstein, and Schrödinger sought a formulation of quantum theory lying unequivocally in the traditional realist approach to physics in which physical properties are independent of observing minds, and did so not simply for the sake of tradition. Heisenberg described Einstein's similar concern as follows.

"He pointed out to me that the very concept of observation was itself problematic. Every observation, so he argued, presupposes that there is an unambiguous connection known to us, between the phenomenon to be observed and the sensation which eventually penetrates into our consciousness. But we can only be sure of this connection, if we know the natural laws by which it is determined. If, however, as is obviously the case in modern atomic physics, these laws have to be called in question, then even the concept of 'observation' loses its clear meaning." ([224], p. 114)

Recall that, in standard quantum mechanics, the observables are Hermitian linear operators on separable Hilbert spaces, which is a rather indirect representation by comparison to the direct representation of physical magnitudes as functions in classical mechanics. A difference between classical and quantum mechanics also arises in that, in the quantum case, the dynamical variables, that is, the observables are constrained by commutation relations, and form a non-Abelian (non-commutative) algebra; in the case of simple position Q and momentum P, for example, one has $[Q, P] = i\hbar$.

In order to enable a logical interpretation of the quantum state $|\psi\rangle$ along the lines of that of classical mechanics, physical magnitudes could be understood via a *value state*, which is a Boolean value (0 or 1) attributed to the system for each pairing of observable and value of that observable. However, according to the Born rule, the Hilbert-space ray for pure quantum states $|\psi\rangle$ instead provides only *probabilities* for all observables at any one time, in particular, to the idempotent observables (projectors) which correspond to propositions [67]; even in the case of pure states, only the proposition corresponding to one *Hilbert-space ray* which takes the value 1, and the propositions corresponding to rays orthogonal to it, which take the value 0, have truth values. The probabilities provided by the Born rule are not all definable as measures over properties on a classical Kolmogorovian probability space, but are well defined as such only in relation to quantities that have been measured and not subsequently disturbed by further measurements of non-commuting quantities [244, 439]. As a result, the probabilities provided by quantum mechanics are best not defined on simple sets but instead on quantum events or, operationally, experimental questions, that is, on pairings of individual observables (defined on Hilbert space) and individual Borel subsets of the real line (associated with values of measurement outcomes). Thus, the sort of probability that appears in quantum mechanics is a *generalized* probability.

As with probability, which is taken up in detail in the following section, it has been suggested that a generalization of standard *logic* is required for the proper description the quantum world [166, 373, 437, 480]. In particular, it has been suggested by Hilary Putnam and others that such a revision might lend greater coherence to physical theory through the adoption of a conception of logical truth under an empiricist approach to logic, within which logic is capable of revision when our knowledge of the world increases as it has with the emergence of quantum theory [366]; Redhead has suggested this view of logic might be called *instrumentalist* [371]. By contrast with the case of probability, however, the position that logic is empirical has come under considerable criticism. One objection arises from the fact that it eliminates the universality of logic in that, for example, human reasoning is already well described by classical propositional logic but is apparently *not* by quantum logic; there then would be, at the very least, different logics in different realms [427].[2] Other problems, as pointed out by Allen Stairs, are that

" 'Quantum Logic' means so many things to so many people that it has almost ceased to be a useful term... There is disagreement as to whether quantum logic has anything to do with *logic*, and even among those who think it does, there is disagreement concerning the nature of logic in general and what quantum logic in particular can do for us in our quest to understand quantum theory... In what I take to be the most interesting version of quantum logic, the word 'logic' is taken very seriously. The aim is to present quantum mechanics as a theory that posits novel relations among events or states of affairs... which are reflected in logical structures of a strongly non-classical character. On this view, logic functions in *explanations:* it is part of an account of the strange behavior which quantum systems exhibit." ([432])

Quantum logic in the broadest sense has a history stretching back to the early 1930's and the work of Garrett Birkhoff and von Neumann, who demonstrated the possibility of connecting the mathematics of lattice theory and Hilbert space [48]. Their manner of associating logical states with quantum systems was to straightforwardly assign binary values to closed linear sub-

---

[2] An extensive but accessible introduction to more contemporary quantum logic can be found in Chapter 7 of [244]. A useful bibliography is [340].

spaces of the Hilbert space of the quantum system as mentioned above.[3] In the context of classical physics, Boolean logic and the traditional understanding of the logical connectives and of negation interpretable in set-theoretic terms can be unproblematically used, with the atomic propositions being those corresponding to a system being in one of the subsets of classical phase space. Again, however, the projection operators representing propositions regarding a quantum system form a non-Boolean algebra. Thus, to construct a quantum logic, one must use different structures, for example, those related to *partial Boolean algebras*.[4]

The quantum logic of von Neumann and Birkhoff, often called the *logic of subspaces*, arises from the set of Hilbert subspaces of the complex Hilbert space $\mathcal{H}$ describing the quantum system of interest, as follows. Each subspace $\bar{h}$ is identified with the operator $P_{\bar{h}}$ that projects onto the subspace. The lattice $\bar{L}(\mathcal{H})$ of closed linear subspaces of a Hilbert space $\mathcal{H}$ is seen to be equivalent to the lattice of projection operators on $\mathcal{H}$. One can define the two operations $\wedge$ (*meet*) and $\vee$ (*join*) acting pairwise on any two projectors P_1 and P_2 by $P_1 \wedge P_2 \equiv P_1 P_2$, $P_1 \vee P_2 \equiv P_1 + P_2 - P_1 P_2$, and identify the zero as the projector $\mathbb{O}$ onto the zero vector $\mathbf{0}$ and the identity as the projector $\mathbb{I}$ onto all of $\mathcal{H}$; $\vee$ corresponds to the linear span, $\wedge$ to intersection. The rays of $\mathcal{H}$ are considered to be the atomic propositions of $\bar{L}(\mathcal{H})$.[5] Compound propositions formed from them correspond to higher-dimensional closed linear subspaces. The conjunction, $\wedge$, is essentially the same as the conjunction of classical logic. The central conclusion of Birkhoff and von Neumann was that

> "one can reasonably expect to find a calculus of propositions which is formally indistinguishable from the calculus of linear subspaces with respect to *set products, linear sums*, and *orthogonal complements*— and resembles the usual calculus of propositions with respect to *and, or*, and *not*." ([48])

However, the disjunction $\vee$ (join) and negation $'$ behave much differently from to set-theoretic disjunction and negation even if, as Putnam claims, these are taken to be synonymous with those of traditional propositional logic. In particular, Putnam has argued that one can reasonably claim that "adopting quantum logic is *not* changing the meanings of the connectives, but merely changing our minds about [the distributive law]" [366].

---

[3] Given this, it is especially important to distinguish quantum logic from the implementation of Boolean logic in quantum information science, which is based on the manipulation of computational basis states by 'quantum logic gates'; *cf.* Section 4.5.

[4] A detailed summary of the elements of Boolean and quantum logic are given in section 5 of the Appendix.

[5] An excellent critical résumé of the quantum-logical approach to quantum theory can be found in [427]. For up-to-date reviews of quantum logic, see [117, 440].

The propositions of quantum logic refer to the state of the physical system at a given time. Their semantic interpretation involves no reference to the preparation or measurement of the system. Following Birkhoff and von Neumann, a quantum logic of events involving a system can be "read off" the Hilbert space that describes it. The inherent algebraic structure of the projection operators on the Hilbert space, that of the partial Boolean algebra, is formed by a family of Boolean algebras $B^{(i)}$ if the following conditions are satisfied.

(1) The set-theoretical intersection, $B^{(i)} \cap B^{(j)} = B^{(k)}$, of two members $B^{(i)}, B^{(j)}$ is a member of the family.

(2) If three elements of the partial Boolean algebra are such that two of them belong to a given member of the family, then there exists a Boolean algebra of which all three are members.

The complement of any element of the partial Boolean algebra is its complement with respect to any of the members of a family to which it belongs; the complement of the projector P_i is the operator $\tilde{P}_i = \mathbb{I} - P_i$ such that $\tilde{P}_i \wedge P_i = \mathbb{O}$ and $\tilde{P}_i \vee P_i = \mathbb{I}$. This complement is then unique and belongs to any family to which the element belongs. Importantly, the matching of truth values with closed linear subspaces of Hilbert space does *not* require that the corresponding propositions are necessarily either true or false; this mathematical correspondence is compatible with the indefiniteness of truth values of statements regarding physical properties inherent in quantum mechanics discussed in Chapter 1.[6] Furthermore, the truth of an elementary quantum proposition is *insufficient* to determine the value of all other propositions.

Again, the central move of the quantum logic approach is to consider the lattice of propositions defined on Hilbert space, as described above, rather than Boolean lattice of propositions defined on classical phase space or, equivalently, to consider the related partial Boolean algebra. One can then attempt to provide a truth-valuation, mapping subspaces onto the truth values of B_2, to arrive at a quantum propositional calculus in a manner resembling that in classical propositional calculus. Such a valuation is subject to an *admissibility criterion*: A truth-valuation is admissible if and only if there is a ray R such that for every subspace S the value is 1 if and only if the former is a subspace of the latter. Another peculiarity, in comparison with traditional logic, is that the logic depends *on the particular quantum system in question*. Furthermore, for Hilbert spaces of dimension greater than two, the valuations are not homomorphisms from the lattice corresponding to the logic of subspaces $\bar{L}(\mathcal{H})$ to B_2; the valuations are *not* truth-functional, in that the values of compound propositions are not determined by their components [371, 427].

---

[6] It is also noteworthy in this regard that Reichenbach introduced a third logical value as a basis on which to approach quantum mechanics ([373], Section 30).

Recall that, most distinctively, the lattice structure of quantum logic is *non-distributive*. As a relatively concrete illustration of this, consider the propositional structure of a quantum two-level system, the case of spin-1/2. The propositions corresponding to spin along a given direction z can be written $L_z = \{0, p_{up}, p_{down}, 1\}$. For another spatial orientation, $\bar{z}$, one has a similar system of propositions $L_{\bar{z}} = \{\bar{0}, \bar{p}_{up}, \bar{p}_{down}, \bar{1}\}$. By identifying the least upper bounds (lub's) and greatest lower bounds (glb's) of these two sets, one obtains the "horizontal sum" $L_z \otimes L_{\bar{z}}$, which can be endowed with a modular orthocomplemented structure. Even in this simple case involving the 'pasting together' of Boolean subalgebras, one obtains a structure differing from a Boolean algebra because the distributive law fails to hold. For example, beginning with the complex proposition $p_{down} \vee (\bar{p}_{down} \wedge \bar{p}'_{down})$ and applying the distributive law, one finds

$$p_{down} \vee (\bar{p}_{down} \wedge \bar{p}'_{down}) = (p_{down} \vee \bar{p}_{down}) \wedge (p_{down} \vee \bar{p}'_{down})$$
$$p_{down} \vee 0 = 1 \wedge 1$$
$$p_{down} = 1,$$

which is erroneous. Moreover, beginning with the complex proposition of the same form as that above, but with conjunction and disjunction interchanged, one can also similarly obtain $p_{down} = 0$ [440].

In order understand the evolution of quantum systems in time, one considers the quantum state space, which was described by Birkhoff and von Neumann as follows.

"[Points in this space] correspond to the so-called 'wave functions,' and hence [phase space] is again a function space—usually assumed to be Hilbert space... the law of propagation is contained... in quantum mechanics, in equations due to Schrödinger. In any case, the law of propagation may be imagined as inducing a steady fluid motion in the phase-space. It may be noted that in quantum mechanics the flow conserves distances (*i.e.* the equations are unitary)." ([48])

Notably, they viewed this as entirely compatible with causality.

"The [phase space] point p_0 associated with [a system] S at time t_0, together with a prescribed mathematical 'law of propagation,' fix the point p_t associated with S at any later time t; this assumption evidently embodies the principle of *mathematical causation*." ([48])

This reflects a key element of von Neumann's interpretation of quantum mechanics, which is discussed in the following chapter. Birkhoff and von Neumann then made the following comments with regard to measurement that touches on the distinction between quantum and classical behavior.

"Now before a phase-space can become imbued with reality, its elements and subsets must be correlated in some way with 'experimental propositions' (which are subsets of different observation-spaces)... There is an obvious way to do this in dynamical systems of the classical type. [*their footnote*: "Like systems idealizing the solar system or projectile motion"] One can measure position and its first time-derivative velocity—and hence momentum—explicitly, and so establish a one-one correspondence which preserves inclusion between subsets of phase-space and subsets of a suitable observation-space." ([48])

They then explicitly distinguished this from classical statistical theory.

"In the kinetic theory of gases and of electromagnetic waves no such simple procedure is possible, but it was imagined for a long time that 'demons' of small enough size could by tracing the motion of each particle... measure quantities corresponding to every coördinate of the phase-space involved. In quantum theory not even this is imagined, and the possibility of predicting in general the readings from measurements on a physical system S from a knowledge of its 'state' is denied; only statistical predictions are always possible. This has been interpreted as a renunciation of the doctrine of predetermination; a thoughtful analysis shows that another and more subtle idea is involved. The central idea is that physical quantities are *related*, but are not all computable from a number of *independent basic* quantities (such as position and velocity)." ([48])

It is noteworthy that Birkhoff and von Neumann considered the requirements of imbuing the space of states with "reality," echoing the vocabulary used by EPR the year before, and that they reflect on causality; one motive for pursuing quantum logic in recent times, as is typical of interpretations of the quantum formalism, has been to remove traditional paradoxes associated with the theory's description of the world while being straightforwardly compatible with 'realism' in the EPR sense.[7] Again, however, the atomic propositions must be attributed definite truth values for this to work. Putnam proposed mapping a new set of propositions $\{\Delta_Q\}$, each associated with the value of each observable Q lying in an Borel subset Δ of its set of possible values, onto corresponding projection operators; these projection operators $P_Q(\Delta)$ are those associated with the subspaces the ranges of which are the subspaces spanned by all eigenvectors $|q_i\rangle$ corresponding to the eigenvalues $q_i \in \Delta$, the set of corresponding measurement outcomes associated with measuring the observable Q. One thereby associates with the operator the proposition 'The state of the system is in the range of $P_Q(\Delta)$' [366]. This is problematical if the state of the system, which Redhead calls the *Putnam state* [371], is not

---

[7] The usage of the term *realism* is taken up in detail in Chapter 3.

in this range and one wishes to assert the value of the elementary proposition in order to retain 'realism.'

The Kochen–Specker theorem, considered in a later section, presents a significant obstacle to straightforwardly providing a truth valuation under these circumstance. In an attempt to enable an admissible truth valuation, Putnam asserted that in quantum logic "every observable Q has a value, but there is no value which it has." Although consistent, this move undermines the intuitions motivating 'realism' in the first place, to such an extent that its appeal is almost entirely lost. The underlying components of the 'realist' understanding of quantum mechanics, are (i) the *value-definiteness thesis*, the idea that every physical magnitude has a definite value at all times [433], and (ii) that measurements reveal those values. These run contrary to the idea objective indefiniteness, which allows a less conflicted description of the behavior of quantum systems. As Richard Healey has argued,

> "The *content* of the claim that every dynamical variable has a precise value is quite obscure once one has adopted quantum logic, given that the standard (classical) inferences can no longer be drawn from it. And, in particular, it is by no means clear that the truth of this claim suffices to permit a naive realist reading of the Born rules, according to which the object system possessed the specific value revealed by a measurement prior to, or independent of, the occurrence of that measurement."([211], p. 22)

Thus, as Peter Mittelstaedt has put it in relation to the historical development of the interpretation of the quantum probability rule itself,

> "[The] original Born interpretation, which was formulated for scattering processes, was... not tenable in the general case... The probabilities must not be related to the system S in state ϕ, since in the preparation [of the quantum state] ϕ the value a_i of an observable A is in general not subjectively unknown but objectively undecided. Instead, one has to interpret the formal expressions $p(\phi, a_i)$ as the probabilities of finding the value a_i after measurement of the observable A of the system S with preparation ϕ. In this improved version, the statistical or Born interpretation is used in the present-day literature.... On the other hand,... the meaning... for an individual system is highly problematic." ([319], p. 41)

2.2 Probability and Quantum Mechanics

From the perspective of logic, one sees that complications arise in quantum mechanics because not all propositions are compatible, that is, the full set of events in quantum mechanics is non-Boolean in a specific way. The probabilities arising in quantum mechanics are probabilities generalizing those of the traditional kind. Nonetheless, conceptions deriving from those of classical probability can still be brought to bear on the generalized probabilities given by the Born rule as a postulate of standard quantum mechanics. In this section, we review the basic conceptions of probability. Two fundamental results that help contextualize and characterize probability within quantum theory, namely, Gleason's theorem and the Kochen-Specker theorem, are then considered in the following section. It is valuable to have all these in mind when later considering various surprising or arguably problematic quantum mechanical situations, that is, the "quantum paradoxes" that have been contemplated.

There are several different conceptions of probability, most significantly for our purposes the classical, relative frequency, propensity, and subjective conceptions. Although there are difficulties associated with each, they typically do not bear directly on the situation in quantum mechanics *per se*.[8] Let us survey these conceptions in the mathematical context of the Kolmogorov axiomatization of the probability calculus in modern form; in particular, the third axiom as presented below is a more general one than that introduced by Kolmogorov but reduces to it when the set of events is finite rather than merely countable.[9] In the Kolmogorov axiomatization, one is given the events $A, B, C, \ldots$ and thus the sample space S of events (the unit event being identified in quantum mechanics with the projector $\mathbb{I}$) defined as their union. The triple (S, F, p), where F is a field of subsets of S, is referred to as a *Kolmogorovian probability space* when the following conditions are satisfied by p, in particular, taking $p(E_i) \in \mathbb{R}$ as the *probability* of the event E_i.

(1) For any set of events $\{E_i\}$: $0 \leq p(E_i) \leq 1$.

(2) $p(S) = 1$.

(3) For any countable sequence of mutually disjoint events $E_1, E_2, \ldots$,
$p(E_1 \cup E_2 \cup \cdots) = \sum_i p(E_i)$ (σ-additivity).

The probability of one event, B, conditional on another, A, is written $p(B|A) = p(B \cap A)/p(A)$. If two events A and B are such that $p(B|A) = p(B)$ and $p(A|B) = p(A)$, then they are *probabilistically independent*. In the context of probability theory, one is primarily interested in *experiments* consisting of a sequence of *trials*, each having an elementary event as an outcome. If the trials are all independent, such a sequence is referred to as a *Bernoulli sequence*. If the trials are such that the probability of each event may *not* be independent of its predecessor but *is* independent of all others of the sequence, it is referred to as a *Markov sequence*.

The *classical conception* of probability, which predates the Kolmogorov axiomatization, arose by abstraction from practical situations in which all outcomes are in some sense equally possible, the probability of any one event

---

[8] For an illuminating summary of difficulties in interpreting probability, from which this survey has greatly benefited, see [203].

[9] In some versions of the subjective interpretation of probability, this axiom is brought into question.

being the fraction of the total number of events that it represents, as in the appearance of a quantity as a sum obtained in the rolling of a pair of fair dice by adding together the value of the two upward faces. The introduction of the principle of indifference, namely, that whenever there exists no evidence that favors one possibility over another possibility they have equal probabilities, helps avoid a circularity in the classical conception of probability. However, it may be problematic to explicate the idea of "equally good evidence" for two uncertain events without making use of probability. There is also an issue of the impact of the *number* of events. In some applications, most importantly here being quantum mechanical ones, probabilities may take irrational values; this conception of probability requires the consideration of an infinite number of events for these values to be defined. In such cases, one may appeal to a generalization of the principle of indifference, in the form of the maximum entropy principle. On this principle, one takes from the set of all distributions consistent with background knowledge the one maximizing the information-theoretic entropy. Although this is viable in the countably infinite case, there is a certain arbitrariness to the procedure that is particularly problematic in the case of uncountably infinite sets of events, because the principle of indifference can be applied in ways that are incompatible with each other.

The *frequency conception* of probability is based on the direct identification of the probability of events with their relative frequency of occurrence in the total set (reference class) of actual events. Advocates of the Bayesian subjective interpretation of probability consider this identification a category mistake; that subjective view has been adopted by advocates of the Radical Bayesian interpretation of quantum mechanics, discussed in the following chapter.[10] A distinguishing element here is the consideration of *actual* outcomes as opposed to possible outcomes. Thus, on this conception, probability is defined *operationally*. This poses an immediate problem in cases where irrational values of probability might be considered necessary because such values clearly cannot exist for finite sets of events, such as *physical* measurements, which clearly cannot ever constitute an infinite class if measurement is to be carried out by agents. This problem is typically avoided by considering this probability as an ideal limit as the number of events becomes infinite, which is counterfactual in character, at some cost to its operational character.

The *propensity conception* of probability, by contrast, takes probability to be a physical disposition or tendency of a situation in the world to provide each kind of outcome, or a limiting relative frequency for each such outcome. This conception is particularly practical when one desires to attribute probabilities to events that by definition can only occur *once*, something quite unnatural for the conceptions considered above. For example, Karl Popper considered the probability of an outcome of a given type to be a propensity of a repeatable experiment to produce the given outcome with just that limiting relative

---

[10] A category mistake is committed when one discusses a matter in terms appropriate only to matters of a significantly different character (*cf.* [386], p. 16).

frequency [360]. Propensity may also derive its meaning from the role it plays in theories of interest, including quantum mechanics. Bernard d'Espagnat has argued that quantum "probabilities are both intrinsic and 'of appearing,' which mean [*sic*], they are 'probabilities to appear to observers.' " ([126], p. 326). Typically, however, those advocating the propensity theory of quantum probabilities typically wish *not* to view them in this way.

The *subjective conception* of probability identifies probability not with any objective property of the world but directly with the degrees of belief of relevant agents about events, subject to chosen constraints such as rationality, which involves at least consistency. Traditionally, the subjective conception of probability assumes that events are definite and that probabilities arise due to the ignorance of subjects. A set of alternatives are considered that are in a certain sense symmetric relative to this ignorance, with the result that probability is uniformly divided over the elements of this typically finite set [479].

The specific subjectivist approach to probability due to de Finetti has been taken on board by 'Radical Bayesian' interpreters of the theory. The approach of de Finetti takes an agent's degree of belief in an event to be the probability p if and only if p units of utility is the price (the so-called 'fair price') the agent would buy or sell a wager that pays one unit of utility if E occurs and 0 otherwise, assuming that there is precisely one such price, an assumption that is often challenged. One considers a 'Dutch book,' which is a series of bets against an agent that the agent considers acceptable; such a series of bets can be avoided by the agent if his subjective probabilities obey the Kolmogorov axioms, that is, are *coherent* [268]. This provides an operational definition of probability. Probabilities are considered to differ categorically from propositions, so that probability assignments are not considered propositions within the theory. One considers an *arbitrary sum* as being the reward of betting on E. Similarly to the situation in other conceptions of probability, one finds that this sum must then be infinitely divisible in principle in order to guarantee full precision of probability measurement; utility must also depend linearly on the sums to avoid dependency of betting on probability assignments. Upon learning new facts, agents probabilities are updated in accordance with Bayes' rule and are dependent on their prior probability assignments.

With these various conceptions of probability in mind, let us now consider a number of central technical results in the foundations of quantum mechanics that have been obtained in relation to quantum mechanical propositions and related probabilities.

2.3 The Completeness of Quantum Mechanics

The traditional approach to physics before the arrival of quantum mechanics was, as seen above, one in which each state of the system attributes a definite value to *all* physical magnitudes, that is, one that fulfilled the value-definiteness condition: each proposition regarding the system, which is of the form "$O \in \Delta$" where O is a quantity describing the physical magnitude and Δ is a Borel subset of the real numbers, is assigned a definite truth value. Statistical states are then given as probability measures $\mu_O : \Delta \to p(O, \mu, \Delta)$ on a phase space specifying the probability that a measurement of the magnitude O will lie in Δ when the system is in the state μ. The Kochen-Specker theorem shows this approach to be impossible within the structure of standard quantum mechanics. The related theorem of Gleason specifies the precise form of the admissible quantum mechanical probability measures as a functional of the quantum state.

Gleason's theorem justifies, *contra* EPR, the claim that the state description of standard quantum mechanics is complete, by identifying the form of all admissible measures with that of standard quantum mechanics [193], as assumed, for example, under the Basic interpretation of the theory discussed in Section 3.2 below. In the process, it eliminates an entire class of conceivable local hidden-variables theories that had been considered to complete the description of physical systems and would have relegated quantum mechanics to the status of a trivially statistical theory and so would have demoted it from the status of a fundamental theory of physics.

To understand the completeness of quantum state descriptions, one can begin by considering a map $p(P_i)$ from sets of quantum projectors $\{P_i\}$ to the real numbers between 0 and 1, $p : P_i \mapsto p(P_i)$, such that $p(\mathbb{O}) = 0$ and $p(\mathbb{I}) = 1$, such that $P_1 P_2 = 0$ implies $p(P_1 + P_2) = p(P_1) + p(P_2)$, taking p to be a *countably additive* probability measure, where $\mathbb{O}$ is the projector onto the zero vector $\mathbf{0}$ and $\mathbb{I}$ projects onto the entirety of the Hilbert space pertaining to the quantum system in question. In the course of his proof, Gleason provided an important *Lemma* ([193][11]): Let $|\phi\rangle$ and $|\psi\rangle$ be two state-vectors in a Hilbert space $\mathcal{H}$ of dimension at least 3, such that for a given system state $\langle P(|\psi\rangle) \rangle = 1$ and $\langle P(|\phi\rangle) \rangle = 0$. Then $|\phi\rangle$ and $|\psi\rangle$ cannot be arbitrarily close to each other. In particular, $\| \, |\phi\rangle - |\psi\rangle \, \| > \frac{1}{2}$.

The central result of Gleason is the following.

Theorem ([193]): All probability measures that can be defined on the lattice of quantum propositions from the quantum statistical operators, that is all quantum probabilities, are of the form $p(P_i) = \mathrm{tr}(\rho P_i)$, for some statistical operator ρ on Hilbert space $\mathcal{H}$, for all $\mathcal{H}$ of dimension greater than two.

Gleason's theorem shows that every probability measure over the set of projectors arises from a quantum state ρ on the Hilbert space of the system. The trace measure assigns to each projector the dimension of its range, which can then be normalized by the dimension of the pertinent (finite-dimensional) Hilbert space. It is thus obtainable by considering ρ to be the maximally mixed state on the space (see [371], Section 1.5). Gleason's theorem shows

---

[11] This version of the lemma is that given by Bell [24].

that the only natural generalization of Kolmogorovian probability functions of the type needed for quantum mechanics is just that appearing in Hilbert-space formulation. The values corresponding to orthogonal projectors thus obey a rule of the type introduced by Born [67] and explicated by Pauli.

The relationship between these results and the putative dispersion-free states, for which projectors take expectation values of only either 0 or 1 under the above mapping, can be understood as follows [24]. The condition $\sum_i \langle P(|\phi_i\rangle)\rangle = 1$ implies that both 0 and 1 occur because (1) there are no other possible values for satisfying the condition and (2) neither alone suffices. But, then, there must be arbitrarily close pairs $|\psi\rangle, |\phi\rangle$ having different expectation values, 0 and 1 respectively; however, such pairs cannot be arbitrarily close, by the above lemma. Therefore, there can be *no* dispersion-free states providing quantum statistics. Accordingly, no hidden variables that parameterize dispersion-free probability measures exist for systems with Hilbert spaces of dimension greater than 2 [24]; because it provides the probability measures definable on the lattice of quantum propositions corresponding to the quantum projectors, the set of quantum states is *complete*. In the context of quantum information theory, this is related to the no-cloning theorem, discussed in Section 4.2; dispersion-free states would enable perfect quantum cloning.

To see more explicitly how this considerations bear on the class of local hidden-variables theories, following Kochen–Specker, consider the complete set of Hermitian self-adjoint operators for the entire set of quantum states of a system with a Hilbert space of dimension greater than two, with the very natural constraint that the algebraic relations of these operators must be reflected in the assigned values and the assignment of real numbers to the operators of quantum mechanics that might be taken to represent the values of the corresponding properties of that system. The Kochen–Specker theorem shows that such an assignment cannot be found for a finite sublattice of quantum propositions [277]. Consider the *value function*, v_ψ connecting an observable O to a value of a physical magnitude O when a system is in a state ψ, as mentioned at the outset of this chapter. It is natural to define $F(O)$, the value associated with $F(O)$ for all functions F, where the mapping from values of O to O is one-to-one and onto. One can then imagine taking $v_\psi(F(O)) = F(v_\psi(O))$, which has the consequence that v_ψ is additive and multiplicative on operators that commute, the latter itself having the consequence that $v_\psi(\mathbb{I})$ for all states ψ so long as there is at least one magnitude O for which $v_\psi(O) = 0$ (*cf.* [247], pp. 191-192). Another consequence of this multiplicativity is that $v_\psi(P_i)$ must be either 0 or 1 for all propositions P_i, which have corresponding projectors P_i. Thus, if one considers a resolution of the identity into a set of projectors $\{P_i\}$, that is, this set is such that $\sum_i P_i = \mathbb{I}$ in an interpretation of quantum properties where *one and only one* of the corresponding magnitudes P_i can take the value 1, one finds that, except for an overly restricted class of properties, no such function exists.

It is worth noting in passing that, because the calculations involved in Gleason's proof require the dispersion-free states to provide relationships between experiments that cannot, as a matter of principle, be made *simultaneously*, as noted by Bell ([22], Ch. 1), it does not preclude an essentially empty hidden-variables theory of the kind that could be produced for *any* theory, that is, if *no* constraints are placed on the relationships between observables ([277]); it remains possible for a set of observables and states assigning probabilities to Borel sets of values of the observables to have a set of hidden variables that are sufficient, by simply considering real functions, one for each observable such that probabilities associated with the Borel sets, for each state—for each observable, a probability given by the hidden variable distribution given by the inverse of the associated real function will suffice.

2.4 Problems with Measurement in Quantum Mechanics

Many have been uncomfortable with the fundamental role of measurement in quantum mechanics. For example, one of Bell's greatest concerns was that it introduces *conceptual imprecision* into physics.

> "If the theory is to apply to anything but idealized laboratory operations, are we not obliged to admit that more or less 'measurement-like' processes are going on more or less all the time more or less everywhere? Is there then ever then a moment when there is no jumping and the Schrödinger equation applies? The concept of 'measurement' becomes so fuzzy that it is quite surprising to have it appearing in physical theory *at the most fundamental level*... does not any *analysis* of measurement require concepts more *fundamental* than measurement? And should not the fundamental theory be about these more fundamental concepts?" ([24], pp. 117-118)

Instead of the term *measurement*, Bell strongly preferred that the term *experiment* be used, because

> "in fact the [former] word has had such a damaging effect on the discussion, that I think it should now be banned altogether in quantum mechanics... the latter word is altogether less misleading." ([25], p. 20)

This concern can be related to the concern about the prominence of observation. Some have accepted the centrality of measurement to quantum mechanics despite the fact that it is seen as a key shortcoming of the theory. Nonetheless, even Wigner, one of the more radical in approach to achieving consistency between measurements and the remainder of quantum theory due to his explicit appeal to observation, understood the discomfort it causes.

"The principal difficulty [of quantum mechanics] is that it elevates
the measurement, that is the observation of a quantity, to the basic
concept of the theory... it seems dangerous to consider the act of
observation, a human act, as the basic one for a theory of inanimate
objects. It is, nevertheless, at least in my opinion, an unavoidable
conclusion. If it is accepted, we have considered the act of observation, a mental act, as the primitive concept of physics..." ([502])

That quantum measurements are not dynamical defined in a way consistent
with the Schrödinger equation but are used to explicate how observables come
to have definite values after not having had them, was well understood by
Wigner, who acknowledged that by accepting measurement as fundamental
to the theory "it may well be said that we explain a riddle by a mystery"
([502], p. 1).

Given that quantum mechanics remains a fundamental theory, something
the results of the previous section support, and that theory ought to be
properly connected with practice, confronting this difficulty is unavoidable,
as Wigner noted, even if his own solution to it may be rejected. The standard approach to measurement in quantum theory is to consider measuring
instruments to be physical objects that can in principle be realized in the
world, say in a laboratory environment, rather than being mere conceptual
crutches. This is particularly important to bear in mind given the prominence
of thought experiments in the arguments pertaining to the foundations of the
theory, which would lose considerable force were they to be absolutely distanced from practice. Indeed, because quantum mechanics is one of our most
fundamental theories and, perhaps, the most empirically successful one in history, it must describe measurement whether it does so alone or together with
other equally fundamental elements of physics, if it is to be considered truly
complete, although, as with indeterminism, there also remains some question
regarding the uniqueness of quantum mechanics in this regard (*cf.* [389], p.
55). As Krips has pointed out, it is

"[not] necessary that [quantum theory] by itself generates [models
for the process of measurement], any more than we require a description of the functioning of devices for measuring charge, say, to
be entirely given from within electromagnetism (although arguably if
[quantum theory] is to be 'complete' then it must provide such models)... [Nonetheless,] one would expect that [quantum theory] places
certain restrictions on how to measure... (just as the laws of electromagnetism do...). These turn out to be quite strong..." ([281], p.
106)

In quantum mechanics, as in classical physics, measurements must be performed for one to find the values of physical magnitudes of a system, which
may have been prepared in an incompletely known state. Measurements require the physical coupling of some apparatus to the system that is the object

of measurement. In quantum mechanics, however, due to (at least one version of the) the Heisenberg relations, it cannot simply be assumed that measurements have no effect on the physical state or its description, as can be consistently assumed in classical mechanics. Recall that the distinctive element of quantum theory most clearly setting it apart from classical statistical mechanics is the superposition principle regarding quantum states, which implies the Heisenberg relations. A measurement result is given as the registration of an appropriate physical magnitude, a *pointer observable*. Pointer-magnitude values can differ from the those of the measurement memory register so long as there is a well defined *pointer function* serving to bring the elements of the two sets of values into one-to-one correspondence. In this way, quantum mechanics specifies the connection between the subject (observer) and observed phenomena to which Einstein referred. However, the Schrödinger dynamics typically fails to specify the result obtained.

The ongoing interest in the interpretation of quantum mechanics is strongly related to the degree of concern over the difficulty of providing a detailed account of measurement, which is far greater than, for example, providing one for classical measurements. Standard quantum mechanics assumes a formal relationship between physical magnitudes and eigenstates, namely, that a system magnitude is attributed a definite value *when and only when* the system is in a state that is an eigenstate-vector of the operator corresponding to that magnitude. Otherwise, at least on the Basic interpretation of the theory, no such precise value need be attributed to the physical magnitude of the system. This formal relationship, already mentioned in Chapter 1, known as the *eigenvalue–eigenstate link* [371], makes explicit the meaning of Postulate I of the standard formalism (*cf.* Appendix). Although some interpretations *do* deny this link, those that do inherit other problems as a result, as shown in the following chapter.[12]

The connection between measurement results and physical magnitudes is made explicit through the *calibration postulate*, namely, the requirement that if a system is in an eigenstate of an operator corresponding to a physical magnitude then a measurement of this magnitude leads *with certainty* to an outcome, indicating that the system is in this eigenstate at the moment the measurement ends [319]. A quantum state is thus typically characterized by its *preparation*—for example, as described by Schrödinger in his summary quoted in the introduction of this chapter—which may be performed in the same way as a measurement and may determine the system's quantum state, allowing the prediction of its future quantum state. In this way, one can view probabilities specified by the quantum state as having an implicitly conditional character. However, the conditional character of quantum probabilities does not automatically render them epistemic in nature. The state may (or may not) predict the outcomes of future measurements with perfect accuracy, depending on whether or not the two projectors corresponding to the prepa-

---

[12] See also, for example, [137], p. 22.

ration and the measurement commute; in the former case, they may, whereas in the latter they don't.

The unusual role apparently played by measurement in quantum mechanics, by comparison with classical mechanics, when straightforwardly applying the theory to even the simplest measurement situation, has been one of the motives for the explicit interpretation of the quantum formalism, as evidenced by the Bohr–Einstein debate. Indeed, there is considered to be a *measurement problem* in quantum theory because, for example, when one takes the Schrödinger evolution to describe the closed system constituted by the measuring apparatus and the system under measurement, including their environments when appropriate, absurdities appear due to the unitary character of this evolution.[13] Such a description predicts a number of different but equally valid measurement outcomes if one assumes measurement outcomes to exist whenever the appropriate one-to-one correlation of measuring apparatus states and object system states occurs and the superposition principle is enforced.[14] The measurement problem is also referred to as the problem of 'the reduction of the wave packet' to the post-measurement eigenstate or of 'the actualization of potentialities,' that is, of the appearance of individual measurement outcomes (eigenvalues). The significance of this under a given interpretation of quantum mechanics ultimately relates to some reliance on the eigenvalue–eigenstate link or on causality.

Comparing the standard quantum predictions with the observed results, one can fairly easily see why such a description of measurement fails (*cf.*, *e.g.*, [87], [371]). Consider a measuring apparatus initially in an eigenstate $|p_0\rangle$ of its pointer magnitude. Take the system measured for a magnitude of interest having corresponding Hermitian operator O with discrete non-degenerate eigenvalues $\{o_j\}$ to be in any eigenstate $|o_i\rangle$ before the measurement process begins. Finally, assume that the latter remains unchanged during the measurement process, so that the physical magnitude measured is that before as well as that after measurement has finished. Measurement should then result in composite-system state transformations $|\Psi_j^{(i)}\rangle \equiv |p_0\rangle|o_j\rangle \to |\Psi_j^{(f)}\rangle \equiv |p_j\rangle|o_j\rangle$, for each value of j that is a possible measurement outcome. However, the system being measured must also be capable of being measured were it instead initially *not* in an eigenstate of O but instead the state $\sum_j a_j|o_j\rangle$, because that is also a state allowed by the superposition principle. Because the temporal evolution operator acting on the composite system of the apparatus subsystem and the measured subsystem—assuming they form a closed system—is unitary, it acts linearly on state-vectors and the formal description of the measurement process involves a transformation of the form $|\Psi\rangle \equiv |p_0\rangle \sum_j a_j|o_j\rangle \to |\Psi'\rangle \equiv \sum_j a_j|p_j\rangle|o_j\rangle$.

---

[13] This is option discussed in detail in the account of the Collapse-Free interpretation of quantum mechanics given in Chapter 3.

[14] This has been seen as both a weakness and a strength in the case of the Collapse-Free approach, *cf.* 3.4.

Thus, in the statistical operator description, $\rho(0) \to \rho'(t) = U(t)\rho(0)\, U(t)^\dagger$, where

$$\rho = P(|\Psi\rangle)\,, \quad \rho' = P\left(\sum_j a_j |p_j\rangle |o_j\rangle\right)\,; \tag{2.1}$$

there is a transition from a pure state to a pure state because unitary transformations preserve purity. However, what one requires of measurement is that the transformation of the overall system result in a final state of the form

$$\rho^{(f)} = \sum_j |a_j|^2 P(|\Psi_j^{(f)}\rangle)\,, \tag{2.2}$$

that is, a mixed state describing a collection of distinct states with probabilities $|a_j|^2$, where a specific definite outcome is obtained for each measurement. Because the composite system evolves into a coherent superposition involving several distinct measuring system states for the ensemble, rather than just one, this does not constitute a good measurement description. Indeed, any unitary evolution, including that given by the Schrödinger equation, predicts that measurement of the quantity O yields neither a definite outcome nor an appropriate mixture. Furthermore, including the pure environmental state in the description makes no difference in this regard.

This failure presents difficulties under a number of interpretations of quantum mechanics. On the Basic interpretation, the Schrödinger cat thought experiment and the Wigner friend experiment, both discussed below, strikingly illustrate the measurement problem, which was to an extent anticipated by the EPR thought experiment. Furthermore, when not introducing state collapse during measurement, again if the above sort of correlation between measuring apparatus and system is taken to constitute a successful measurement as is naturally assumed, for example, in the absence of the introduction of some additional psychophysical process *à la* Wigner, an exponential number of differing sequences of outcomes will arise when *sequences* of measurements are performed.

2.5 Elements of Quantum Measurement Theory

The quantum measurement problem is the result of considering the measurement process to take place in the same way as does any other quantum physical process, as discussed by Bell in the quote provided in the previous section. A natural way of avoiding the problem while continuing to treat measurement devices as quantum systems is to make a distinction between measurements and other physical processes. Measurement has been viewed via two 'stages.'

The first stage is that of the system–apparatus quantum correlation discussed in the previous section, sometimes called *pre-measurement*, which, although necessary, is in itself insufficient for a successful measurement, as just seen. The second stage is a change of state of the measured system, the stage

in which a measurement outcome appears. Von Neumann explained the need
for a two-stage description as follows.

"Why then do we need the special process 1. for the measurement?
This reason is this: In the measurement we cannot observe the system
S by itself, but must rather investigate the system **S+M**, in order
to obtain (numerically) its interaction with the measuring apparatus
M. The theory of measurement is a statement concerning **S+M**, and
should describe how the state of **S** is related to certain properties of
the state of **M** (namely, the positions of a pointer, since the observer
reads these). Moreover, it is rather arbitrary whether or not one
includes the observer in **M**, and replaces the relation between the **S**
state and the pointer positions in **M** by the relations of this state
and the chemical changes in the observer's eye or even the brain (*i.e.*
to that which he has 'seen' or 'perceived')." ([477], Section V.2)

In von Neumann's formal explication of this process, the first stage takes place
entirely in accordance with the usual unitary Schrödinger equation (process
2.), whereas the second stage involves a non-unitary change of state (process **1.**); the apparatus includes the physical system from which measurement
outcome data can be obtained, subject to the awareness of the measuring
agent.

The discontinuous change of quantum state in the second stage, associated
with the measurement outcome coming to be known, accords with the *the repeatability hypothesis*, that an immediate repetition of the measurement with
certainty yields the same result as the initial measurement. This second stage
was given a specific form, now known as the (original) *projection postulate*,
used by Dirac, Heisenberg, Pauli, and others beginning in the late 1920s. The
name *projection postulate* for the prescription was first given by Henry Margenau in 1958 [310]; before then, it was typically referred to as a *quantum
jump*. Dirac described quantum jumping and corresponding state change, as
follows. "[A] measurement always causes the system to jump into an eigenstate
of the dynamical variable being measured" [139]. That is, there is a "quantum
jump" to the state corresponding to the outcome.[15] This description of state
change is sometimes referred to as the *collapse of the wave-function*, as if it
were a lawful dynamical process rather than an instantaneous indeterministic
change.

One typically finds differences in the descriptions of measurement among
interpretations of the theory. For example, on the Copenhagen interpretation, discussed in Section 3.3 below, as articulated by Bohr in particular, the
measurement apparatus and process need not necessarily be given an explicit
quantum mechanical description as above but must only be capable of description in *classical* terms (*cf.* [322]). Heisenberg described the occurrence of a

---

[15] The projection postulate is a central element of the Basic interpretation of quantum mechanics, *cf.* Section 3.2.

quantum jump in a way that emphasizes the subjective character of the observation associated with (selective) measurement. "[O]bservation itself changes the probability function discontinuously; it selects of all the possible events the actual one that has taken place" ([219], p. 54). Fellow *Copenhagener* Pauli viewed the need for a random projection as arising naturally from the fact that the interaction between the measuring system and the measured system is "in many respects intrinsically uncontrollable," in particular in relation to the idea, rejected by Heisenberg in response to Bohr's cautionary comments, that the uncertainty relations might be similarly understood, as discussed in Section 1.2 [338].

On the Collapse-Free interpretation of the theory, the distinction between the above two sorts of ("relative state") change is not fundamental. Instead, it is understood (in most versions) to result from the emergence of different observational perspectives. On it, measurement is entirely described by the Schrödinger evolution acting in the tensor-product space for the composite system formed by the measured system, the measuring system, and the entire environment of the two, within a superposition of states containing correlations between all of these systems formally similar to that of Equation 2.6; ultimately, one considers this to involve a unique 'wave-function of the universe,' an elaborate superposition state that never 'collapses.' Classical phenomena are also sometimes viewed as naturally emerging in it due to the pervasive effect of quantum state decoherence. Those portions of the resulting ramifying set of situations in which different sequences of measurement outcomes are obtained by measurers are assumed in some way to be inaccessible to each other [319, 410]. This interpretation thus represents an attempt to avoid the quantum measurement problem at the level of the entire universe rather than invoking physical state projection at the fundamental level. The Collapse-Free treatment was first carefully investigated by Everett with the support of John A. Wheeler [135], although it was eventually rejected by the latter. The various versions of the resulting Collapse-Free interpretation and its measurement-related problems are discussed in Section 3.5.

In von Neumann's statistical operator formulation, the change of state during measurement of a quantity, typically one not commuting with that of the preparation, is expressed explicitly by the rule that, when subject to measurement, a quantum system initially in a pure state $P(|\psi\rangle)$ evolves, typically non-unitarily, into a mixed state ρ':

$$P(|\psi\rangle) \longrightarrow \rho' = \sum_i \left(\langle \psi_i | P(|\psi\rangle) | \psi_i \rangle \right) P(|\psi_i\rangle) , \tag{2.3}$$

where the projectors $P(|\psi_i\rangle)$ onto the eigenvectors $|\psi_i\rangle$ sum to the identity. The weights $\langle \psi_i | P(|\psi\rangle) | \psi_i \rangle = w_k$ sum to unity and correspond to probabilities that the values of the physical magnitude being measured are found to be those of the subensembles corresponding to the projectors $P(|\psi_i\rangle)$, after the system was prepared in the state $|\psi\rangle$; see Figure 2.1. The state is thus no longer coherent in the *measurement basis* $\{|\psi_i\rangle\}$.

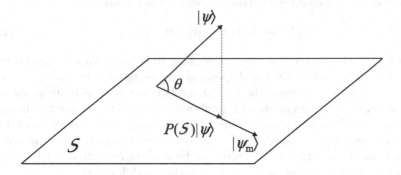

Fig. 2.1. Projection of a quantum state-vector $|\psi\rangle$ into a vector subspace $\mathcal{S}$ by a projector $P(\mathcal{S})$. The projection of $|\psi\rangle$ onto a ray corresponding to $|\psi_m\rangle$, with which it makes an angle θ, is shown here; the probability for this transition to occur is $\cos^2\theta$. A von Neumann measurement corresponds to the set of possible such projections onto a complete orthogonal set rays of the Hilbert space being measured. In a Neumann-Lüders measurement, the projections, then denoted by P_k rather than $P(|\psi_k\rangle)$, are onto subspaces that are not necessarily rays as shown here.

Projectors corresponding to any pair of possible outcomes i and j are related by $P(|\psi_i\rangle)P(|\psi_j\rangle) = \delta_{ij}P(|\psi_i\rangle)$; for each possible outcome, the same subspace would projected to were the measurement immediately repeated, in accordance with the repeatability hypothesis. In an arbitrary basis $\{|\alpha_i\rangle\}$, the matrix elements of the final statistical operator are

$$[\rho']_{ij} = \sum_k \langle\alpha_i|\psi_k\rangle\langle\psi_k|\psi\rangle\langle\psi|\psi_k\rangle\langle\psi_k|\alpha_j\rangle = \sum_k w_k[\rho'_k]_{ij} \,, \qquad (2.4)$$

where $w_k = |\langle\psi_k|\psi\rangle|^2$ and $[\rho'_k]_{ij} = \langle\alpha_i|P(|\psi_k\rangle)|\alpha_j\rangle$, exhibiting the fact that, in general, von Neumann's Process 1 takes pure states to mixtures described by the weights w_k. That is, these probabilities can be understood as standard probabilities. Whenever the state is measured in a basis different from the one in which it was prepared, that is, in general, Process 1 gives rise to a non-unitary change in state and is irreversible [477]. When a subensemble, corresponding to a given value of k, constituting a proportion w_k of the total normalized ideal ensemble, is then also *selected*—in the case of an individual system by its being actualized, which happens with probability w_k—one has $P(|\psi\rangle) \rightarrow P_k|\psi\rangle$, the result generally having norm $w_k \neq 1$. The resulting ensemble state must then also be renormalized through division by w_k, and the state may be describable by the statistical operator $P(|\psi_k\rangle)$. Process 1 together with pure subensemble selection constitutes a maximal selective measurement.

By contrast, during the continuous Schrödinger evolution,

$$\rho(0) \to \rho'(t) = U(t)\rho(0)\,U(t)^{\dagger} \,, \tag{2.5}$$

where $U(t)$ is the unitary operator describing that 'automatic' temporal evolution, the purity of the statistical operator always remains unchanged. As von Neumann emphasized, when the interaction between a measurement apparatus and a system being measured is analyzed, the unitarity of the Schrödinger equation describing the state evolution of the composite system formed by these two systems provides consistency between alternative descriptions of system behavior in which the division (or *Heisenberg Schnitt*, or *cut*) between measuring system and measured system is chosen differently.

Schrödinger vigorously objected to the introduction of Process 1, due to its evident inconsistency with the usual Process 2, in that

> "any *measurement* suspends the law that otherwise governing the continuous time-dependence of the ψ-function and brings about in it a quite different change, not governed by any laws, but rather dictated by the result of the measurement. But laws differing from the usual ones cannot apply during a measurement, for objectively viewed it is a natural process like any other, and it cannot interrupt the orderly course of natural events." ([394])

Therefore, Schrödinger argued, 'wave-function collapse' could not be a dynamical process; his objection is interpretational and distinguishes Schrödinger's particular views as to how quantum mechanics ought ultimately to be articulated, although he, like Einstein, never arrived at a full interpretation of quantum mechanics with which he was satisfied (*cf.* [49]).

A selective measurement of an observable is said to be *maximal* (or *complete*) when it provides fully distinct values for the quantity measured, so that no further knowledge of its preparation can be obtained by further measurement of the observable. For such measurements, the projection as described by von Neumann's Process 1 in its original form is entirely satisfactory. However, if instead the measurement performed is capable of discriminating only sets of values, the measurement is said to be *non-maximal*; in that case, it provides incomplete knowledge of the observable. Consider, for example, the measurement of a spin-1 system (or *qutrit*), which is a quantum system possessing a trivalent observable O with eigenvalues $o_i = -1, 0, 1$; a maximal measurement will have three possible outcomes, one for each of the possible values. By contrast, a measurement with only two outcomes, say "$-$" for system observable values -1 or 0, and "$+$" for system observable value $+1$, is a non-maximal measurement. An example situation wherein the latter is realized is a (imperfect) Stern–Gerlach type apparatus acting on a spin-1 system such that a particle with z-spin $+1\hbar$ enters a distinct spatial beam downstream from the magnet but particles with spins $0\hbar$ or $-1\hbar$ are not separated and enter a common, second beam.

If the outcome for a measurement of an observable corresponding to a sub-space of finite dimension *greater than one*, then the von Neumann projection postulate, including subensemble selection and renormalization, prescribes the process

$$\rho \longrightarrow \rho' = P_k \,, \tag{2.6}$$

where the projector is written P_k, rather than $P(|\psi_k\rangle)$, because the projection is made onto a subspace of dimension greater than one. However, this rule dictates a system state after measurement that is *independent* of the details of the state before the measurement, beyond those pertinent to the measurement outcome itself, as can be seen mathematically in that state ρ associated with its prepration does not appear in the description of the resulting state ρ'. Thus, the von Neumann prescription fails to maintain the distinction between initially pure states and initially mixed states, and so fails to preserve coherence of pure states in non-maximal measurements, which is not necessarily lost because state decoherence does not always occur in non-maximal measurements. Nonetheless, the original projection rule can be easily and naturally adjusted to characterize more precisely measurements of physical magnitudes whose eigenvectors have degenerate, that is, non-unique eigenvalues. The prescription for describing the state-change as a result of a selective quantum measurement that maintains the distinction between the two sorts of preparation in such cases is the *Lüders projection* (*Lüders rule*),

$$\rho \longrightarrow \rho' = \frac{P_k \rho P_k}{\text{tr}(\rho P_k)}. \tag{2.7}$$

By contrast to the original projection rule, the Lüders prescription of state after measurement clearly *is* dependent on the state of the system before measurement; in particular, the values of successive measurements under this rule will coincide when another measurement is made between successive measurements of O of an observable *compatible with* O in the sense that the corresponding operators commute [273, 305]. That is, under the above rule, first introduced by Gerhard Lüders, if one prepares two ensembles of systems in the state ρ, the first being measured for some observable that is compatible with O and the second having O measured *first*, the relative frequencies of the values of the compatible observable for those two cases are the same and yield pure subensembles from pure ensembles. Furthermore, this is the *only* projection rule for which this is true [432], as shown in the next section. This rule is now commonly considered to be the appropriate general description of precise measurements in the standard theory. It can also be viewed as a consequence of Feynman's rules for computing quantum probability amplitudes based on the indistinguishability of processes [427], discussed in Section 3.8.

An important aspect of the approach to measurement of von Neumann is that physical theories are required to accommodate a physical correlate to subjective perception. This is the principle of *psychophysical parallelism*:

"it is a fundamental requirement of the scientific viewpoint—the so-called principle of the psycho-physical parallelism—that it must be possible so to describe the extraphysical process of the subjective perception as if it were in reality in the physical world—*i.e.*, to assign to its parts equivalent physical processes in the objective environment, in ordinary space." ([477], p. 418-419)

He considered chains of physical systems connecting human observers with physical phenomena, as in the situation of Schrödinger's cat discussed below, concluding that one "must always divide the world into two parts, the one being the observed system, the other the observer" (Heisenberg's *Schnitt*), despite the fact that, mathematically, this inclusion can be shown to be arbitrary. That is, in quantum mechanics, one ultimately must be able to consider a bipartite decomposition of composite system involved in a measurement into *observed system* and *observer system*, even if the former lies within a human observer's body (*cf.* [477], p. 419).

In relation to this principle, von Neumann stated that

"[Bohr] was the first to point out that the dual description [of measurement into the usual evolution and state projection] which is necessitated by the quantum mechanical description of nature is fully justified by the physical nature of things that it may be connected with the principle of the psycho-physical parallelism." ([477], p. 420)

This requires only that the *abstraktes Ich* ('abstract I') be associated with one component of the division and that the object of its attention be associated with the other. In the case of Schrödinger's cat, if cats have consciousness of the relevant sort humans have for observation, a conscious cat would certainly be aware that it is alive whenever it is so. If one is concerned primarily with preserving the appearances, then there being two distinct state-evolution processes in the standard formulation of quantum mechanics is not a grave concern; Process 1 comes into play only when and as soon as the observer's interaction with the observed occurs and is consistent with its observations.

However, there remains the issue that "the principle of the psycho-physical parallelism is violated, so long as it is not shown that the boundary between the observed system and the observer can be displaced arbitrarily..." ([477], p. 421). To address this, von Neumann also explicitly demonstrated that changing the boundary between the above two portions of the world has no affect on quantum mechanical predictions, that is, that "the boundary between the two parts is arbitrary to a very large extent" ([477], p. 420), although he did not demonstrate that the stage of the progressing chain of measurement subprocesses from the system to the observer's brain at which the application of (the measurement) Process 1 occurs *in all cases* does not create inconsistency in quantum predictions. In any event, there is no evidence that the superposition or projection of 'states of consciousness' can occur.

This point was considered by Wigner, who argued that to be adequate quantum mechanics must at least provide a prescription for the precise *physical* circumstances in which Process 1 should take place [499], if the quantum measurement problem is truly to be resolved. Wigner identified six views regarding the relevance of Process 1 to this problem: Everett's ("there is no need to assume a reduction"), Fock's ("measuring instruments must be described classically"), Ludwig's ("quantum mechanics does not apply to macroscopic systems"), the London—Bauer elaboration [302] following von Neumann's above (adding a collapse postulate), the view that quantum mechanics never describes individual events (which Wigner considers ultimately solipsistic on the basis of the Friend thought experiment discussed in Section 2.7), and Zeh's ("the state of isolation [of the measuring instrument and measured system] is very difficult to maintain... their state-vector... will soon go over into a mixture")—*cf.* [502], p. 1. As mentioned above, Wigner ultimately held the view that the *act of observation* by the conscious agent making a measurement plays a fundamental role in inducing a wave-funtion projection, through a process now know as the *von Neumann–London–Bauer collapse*.

The quantum-logical approach discussed in Section 2.1, assuming state projection takes place when a measurement outcome is registered by a conscious agent, has been used in the attempt to solve the measurement problem, in the sense of explaining how measurements can have objectively definite outcomes and still accord with quantum equations of motion. The idea is that this problem will not arise if one assumes value-definiteness at *all* times rather than only when the quantum system is in an eigenstate of the observable of interest: if both the measuring apparatus and the measured system have definite values for all physical magnitudes, then the outcomes would naturally arise despite the existence of the joint superposition of Equation 2.1. However, as already noted, the value-definiteness thesis is highly problematic.

Some have argued that the measurement problem is artificial, that is, "it is often held that the restrictions placed on the measurement process from within [quantum theory] are *too* strong in that they impose paradoxical requirements" ([281], p. 107). However, as Shimony has pointed out,

> "[it] has often been claimed that [the measurement problem] is specious, arising from a narrow or inadequate representation of the measurement process... however... the problem is a fundamental anomaly, which cannot be lightly dismissed. Some proposals for solving this problem [postulate] nonlinear or stochastic modifications of quantum mechanics... Our conclusion is that the main conceptual innovations of quantum mechanics [objective indefiniteness, objective chance, objective probability, potentiality, entanglement, and quantum non-locality] are probably embedded permanently in physical theory, but that some further radical innovation will probably have to be made." ([410], p. 373)

Whatever the nature of this innovation, it appears increasingly likely that it will be related to quantum entanglement and have some connection to information theory.[16] Because the following section is somewhat technical, one may prefer on a first reading to proceed directly to Section 2.7.

2.6 Advances in Quantum Measurement Theory

An important result obtained by considering quantum mechanics from the perspective of logic is the demonstration, mentioned above, that Lüders' rule for the state after quantum measurement is the *only one* that prescribes the correct generalization of conditional probability to the quantum realm, in the sense that it provides the classical rule for conditional probability when the two operators related to the pertinent events commute [86, 244, 432].

To see how the Lüders rule is naturally picked out, consider a generalized probability function q on the set of subspaces of Hilbert space $\mathcal{H}$. Because any such function is additive over orthogonal subspaces, it is defined by its probability assignments to the one-dimensional subspaces of $\mathcal{H}$. In addition, Gleason's theorem shows that q is provided via a density operator, ρ_q. Finally, if such a q assigns the value 1 for a projector Q, then it assigns the value 0 to projectors onto rays in the *complement* of the subspace onto which Q projects so that $\rho_q|v\rangle = \mathbf{0}$ for all (not necessarily normalized) vectors along such rays. For any such ray $\bar{v}$, $q(\bar{v}) = \mathrm{tr}(\rho_q|v\rangle\langle v|) = \mathrm{tr}(\rho_q(Q+(I-Q))|v\rangle\langle v|) = \mathrm{tr}(\rho_q Q|v\rangle\langle v|)$, for all $|v\rangle \in \bar{v}$, because of the linearity of the trace and because $(I - Q)|v\rangle = \mathbf{0}$ by the definition of $\bar{v}$. Thus, $q(\bar{v}) = |a|^2 q(|u\rangle\langle u|)$ for some $a \in \mathbb{C}$, where $|u\rangle$ is the normalized vector along the ray, so that *any* q such that $\rho_q = 1$ is fully specified by the values it assigns to vectors in the ray onto which Q projects. Hence, for any generalized probability function p on the subsets of $\mathcal{H}$, there is a unique q on those subsets such that for all subspaces of the subspace onto which Q projects one has $q(P) = p(P)/p(Q)$ and in turn q is uniquely represented by ρ_q. Note then that

$$\frac{QPQ}{\mathrm{tr}(QPQ)} \tag{2.8}$$

is a statistical operator, so that one can write

$$\rho_q = \frac{QPQ}{\mathrm{tr}(QPQ)} . \tag{2.9}$$

Therefore, Lüders' rule provides the *unique* generalized probability function q such that, for all projectors into subspaces of the space onto which Q projects, $q(P) = p(P)/p(q)$. This uniqueness result bolsters standard quantum measurement theory, despite the presence of the quantum measurement problem

---

[16] One sophisticated move along these lines is discussed later in Section 3.8.

which might be addressed by an innovative interpretation of the quantum formalism.

One can further advance the description of processes within quantum mechanics along similar lines by continuing to elaborate and generalize the standard theory of quantum measurement, keeping in mind the basic requirements of probability theory. Consider for a moment quantum measurement in a broad mathematical setting where the spectrum of the Hermitian operator O representing a physical magnitude may be continuous, so that a measurement might find its value within a Borel set $\Delta \in \mathbb{R}$ and leave the state of the system with support (O, Δ) with respect to O. A projector $P_O(\Delta)$ from the spectral decomposition of O might describe the quantum mechanically maximally specified state of such a system. Although there are conceptual difficulties with such a description, unsharp measurements do provide a well-defined way of proceeding [95, 426].

For simplicity, let us again consider measurements in the more straightforward case of discrete spectra. *Unsharp measurements* are the class of quantum operations that are described by (normalized) positive-operator-valued measures (POVMs) [118]. Given a nonempty set S and a σ-algebra Σ of its subsets X_m, a *positive-operator-valued measure* E is a collection of operators $\{E(X_m)\}$ satisfying the following conditions.[17]

(i) *Positivity*: $E(X_m) \geq E(\emptyset)$, for all $X_m \in \Sigma$.

(ii) *Additivity*: for all countable sequences of disjoint sets X_m in Σ,

$$E(\cup_m X_m) = \sum_m E(X_m) . \qquad (2.10)$$

(iii) *Completeness*: $E(S) = \mathbb{I}$.

If the value space (S, Σ) of a POVM E is a subspace of the real Borel space $(\mathbb{R}, \mathcal{B}(\mathbb{R}))$, then E provides a unique Hermitian operator on $\mathcal{H}$, namely $\int_{\mathbb{R}} \text{Id } dE$, where Id is the identity map. The positive operators $E(X_m)$ in the range of a POVM are referred to as *effects*, the expectation values of which provide the quantum probabilities.

Each quantum state ρ induces an expectation functional on $\mathcal{L}(\mathcal{H})$, the space of linear operators on the Hilbert space $\mathcal{H}$. This provides well defined probabilities because the effects are bounded by $\mathbb{O}$ and $\mathbb{I}$, so that the ranges of the effect spectra are restricted to lie in the closed unit interval, due to the positivity and normalization of the POVM; the operator ordering $\leq$ on

---

[17] A *Borel σ-algebra* is the σ-algebra generated by the open intervals (or the closed intervals) on a topological space—for example, in $\mathbb{R}$—which are the *Borel sets*. The set S is often a standard measurable space, that is, a Borel subset of a complete separable metric space. Because such spaces of each cardinality are isomorphic, they are all measure-theoretically equivalent to Borel subsets of the real line, $\mathbb{R}$. The sequences here are taken to converge in the weak operator topology on $\mathcal{L}(\mathcal{H})$ [95].

the effects has the zero operator and identity as its upper and lower bounds. Only if the Hilbert space in question is $\mathbb{C}$ do the effects constitute a lattice, as described in the Appendix A.5; a complementation $^\perp$ defined by $E \equiv 1 - E$ exists that satisfies $(E^\perp)^\perp = E$ and reverses the operator order but is not an orthocomplementation, so that law of the excluded middle does not hold, however.[18] POVMs are thus the natural correspondents of standard probability measures, described in Section 2.2, in the operator space of quantum mechanics.

The probability of outcome m upon a generalized measurement of a pure state $P(|\psi\rangle)$ is given by

$$p(m) = \langle \psi | E(X_m) | \psi \rangle = \mathrm{tr}\big((|\psi\rangle\langle\psi|)E(X_m)\big) . \qquad (2.11)$$

When the state is mixed, this probability is given by

$$p(m) = \mathrm{tr}\big(\rho E(X_m)\big) \qquad (2.12)$$

(*cf.* Section 2.2). The output of a POVM measurement of the initial state is exhibited by post-measurement states and corresponding outcome probabilities $p(m)$. The post-measurement state ρ'_m of a system initially described by a statistical operator ρ under a POVM $\{E(X_m)\}$ is often taken to be

$$\rho'_m = \frac{M_m \rho M_m^\dagger}{\mathrm{tr}\big(M_m \rho M_m^\dagger\big)} , \qquad (2.13)$$

where each of the $E(X_m)$ can be written $M_m^\dagger M_m$, M_m being called a *measurement decomposition operator* (*cf.* [324]); in the special case that the M_m are projectors, this expression coincides with the Lüders–von Neumann measurement rule given by Equation 2.7—this can be seen by recalling that projectors are Hermitian and idempotent.

When, and only when, the measurement operators M_m are *projectors*—so that the POVM is a *projection-valued measure* (PVM)—are they *identical* to decomposition operators $E(X_m)$, in which case they are also multiplicative, that is, $E(X_m \cap X_n) = E(X_m)E(X_n)$ for all countable subsets of the corresponding set—equivalently, $E(X_m)^2 = E(X_m)$. When providing positive outcomes, POVM elements allow one to eliminate quantum states from consideration as describing the measured system.[19]

---

[18] Recall, as noted above in the first section of this chapter, the projection operators *do* form an orthocomplement lattice with just this order and complement. The lattice of projection operators $\bar{L}(\mathcal{H})$ has the *sharp properties* as its elements.

[19] An example of a POVM used in this way is the following [32]. Given the two projectors $P(\neg|\phi\rangle) \equiv \mathbb{I} - P(|\phi\rangle)$ and $P(\neg|\phi'\rangle) \equiv \mathbb{I} - P(|\phi'\rangle)$, where $\langle\phi|\phi'\rangle = \sin 2\theta$, one can construct a POVM $\{E_m\}$ with the elements $E_1 = P(\neg|\phi\rangle)/(1+|\langle\phi|\phi'\rangle|)$, $E_2 = P(\neg|\phi'\rangle)/(1 + |\langle\phi|\phi'\rangle|)$, $E_3 = \mathbb{I} - (E_1 + E_2)$. POVM measurements using $\{E_1, E_2, E_3\}$, for example, are more efficient for quantum key distribution and quantum eavesdropping than traditional measurements described by the projectors $\{P(\neg|\phi\rangle), P(\neg|\phi'\rangle)\}$. Similarly, POVMs sometimes allow quantum state tomography to be performed with improved efficiency.

The effects form a convex subset of $\mathcal{L}(\mathcal{H})$, the extremal elements of this subset being the projection operators. A collection of effects is said to be *coexistent* if the union of their ranges is contained within the range of a POVM. Any two quantum observables E_1 and E_2 are representable as PVMs on $(\mathbb{R}, \mathcal{B}(\mathbb{R}))$ exactly when $[E_1, E_2] = \mathbb{O}$, following from results of von Neumann for Hermitian operators. For POVMs, however, commutativity remains sufficient but is not necessary for coexistence [95]. The use of POVMs, that is, unsharp measurements thus allows one to circumvent the restriction of commutativity on measurements of noncommuting observables by including *unsharp properties*. A *regular effect* is an effect with spectrum both above and below $\frac{1}{2}$. One can define *properties* in general by the following set of conditions, given an effect A.

(i) There exists a property $A^\perp$;

(ii) There exist states ρ and ρ' such that both $\mathrm{tr}(A\rho) > \frac{1}{2}$ and $\mathrm{tr}(A\rho') > \frac{1}{2}$;

(iii) If A is regular, for any effect B below A and $A^\perp$, $2B \leq A + A^\perp = \mathbb{I}$.
   (This renders $\perp$ an orthocomplementation for the regular effects.)

The set of properties $\mathcal{E}_p(\mathcal{H}) = \{A \in \mathcal{E}(\mathcal{H}) | A \nleq \frac{1}{2}\mathbb{I}, A \ngeq \frac{1}{2}\mathbb{I}\} \cup \{\mathbb{O}, \mathbb{I}\}$ satisfies these conditions. The set of *unsharp properties* is then $\mathcal{E}_u(\mathcal{H}) = \mathcal{E}(\mathcal{H})_p/L(\mathcal{H})$. A POVM is an *unsharp observable* if there exists an unsharp property in its range [95]. Coexistent observables are those that can be measured simultaneously in a common measurement arrangement; when two observables are coexistent, there exists an observable the statistics of which contain those of both observables, known as the *joint observable*—typically, the two observables are recoverable as marginals of a joint distribution on the product of the corresponding two outcome spaces.

The technical advances in the theory of quantum measurement described in this section are formidable. However, beyond the first of the above, their effect on the quantum mechanical world view remains unclear in the absence of an interpretation of quantum mechanics that involves a novel theory of measurement that exploits them. In any event, the importance of measurement in quantum mechanics was recognized early on in the history of the theory; quantum measurement theory makes predictions that threaten basic elements of the traditional understanding of the physical world. We now consider two thought experiments illustrating this.

2.7 Schrödinger's Cat and Wigner's Friend

Two challenging thought experiments relating to the question of the sharpness of measurements and the experiences of observers were provided by Schrödinger and Wigner. Introduced in the decades following the formalization of quantum theory, they helped illustrate the surprising character of its predictions and continue to do so today. They suggest more than ever that radical innovations are needed to come to terms with the implications of the theory. As will be seen in the following chapter, the interpretations of quantum mechanics that have been offered since the introduction of these thought experiments continue to leave much to be desired, assuming that quantum mechanics remains one of our fundamental theories. As a result, these experiments remain important examples with respect to which one can compare new interpretations and consider the relevance of various physical effects, such as state decoherence. In particular and foremost among the thought experiments that followed the EPR experiment considered in the last chapter are those of Schrödinger's cat and Wigner's friend; the latter is essentially an extension of the former by the inclusion of an additional human observer.

Quantum mechanics allows one to conceive of situations clearly in contradiction with common sense, resulting from the application of the superposition principle in the macroscopic realm. Such an application is desirable given that it has the potential to serve as our most fundamental mechanical theory when proper care is taken in the relativistic case. Schrödinger conceived a number of physical examples, including what he called the "ridiculous case" of the now infamous cat in a box that appears both alive and dead, in which all matter is described quantum mechanically, in order to point out as starkly as possible the difficulties posed by standard quantum theory within a realist world view, such as that of common sense. Indeed, one of the characteristics of standard quantum mechanics is that both the measurement apparatus and the system being measured are treated exactly the same way as any other quantum mechanical object is treated.[20]

Before considering these two thought experiments in detail, it is helpful to have in mind at least one more recent example, beyond EPR's early reality criterion, of how *realism*, which is discussed further in the next chapter, has been related to quantum mechanics. A clear such example has been provided by Krips, who spelled out principles cohering with a realist world view in terms of the elements Q of the set of physical quantities as follows. "(Det Q) Physical quantities always have determinate values, (Pass Q) The measured value of Q in [state] S at [time] t = the value possessed by Q in S at t, and (NDQ) The value possessed by the measured quantity just after measurement = the value registered by the measurement, *i.e.* the measured value" ([281], Index of principles). This amounts to the conjunction of the value-definiteness thesis, a principle of faithful measurement, and the repeatability hypothesis. Although, as will be seen later, this involves a very strong form of realism, it is the sort of realism that Schrödinger appears to have had in mind.

---

[20] Note, however, that the Copenhagen interpretation can finesse this point by appealing to classical mechanics as an equally fundamental theory at the macroscopic scale; arguably, Bohr equivocated in this regard in that in one debate with Einstein he applied the uncertainty principle to at least one macroscopic measuring apparatus (*cf.* Section 3.3).

Roger Penrose has explained the difficulty posed by these experiments, particularly for the realist, as follows.

"Why, then, is there such reluctance to accept the state-vector as describing an actual physical reality?... 'Schrödinger's cat' (or some essential equivalent such as 'Wigner's friend,' etc.) It seems hard to believe in any actual 'reality' at the level of a cat which requires that such states of death and life can co-exist. The situation is, of course, more puzzling for Wigner's friend... According to *strict quantum mechanics*, it seems to me, ψ_t—representing some complex linear combination of a live cat and a dead one—should still have to provide an objective description of a reality [in the box]." ([344], p. 132)

Schrödinger himself had at one point hoped that the complex-square of the state amplitude of the wave-function could provide a complete direct physical description of quantum systems in ordinary three-dimensional space. On such an understanding of the wave-function, physical magnitudes in quantum mechanics are "smeared out," something which might be viewed as unobjectionable at scales below and up to the atomic scale (*cf.* [49], pp. 1-2). However, he found such a description objectionable at the truly macroscopic scale, such as that of a cat, because it contradicts experience.

Schrödinger's experiment is the following.

"A cat is penned up in a steel chamber, along with the following diabolical device (which must be secured against direct interference by the cat): in a Geiger counter there is a tiny bit of radioactive substance, *so* small, that *perhaps* in the course of one hour one of the atoms decays, but also, with equal probability, perhaps none; if it happens, the counter tube discharges and through a relay releases a hammer which shatters a small flask of hydrocyanic acid. If one has left this entire system to itself for an hour, one would say that the cat still lives *if* meanwhile no atom has decayed. The first atomic decay would have poisoned it. The ψ-function of the entire system would express this by having in it the living and the dead cat (pardon the expression) mixed or smeared out in equal parts." ([394])

In such an arrangement, under the standard quantum mechanical description of measurement, the answer to the question of whether the cat is alive or dead can be viewed as *indefinite* until the box is opened by an experimenter and the cat is observed. In von Neumann's conception, at the point of observation the observed system is the object of the attention of the *abstraktes Ich*, the essence of the observer, occuring in parallel with a state projection that accords with what is observed ([477], p. 421); until then, one might, at best, only 'approximately' (in some sense [14]) characterize the cat's vital status. The purpose of the example is to point out the apparent inadequacy of the quantum-mechanical characterization of this situation as expressed in his summary quoted at the end of the introduction to this chapter: Any cat is always either alive or dead, whereas quantum mechanics (apparently) does not describe the situation in that way, an apparently clear failure of the theory in relation to the measurement problem described in Section 2.4.

Although this thought experiment is easily visualized, there are reasons for doubting its efficacy. For example, one can question the very idea of applying quantum mechanics to entire functioning *biological* organisms, all questions of vitalism aside, as Heisenberg did.

> "Logically, it may be that the difference between the two statements: 'The cell is alive' or 'the cell is dead' cannot be replaced by a quantum theoretical statement about the state (certainly a mixture of many states) of the system." ([434])

It is unclear that 'alive' and 'dead' can be immediately reduced to well-defined mechanical quantities. If this theoretical step cannot be made, the example is a non-starter. Moreover, no rigorous experiment measuring such a biological quantum superposition has been performed.[21] This appears to be difficult precisely because it is unclear what *quantum physical* magnitude one would measure. Physicalists are responsible for providing a careful account of how this should be done in practice. If this example is to be taken seriously as an objection to the completeness of quantum theory such an account is clearly required. Nonetheless, the thought experiment retains considerable force in that the vital signs of any organism are clearly physical. Furthermore, Einstein provided a simpler and less easy to criticize example of the same basic sort which avoids the question of applying quantum theory directly to biology in this way: He asked simply whether it makes sense that his *bed* would jump into a definite state only when he or another observer enters the bedroom and looks at the bed. Einstein's example avoids the difficulties presented by involving the vital state of the cat as a physical observable but without the shock value of the consideration of the death of a potential pet.[22]

The Schrödinger cat experiment has been more effectively criticized on the basis that it assumes that the collection of systems from the radioactive sample to the observer who opens the box constitutes a genuinely *closed* system, which it is clearly not in that, for example, the cat also must breathe air to live (*cf.* [391]); Heisenberg noted that the state of this collection will, *de facto*, be a mixed one. However, one can in principle straightforwardly include *all* pertinent objects in any way interacting with these systems without changing its fundamental implication, assuming again that cat biology and all the necessary air molecules are in the domain of quantum theory. Because the entangled state of the composite system in the 'chain' of interactions from the

---

[21] However, work of Zeilinger and co-workers demonstrating the interference of C_{60} molecules, with the distant goal of similarly using *viruses*, shows that the avenue for the experimental investigation of such examples remains open [9].

[22] This example was offered in a personal communication with Putnam, recounted in [367].

radionuclide to the cat and its environment involves a 'fuzzy' micro-physical object, the quantum-mechanical description of the cat will be 'fuzzy' as well. The resulting state of all objects in this chain of interaction will involve the (at least pairwise) entanglement of systems as they interact. This contradicts common sense and what is observed in the natural world, at least at the level of *cats*, in that a heart monitor on a cat records it as either alive or dead and not *both* and not in a superposition of two such states, not to mention one of the cat together with all the physical elements of the chain of measurement.

A further element of paradox arising from the quantum mechanical description of this measurement situation, in Schrödinger's view, is the following.

"It is rather discomforting that the theory should allow a system to be steered or piloted into one or the other type of state at the experimenter's mercy in spite of his having no access to it," ([394], p. 556)

as is the case when one considers the effect of measurement on entangled systems, as strikingly pointed out by delayed-choice versions of the 'quantum eraser' experiment discussed in Chapter 1. That is, the state of the cat in this example appears to be subject to the choice of measurement of one looking in the box. Nonetheless, von Neumann, in his theory of measurement, provided a demonstration that the changing of boundary between the observing and observed portions of the universe has no essential effect on quantum mechanical predictions and places no requirement to the effect that the application of (the measurement) Process 1 occurs at a specific point in the process of measurement, except to the extent required by psychophysical parallelism.[23]

Wigner argued that an objective prescription for the physical stage at which Process 1 takes place, which is not provided by quantum theory, is needed for a complete characterization of measurement [499]. If the cat's consciousness of its situation were considered part of a self-measurement on its part, this would certainly pertain. More than a quarter century after the introduction of Schrödinger's cat experiment, Wigner offered another thought experiment to probe this question, along the following lines. Consider a system S that flashes when in one state, ψ_1, and does not flash if in an orthogonal one, ψ_2. The friend, F, observing the flash will have corresponding states χ_i, in respective cases. Thus, the bipartite composite system SF will have corresponding states $\psi_i \otimes \chi_i$. If the system, much like Schrödinger's cat, is in a superposition state $\alpha\psi_1 + \beta\psi_2$, the linearity of quantum mechanics dictates that the state of SF must be $\alpha\psi_1 \otimes \chi_1 + \beta\psi_2 \otimes \chi_2$. The probability of the friend seeing the flash will accordingly be $|\alpha|^2$, and that of not seeing the flash will be $|\beta|^2$. In order to provide a good answer to Wigner's query to the friend about his observation, the friend must receive a measurement result when observing S that accords with his answer.

---

[23] To make this connection more specific, a prescription for coordinating psychological or subjective time with physical time might be required.

Wigner noted that, "So long as I maintain my privileged position as ultimate observer," there will be no logical inconsistency with von Neumann's theory of measurement [499]. However, after the experiment, if

"I ask my friend, 'What did you feel about the flash before I asked you?' He will answer, 'I told you already, I did [did not] see a flash,' as the case may be. In other words, the question whether he did or did not see a flash was already decided in his mind, before I asked him. If we accept this, we are driven to the conclusion that the proper wave function immediately after the interaction of the friend and object was already either $\psi_1 \times \chi_1$ or $\psi_2 \times \chi_2$ and not the linear combination $\alpha(\psi_1 \times \chi_1) + \beta(\psi_2 \times \chi_2)$. This is a contradiction because the state described by the wave function $\alpha(\psi_1 \times \chi_1) + \beta(\psi_2 \times \chi_2)$ describes a state that has properties which neither $\psi_1 \times \chi_1$, nor $\psi_2 \times \chi_2$ has. If we substitute for 'friend' some simple physical apparatus, such as an atom which may or may not be excited by the light-flash, this difference has observable effects and *there is no doubt that* $\alpha(\psi_1 \times \chi_1) + \beta(\psi_2 \times \chi_2)$ *describes the properties of the joint system correctly,* [whereas] *the assumption that the wave function is either* $\psi_1 \times \chi_1$ *or* $\psi_2 \times \chi_2$ *does not.* If the atom is replaced by a conscious being the wave function, $\alpha(\psi_1 \times \chi_1) + \beta(\psi_2 \times \chi_2)$... appears absurd because it implies that my friend was in a state of suspended animation before he answered my question." ([499], p. 293)

That is, if the friend is to have a specific result before being asked, the joint state cannot have been the superposition state in the basis defined by the definite results. From this, Wigner concluded that the result must have become determinate at the moment the friend made his measurement *as a result of his being conscious* and that (i) "it follows that the being with a consciousness must have a different role in quantum mechanics than the inanimate measuring device" and (ii) "the quantum mechanical equation of motion cannot be linear" ([499], p. 294). He argued that the alternative denies the friend consciousness and is tantamount to a form of solipsism. Furthermore, as he put it,

"In von Neumann's view... the content of [the observer's] mind is not obtainable by means of the laws of the theory. It is either that these laws do not apply to the functioning of the mind (whatever that word means) or that the conscious content of the mind is not uniquely given by its state vector, *i.e.* by the quantity which quantum mechanics uses for the description of all objects." ([502], pp. 122-123)

Thus, Wigner believed that the problem is to be solved by assenting to a special role for consciousness in quantum theory. However, most physicists view this solution as even more objectionable than the original problem, exactly because of its explicit appeal at the fundamental physical level to processes that are not fully physical.

3

Interpretations of Quantum Mechanics

Quantum mechanics was initially formulated in what appeared to be two fundamentally distinct ways, as *Matrix mechanics* with operator matrices satisfying $[Q_i, P_j] = i\hbar\delta_{ij}$, and *Wave mechanics* with the function ψ satisfying $i\hbar(\partial\psi(\mathbf{q})/\partial t) = H(\mathbf{q}, \mathbf{p})\psi(\mathbf{q})$, the former developed by Heisenberg, Born and Jordan [73, 74], and the latter by Schrödinger [393]. Dirac, Jordan, Pauli, and Schrödinger subsequently provided arguments for the equivalence of these two approaches. However, the Dirac–Jordan equivalence proof made use of the Dirac δ 'function,' which is not well defined as a function because it takes an infinite value at a single point although it can be given a proper definition as distribution (or "improper function"). Von Neumann finally rigorously proved the equivalence and derived the hydrogen atom energy eigenvalue spectrum by making use of Hilbert space, a separable complete vector space with an inner product and a countable, potentially infinite basis (*cf.* [282] and [281], Appendix 4), capturing the theory's mathematical essence [473–477]. Much later, exploring some ideas of Dirac involving the Lagrangian and action [139], Feynman also produced a third, mathematically equivalent formulation of the theory [168].

While quantum theory was being developed and such mathematical details were being worked out, new world views were also considered by its founders, as they engaged in ongoing discussions and debates and interacted within the wider physics community. As Rudolf Peierls put it,

> "[The] general community of physicists had to go through the same arguments, and to face the same difficulties, which result from the fact that our intuition is formed as a result of our every day experience on a scale on which neither the relativistic refinements nor the quantum effects are noticeable. Most of them learned to accept the new ideas, and to understand that they are an essential part of the logic of quantum mechanics." ([342], p. 25)

G. Jaeger, *Entanglement, Information, and the Interpretation*   95
of Quantum Mechanics, The Frontiers Collection,
DOI 10.1007/978-3-540-92128-8_3, © Springer-Verlag Berlin Heidelberg 2009

In the *Mathematische Grundlagen der Physik*, where the standard Hilbert-space formalism was laid out in mathematically elegant form, von Neumann included the quantum mechanical theory of measurement outlined in the previous chapter, which he viewed as an essential element of that formulation. In addition, he raised the issue of interpreting the theory, which he also viewed as an essential activity.

> "The object of this book is to present the new quantum mechanics in a unified representation which, so far as it is possible and useful, is mathematically rigorous...In particular, the difficult problems of interpretation, many of which are even now not fully resolved, will be investigated in detail. In this context the relation of quantum mechanics to statistics and to the classical statistical mechanics is of special importance." ([477], p. vii)

In particular, the measurement theory provided in that work involves the introduction of specific elements of an interpretative character, for example, the explication of the relation between probability and known measurement outcomes and the assumption of psychophysical parallelism. Von Neumann's interpretation, which is sufficiently close to that of Dirac to be considered the same, is here called the *Basic interpretation*.

There are differing views as to exactly what constitutes an interpretation of quantum mechanics. Let us consider several characterizations. Mittelstaedt has required any interpretation to satisfy several requirements.

> "Any interpretation of quantum theory should provide interrelations between the theoretical expressions of the theory and possible experimental outcomes. In particular, any interpretation of quantum mechanics has to clarify which are the theoretical terms that correspond to measurable quantities and whether there are limitations of the measurability [*sic*]...Another essential problem is the question of what kind of experimental results could correspond to the Sch[r]ödinger wavefunction, which turns out to be a very important theoretical entity." ([319], p. 1)

Paul Teller has described an interpretation as follows.

> "I take an interpretation to be a relevant similarity relation hypothesized to hold between a model and the aspects of actual things that the model is intended to characterize." ([445], p. 5)

Bub's view of interpretation differs from both of these. He sees it as

> "an account that shows in what respects the theory is related to preceding theories." ([85], p. 143)

As these examples illustrate, there has been a tendency, for better or worse, to inject philosophical biases or concerns into the very conception of *interpretation*, which is largely avoided in this chapter.

Several different interpretations of quantum mechanics are considered here. A cluster of further recent efforts relating to quantum information are considered in the following chapter. Interpretations can be characterized by a set of more or less simply stated theses. The relationship between quantum mechanics and statistics typically differs from one interpretation to another, as does the suggested relationship between the formalism and the world including conscious agents. Hence, more generally than the characterizations above, the *interpretation of quantum mechanics* is taken here primarily to mean the specification of the epistemic, metaphysical, and operational significance of theory, most importantly of its state operator ρ or pure state-vector $|\psi\rangle$. This sense of the term *interpretation* applies to all the 'interpretations' that have been identified, except for those involving only some modification of the dynamics, that is, the theory itself. It also differs from that wherein interpretation is constituted almost entirely by the specification of an ontology, that is, existents, basic categories, and their relationships—for example, Reichenbach's sense [373]. Often, in a given interpretation, only states $|\psi\rangle$ representable as projectors $\rho_{\text{pure}} = P(|\psi\rangle) \equiv |\psi\rangle\langle\psi|$ have been considered; here ρ, which encompasses all states including the 'mixed' states, is of equal interest.

The explicit consideration of such interpretation of the quantum formalism can be historically traced back at least to Einstein's consideration at the 1927 Solvay conference ([303], p. 253) of two alternative understandings of quantum theory, which he called "interpretation I" (his own proto-interpretation, in which the quantum description is an *incomplete* one for the specification of state for individual systems) and "interpretation II" (Bohr's interpretation, in which the quantum description is understood to be as *complete* a description of quantum phenomena as can be given). This distinction reflects Einstein's philosophical preoccupations and a fundamental disagreement with Bohr. A third early interpretation, first given in Dirac's *The Principles of Quantum Mechanics* and later more rigorously developed in von Neumann's *Grundlagen*, is considered here first, before others falling more neatly into one of the two options of Einstein's early dichotomy. Despite subtle differences between the presentations of von Neumann and Dirac, these two works provided the same first completed interpretation of the theory, which is that primarily used today, even after the undeniable impact of Bohr's "Copenhagen interpretation." Bub refers to it as that following "von Neumann's basic approach" to the measurement problem [84], which underlines that fact that the interpretation of quantum mechanics and quantum measurement theory are connected, with elements of the two deeply influencing each other.

There have been a large number of physicists, including Peierls and perhaps the majority of physicists during certain periods, who have viewed themselves as using the Copenhagen interpretation, which is commonly identified with Bohr, Heisenberg, Jordan and Pauli, after it was fully expounded and as Einstein's "interpretation I" fell into disfavor. Henry Stapp has characterized the development of that interpretation as follows.

"This radical concept, called the Copenhagen interpretation, was bitterly debated at first but became during the 30's the orthodox interpretation of quantum theory, nominally accepted by almost all textbooks and practical workers in the field." ([434])

Although this is an oversimplification of the historical situation likely the result of identifying the Basic and Copenhagen interpretations, which differ in fundamental respects despite shared concerns surrounding measurement, it is true that interpretations of quantum mechanics have often been similarly divided into the *orthodox* and the *unorthodox*, where the former term refers to the Copenhagen interpretation and later 'extensions' thereof.

More helpfully, Mittelstaedt has identified, in addition to the Copenhagen interpretation, three classes of interpretation that he has identified as "probably the most important."

(1) the *Minimal interpretation*, which "does not assume that measuring instruments are macroscopic bodies subject to the laws of classical physics. Instead,...[they] are considered proper quantum systems...with respect to measuring instruments[, it] replaces Bohr's position with von Neumann's approach" but on the other hand "refers to observed data only...merely the values of a 'pointer' of a measurement apparatus" ([319], p. 9).

(2) the *Realist interpretation*, which is similar to the minimal interpretation but "is concerned not only with measurement outcomes but also with the properties of an individual system" ([319], p. 12).

(3) the '*Many worlds*' *interpretation* which, like the previous two, considers quantum mechanics to be universal but "avoids any additional assumption that goes beyond the pure formalism, even the very few weak assumptions that are made in the minimal interpretation" ([319], p. 14).

As is done here, Mittelstaedt declines to consider the modal interpretation(s), primarily on the grounds that "there has been no general agreement about the value, the usefulness, and the philosophical implications of these approaches" ([319], p. 8). The characterization of (3) includes several versions of an interpretation identified in this book with the Collapse-Free interpretation beyond the example mentioned by name there.

Despite claims to the contrary, interpretations of quantum mechanics, including versions of the Collapse-Free interpretation do explicitly or implicitly involve elements that are essentially philosophical, even though many physicists have wishfully thought that quantum mechanics be considered 'self-interpreting' and entirely independent of philosophy. A particularly strong statement against explicit interpretation of the theory, perhaps targeting class (2), was made by Léon Rosenfeld who, while advocating the Copenhagen terminology and method of approaching the quantum formalism, viewed the need for interpretation as arising from a pseudo-problem. "Historically, the false problem ('scheinproblem') of 'interpreting the formalism' appears as a short-lived decay-product of the mechanistic philosophy of the nineteenth century" ([278], p. 41). For him, there was only *one* interpretation, namely, Bohr's.

> "We are not faced with a matter of choice between two possible languages or two possible interpretations, but with a rational language intimately connected with the formalism and adapted to it, on the one hand, and with rather wild, metaphysical speculations... on the other." ([278], p.107)

This sentiment continues to be expressed by others who, revealingly, do *not* subscribe to the Copenhagen interpretation (*cf.* [376–379]). A number of those who have similarly objected to the consideration of 'metaphysical' issues relating to quantum mechanics have been advocates of the Copenhagen interpretation who were influenced by positivism. Bohr has very often been viewed as more a philosopher than a physicist, as Heisenberg saw him.

> "Bohr's view of his theory was much more sceptical than that of many other physicists... his insight into the structure of the theory was not a result of a mathematical analysis of the basic assumptions, but rather an intense occupation with the actual phenomena, such that it was possible for him to sense the relationship intuitively rather than formally. Thus I understood: knowledge of nature was primarily obtained in this way, and only as the next step can one succeed in fixing one's knowledge in mathematical form and subjecting it to complete rational analysis. Bohr was primarily a philosopher, not a physicist, but he understood that natural philosophy in our day and age carries weight only if its every detail can be subjected to the inexorable test of experiment." ([221]; quoted in [462], pp. 21-22)

It would be a mistake to characterize Einstein's "interpretation I," which is most closely related to what is called the *Naive interpretation*, as involving "wild, metaphysical speculation," no matter how problematic it is; following the Mittelstaedt classification, it most closely resembles the "Minimal interpretation." However, *contra* Mittelstaedt, it is legitimate to view the ontological commitment of the Many-worlds variant of the *Collapse-Free interpretation* as extreme and speculative.[1] Many of those wishing to preserve the traditional approach to interpreting physical theory do find themselves needing to postulate the existence of unobserved (even *unobservable*) entities. All of this shows that it behooves one to carefully consider rather than to reject the consideration of philosophical questions relating to quantum mechanics. Later interpretations have been mainly motivated by a desire both to achieve conceptual clarity and to avoid apparently unresolved problems.

A more recent illustration of the desire to consider the quantum formalism 'self-interpreting' is the information-focused approach vigorously advocated

---

[1] This is also a reason for *not* identifying, as has recently been suggested, as part of some "new orthodoxy" flowing directly from the Copenhagen approach.

by Chris Fuchs and Asher Peres in a year 2000 article in *Physics Today*, wherein a polemic was offered for an " 'interpretation without interpretation' for quantum mechanics."

> "The thread common to all the nonstandard 'interpretations' is the desire to create a new theory with features that correspond to some reality independent of our potential experiments. But, trying to fulfill a classical world view by encumbering quantum mechanics with hidden variables, multiple worlds, consistency rules, or spontaneous 'collapse', without any improvement in its predictive power, only gives the illusion of a better understanding. Contrary to those desires, quantum theory does *not* describe physical reality. What it does is provide an algorithm for computing *probabilities* for the macroscopic events ('detector clicks') that are the consequence of our experimental interventions. This strict definition of the scope of quantum theory is the only interpretation ever needed, whether by experimenters or theorists." ([182])

Leaving aside the point that this conflates theory interpretation and theory modification, it mistakenly attributes to all 'non-standard' interpretations a "classical world view;" the intended target appears to be the realist approach to theory interpretation, which requires a non-trivial degree of objectivity that has been viewed as something inappropriately classical and encumbering by *positivist* advocates of the Copenhagen interpretation. The criticism appears very similar to Rosenfeld's, particularly in its empiricism. However, it is now being made by such advocates of a new interpretation, known as the *Radical Bayesian interpretation*, which revolves around the Bayesian subjectivist understanding of *probability* rather than positivism.

The fact that Radical Bayesianism has appeared in the era of quantum information science is not accidental; it explicitly interprets quantum theory almost entirely as a theory of *information* rather than of physical objects except for, as Fuchs puts it, a *"Zing!"*, a ghostly remnant of the physical world. A strong motivation for reconsidering the interpretation of quantum mechanics is to avoid paradox. Indeed, like many of those previously critical of existing non-empiricist interpretations of quantum mechanics, Fuchs and Peres claimed that

> "attributing reality to quantum states leads to a host of 'quantum paradoxes.' These are due solely to an incorrect interpretation of quantum mechanics." ([182])

Of course, many apparent difficulties could be removed by simply disregarding the need to provide naturalist explanations of phenomena. However, as shown below, this too comes at the cost of a loss of genuine physical insight.

No matter how persistent has been the anti-realist temptation, anti-realism has been consistently rejected by the overwhelming majority of physicists, because of its anemic character vis-à-vis explanation, beginning with those who formulated the Basic interpretation, the most 'standard' of interpretations, and the other 'standard' interpretation, the Copenhagen interpretation, which is not truly anti-realist in character, despite various revisionist claims considered below. The next chapter focuses more conservatively on the established relationship between quantum mechanics and information, which often differs from that of the radical Bayesian interpretation. In that chapter, other informational approaches to interpreting quantum mechanics and the ultimate reasons why such an approach has appeared in this era are made clear. Other exotic proto-interpretations are also examined there. It is shown that confusion arises in all of them from a lack of care as to just what is required of a fundamental physical theory and confusion as to the relationship between information and the physical signals that can be used to communicate it.

As Josef Jauch correctly emphasized, specialization in research has often been detrimental to the understanding of foundational questions in quantum theory.

"[T]he pragmatic tendency of modern research has often obscured the difference between knowing the usage of a language and understanding the meaning of the concepts." ([258], p. v)

This assessment is increasingly accurate as regards modern quantum theory after the achievements of its founders, many of whom were remarkably well educated beyond physics proper. However, the failures of interpretation are not entirely attributable to the technical character of physics. The formalistic character of the twentieth-century philosophy to which physicists have been at least indirectly exposed is partly responsible. In his *The philosophy of quantum mechanics*, a comprehensive historical survey of the history of quantum mechanics and its foundations, Jammer remarked that

"what it means to *interpret* this formalism... is not by any means a simple question. In fact, ... just as physicists disagree on what *is* the correct interpretation of quantum mechanics, philosophers of science disagree what it *means* to interpret such a theory. If for mathematical theories the problem of interpretation, usually solved by applying the language of model theory (in the technical sense) requires a conceptually quite elaborate apparatus, then for empirical theories—which differ from the former not so much in syntax as in semantics—the problem is considerably more difficult..."([256], p. 10)

Because philosophical matters pertain to much of this confusion, let us attempt to inoculate ourselves by briefly considering some of the philosophical matters pertaining to the interpretation of quantum mechanics. This will also help us better appreciate why some physicists might be so critical of theory interpretation, beyond the simple view that interpretation itself is responsible for the existence of the well known 'paradoxes' that result when considering the counterintuitive behavior predicted by quantum mechanics.

Bohr has often been incorrectly identified as a positivist.[2] Perhaps under the influence of such portrayals, some physicists have adopted the position that theories are self-interpreting and need not describe an external world. Instrumentalism, the position that theories are to serve as mere instruments for the prediction of observations, can be seen as having strongly influenced the Radical Bayesians; it supports their denial that the quantum state on the one hand correctly predicts quantum measurement outcomes but on the other does not describe the external world. As a more sophisticated example, consider the anti-realist position of Roland Omnès in *The interpretation of quantum mechanics*, which offers a "consistent histories" interpretation of quantum mechanics in a version of the theory in which wave function 'collapse' does not take place. In his book, Omnès defines the term *interpretation* as

> "a translation of the language of facts into the formal language of a theory... in reexpressing the phenomena and the data within the conceptual framework of the theory." ([329], p. 98)

Similarly to the above attempt at "interpretation without interpretation," Omnès claims to

> "reject a priori all kinds of philosophical criteria such as, for instance, demanding that the theory should be an exact description of reality, a complete explanation of it, or its intimate knowledge." ([329], p. 98)

He further claims that quantum mechanics is a theory that "provides its own interpretation," which he considers "*the* interpretation" of the theory but, again, one differing from Rosenfeld's, that is from the Copenhagen interpretation, *and* from the Radical Bayesian interpretation.

Omnès views any well defined such interpretation as "an intrinsic part of the theory" that should satisfy two criteria, consistency and agreement with experiment. He also suggests that an interpretation should be complete, by which he means "that it gives explicit predictions, whatever experiment is being considered," and that it "must be totally explicit in its description of phenomena" and include

> "a clear understanding of the status of determinism and a derivation of its empirical existence, whatever its limitations might be." ([329], p. 99)

On these grounds, for example, Omnès rejects the traditional Copenhagen interpretation, viewing its consistency highly questionable. Further, he views it as incomplete in that, *contra* Bohr, it "does not offer a theory of phenomena." Nonetheless, he believes that a related interpretation can be obtained that satisfies the above criteria and "confirms the Copenhagen empirical prescriptions... while removing the apparent contradictions they contained" ([329], p.

---

[2] *Cf.* [233], p. 14.

100). The latter involves the rejection of some Copenhagen theses, the reformulation of some, and the placing of limitations on others.

Redhead has made the following note regarding the anemic character of this sort of approach.

> "Suffice it to point out that the naive distinction between observation and theoretical terms, and the inability of [an instrumentalist] account to make credible the successful novel predictions in which major theoretical advances such as QM abound." ([371], pp. 44-45)

Any attempt to deny that physics describes at least *some* aspect of the physical world beyond raw data fails because all physical theories must have *some* degree of realist commitment to be about the *physics* in the sense of the term regularly used in science. Any attempt to "reject a priori all kinds of philosophical criteria" is untenable when interpreting quantum mechanics or indeed *any* physical theory, because that is internally inconsistent. Interpretation explicitly involves the consideration of issues beyond the bounds of physics proper; epistemology and metaphysics are not subsumed by physics. It is obviously impossible to achieve a "clear understanding of the status of determinism and a derivation of empirical existence" in quantum mechanics as, for example, Omnès wishes without bringing philosophical criteria to bear. Indeed, it is impossible in the case of *any* physical theory, even when this activity is relatively straightforward, as it is often argued to be in the case of classical mechanics. The tactic might be useful when first *constructing* a theory, in order to avoid hindersome presumptions, but to reject *all* philosophical criteria when interpreting a theory is to reject of one of the fundamental activities of science, namely, arriving at a conception of the external world. As Redhead puts the matter, theories 'interpreted' in the above ways "simply do not contribute to our *understanding* of the natural world" ([371], p. 45).

To aid better the interpretation of quantum mechanics, consider Healey's minimal desiderata, including the *sine qua non* of explanatory power.

> "A satisfactory interpretation. . . would provide a way of understanding the central notions of the theory which permits a clear and exact statement of its key principles. It would include a demonstration that, with this understanding, quantum mechanics is a consistent, empirically adequate, and explanatorily powerful theory. And it would give convincing and natural resolutions of the 'paradoxes'." ([211])

The paradoxes Healey has in mind are those of the sort discussed in the previous chapter. An interpretation of the theory need neither reject all philosophical criteria nor require the rejection of an independent external world the physics of which those who formulated it intended to capture in the first place. A nuanced understanding of the Basic interpretation that includes objective indefiniteness can currently be seen to be the most satisfactory according to these criteria.

3.1 Interpretation and Metaphysics

The revolutionary nature of quantum theory and the conceptual innovations it requires, particularly in light of the experimental disconfirmation of local hidden-variables theories by the observation of violations of Bell-type inequalities, have led physicists to seek new ways of interpreting quantum mechanics, for better or worse, including anti-realist moves that deny that there is an objective universe and realist moves that postulate inaccessible universes.

There are metaphysical, epistemological, and theoretical aspects of most issues in the foundations of quantum mechanics which are typically interrelated in any given case, as seen above regarding the Heisenberg relations and quantum logic. The adjective *realist* has been applied to each of these aspects. *Metaphysical realism*, particularly in the form of scientific realism, is the position most naturally adopted by physicists; it posits the existence, independently of the mental, of the objects to which they devote their studies. However, like *interpretation*, *realism* is a term that has been variously understood and often used with insufficient care in the physics literature. This is particularly so in the context of foundational questions in quantum theory, most often in relation to quantum non-locality [314, 327]. Usage of the term in that regard is typically vague or improper, which has led to the claim that quantum mechanics conflicts with 'local realism.' Whether one agrees with that claim depends on how one understands *local* and *reality*.

The prominence of the term *reality* in relation to quantum physics can be traced at least as far back as the original EPR paper and its 'reality criterion.' Bohr used the term with similar prominence in *Quantum mechanics and physical reality*, his follow-up to the EPR paper which reiterated its title, *Can quantum-mechanical description of physical reality be considered complete?* Bell made similar use of the term in essays, as in *Bertlmann's socks and the nature of reality*. These opened the way toward increasing attention to the philosophical issues arising from quantum mechanics and its explicit interpretation. Because, unlike the locality requirements which they also addressed, the analysis of the relationship of theory to reality is not physics proper, these articles forced physicist readers to confront unfamiliar questions without providing sufficient context for them to grasp fully their exact meaning and full significance. Predictably, this led to a variety of understandings of the *reality* and related terms.

Margenau identified the issue of manifold meanings early on, in his own follow-up article to the EPR paper, in which he made the following comment.

> "The discussion of a recent paper by Einstein, Podolski and Rosen has brought to light an interesting divergence of opinions as to the meaning of reality." ([308])

Accordingly, Heisenberg cautioned physicists against making metaphysical moves, explicitly or implicitly, through the imprecise use of terms. In particular, in relation to

"the statement that besides our world there exists another world...One should be especially careful in using the words 'reality,' 'actually,' etc., ..." ([219], p. 15)

Born later remarked, more strongly than Margenau, that the

"concept of reality is too much connected with emotions to allow a generally acceptable definition. " ([68], p. 103)

D'Espagnat has clearly characterized two of the most famous of realist views in quantum physics, those of Einstein and Bell, as resting on common sense, because they are based, for example, on such arguments as

"...nobody really succeeds in believing that the moon does not exist when not looked at. Some physicists, and not the lesser ones (in his mature years Einstein was, as it seems, quite avowedly one of them) are fully convinced by such commonsense arguments. Similarly, John Bell considered them to be decisive even after he discovered his theorem...and was content, in view of the said discovery, to hold this reality to be nonlocal." ([126], p. 197)

Philosophers, in general, have been less tentative and more careful and precise than physicists have in their use of reality and of related concepts in understanding physics, as well as in their characterization of the positions of founders of quantum mechanics; for them, metaphysics, theories of reference, and semantics are central territories. For example, Krips takes proper care in his use of realism and anti-realism; at the outset of his book, *The metaphysics of quantum theory*, in relation to the case of the Copenhagen interpretation, he has explained that

"Bohr was anti-realist in his attitude toward the formalism...he took the state-vectors of systems to be purely heuristic devices. Moreover Heisenberg was anti-realist in respect of his endorsement...of verificationist principles...But...neither...were anti-realists in the metaphysical sense of denying the existence of an objective external reality...nor did they eschew the scientific realist's commitment to describing that reality within science...In short the disagreement between Bohr and Heisenberg on the one hand and Einstein on the other was not a disagreement over metaphysical realism or indeed scientific realism. Rather it was a disagreement about the terms in which external reality was to be described." ([281], p. 1)

Historians of science are also typically fully aware of all the meanings of the terms involved. For example, Einstein's position regarding reality has been articulated well by Jammer, as follows.

"In 1936 Einstein published his *credo* concerning the philosophy of physics in an essay 'Physics and Reality,' which started with the remark that science is but a refinement of everyday thinking and showed how the ordinary conception of a 'real external world' leads the scientist to the formation of the concept of bodily objects by taking, out of the multitude of his sense experiences, certain repeatedly occurring complexes of sense impressions. From the logical point of view, this concept of bodily object is not to be identified with the totality of sense impressions but is 'an arbitrary creation of the human mind.' Although we form this concept originally on the basis of sense impressions, we attribute it—and this is the second step in building up 'reality'—a significance which is to a high degree independent of sense impressions and we thereby raise its status to that of an object of 'real existence.'... That the totality of sense impressions can be put in order was for Einstein a fact 'which leaves us in awe, for we shall never understand it.' The eternal mystery of the world, he declared, 'is its comprehensibility.' " ([256], p. 230)

Metaphysics plays an important role in the interpretation of any fundamental physical theory. It is, therefore, perilous *not* to engage in it, at least implicitly, as Bohr, Einstein, Heisenberg, Schrödinger and other founders of quantum theory managed to do with alacrity. Nonetheless, among both physicists and philosophers, the greatest error in the pursuit of a well defined interpretation has been *naive realism*, in the sense of interpreting the wave-function as directly representing reality, in particular, directly representing individual objects, which is sometimes incorrectly attributed to Einstein, as opposed to a subtler realism of the kind he espoused. Schrödinger, with whom Einstein was in written correspondence and sympathetic in regard to metaphysical issues in quantum theory, offered the following comments in 1935 in regard to the relationship between the wave-function and reality.

"For each measurement one is required to ascribe to the ψ-function (= the prediction-catalog) a characteristic, quite sudden change, which *depends on the measurement result obtained, and so cannot be foreseen*; from which alone it is already quite clear that this second kind of change of the ψ-function has nothing whatever in common with its orderly development *between* two measurements. The abrupt change by measurement... is the most interesting point of the entire theory. It is precisely *the* point that demands the break with naive realism. For *this* reason one can *not* put the ψ-function directly in place of the model or of the physical thing. And indeed not because one might never dare impute abrupt unforeseen changes to a physical thing or to a model, but because in the realism point of view observation is a natural process like any other and cannot *per se* bring about an interruption of the orderly flow of natural events." ([394])

As Omnès has simply put the matter, "nobody will ever see a wave function when looking at a car or a chair" ([330], p. xi).

Again, however, criticism prompted by metaphysical excess has often been incorrectly directed toward *any* metaphysical contemplation rather than toward the specific offending moves. This was clearly noted by Bell, who described the historical significance of this attitude for the development of quantum physics itself when he said that quantum phenomena

> "made physicists despair of finding any consistent space-time picture of what goes on on the atomic and subatomic scale. Making a virtue of necessity, and influenced by positivistic and instrumentalist philosophies, many came to hold that not only is it difficult to find a coherent picture but it is wrong to look for one – if not actually immoral then certainly unprofessional." ([24], p. 142)

To understand better why such reactions are inappropriate, it is helpful to know precisely how philosophers understand metaphysical realism.[3] Popper explicated the realist metaphysical position as follows.

> "[T]hat this world exists independently of ourselves; that it existed before life existed, according to our best hypotheses; and that it will continue to exist, for all we know, long after we have all been swept away." ([361], p. 2)

He provided the following simple argument for the realist position,

> "[T]heories are our own inventions, our own ideas...But some of these ideas are so bold that they can clash with reality: they are the testable theories of science. And when they clash, then we know there is a reality: something that can inform us that our ideas are mistaken. And this is why the realist is right. (Incidentally, this kind of information—the rejection of our theories by reality is the only information we can obtain from reality: all else is our own making. This explains why our theories are coloured by our human point of view, but less and less distorted by it as our search goes on.)" ([361], p. 3)

This is essentially the way, for example, Bell himself understood the position, which he also advocated.[4]

A more precise, technical definition of metaphysical realism is the following, given in the following strong form by Michael Devitt.

---

[3] The reader may also find it useful to consult d'Espagnat's description of various forms of realism in relation to physics [126], pp. 24-30.

[4] See, for example, Bell's comments on the Collapse-Free interpretation shown in Section 3.5.

"Tokens of the great majority of current common-sense, and scientific, physical types objectively exist independently of the mental."
([133])

The distinction between *tokens* and *types* is that a type is a general sort of thing, as in "*the* ape," whereas a token is a particular concrete instance of a general sort of thing, as in "*this* particular ape"; types are considered abstract and unique in character whereas tokens are concrete and particular, say composed of a specific sort of material (*cf.*, *e.g.*, [488]). The token–type distinction is of particular importance where logical and linguistic considerations come into play. With this understood, one can go on to consider other and subtler matters, such as what properties tokens are expected to have; in common sense and fairly mature sciences, these characteristics are known well enough for our purposes. As noted above, considerable attention was also paid by philosophers to anti-realism during much of the twentieth century perhaps more than to articulations of the forms of realism to which it is opposed (*cf.* [145, 146]). A rather crude but direct characterization of anti-realism and its relation to quantum mechanics has been given by Peter Gibbins.

"Philosophers have another motive, one which is inspired by their intellectual cussedness. Quantum mechanics is most easily interpreted *antirealistically*, that is, as a theory which, though it works, does not describe the way the world is." ([187], p. ix)

With this characterization, Gibbins could equally well be describing the related position of instrumentalism, namely, that scientific theories need only serve as mere tools for making predictions of future observations, so that believing them to be valid in no way obligates one to the *existence* of the entities to which they refer. Although the detailed consideration of anti-realism is philosophically important, the issues that such detailed consideration involves are not especially pertinent to understanding quantum theory over and above understanding other physical theories and will not be delved into further here.[5]

As noted at various points above, physicists have paid far more attention to the option of instrumentalism, which for obvious reasons comes far less naturally to them, than to the much stronger position of metaphysical anti-realism. Another, closely related position is scientific empiricism, such as described by Bas van Fraassen in the context of a discussion of the implications of Bell's theorem.

"I wish merely to be agnostic about the existence of the unobservable aspects of the world described by science. . ." ([464], p. 72)

---

[5] The interested reader may wish to consult Peter Forrest's book *Quantum metaphysics* for a description of the pertinence of anti-realist thought, as well as Peircean realism, to the interpretation of quantum mechanics [177].

Van Fraassen more fully explicates this position as follows.

"To be an empiricist is to withhold belief in anything that goes beyond the actual, observable phenomena, and to recognize no objective modality in nature. To develop an empiricist account of science is to depict it as involving a search for truth only about the empirical world, about what is actual and observable...it must involve throughout a resolute rejection of the demand for an explanation of the regularities in the observable course of nature, by means of truths concerning a reality beyond what is actual and observable, as a demand which plays no role in the scientific enterprise." ([463], pp. 202-203)

It is interesting here to note that Schrödinger, in his criticism of naive realism, comment that

"The classical method of the precise model... is [perhaps] based on the belief that *somehow* the initial state *really* determines uniquely the subsequent events, or that a *complete exactness* would permit predictive calculation of outcomes of all experiments with complete exactness. Perhaps on the other hand this belief is based on the method. But it is quite probable... that 'complete model' is a contradiction in terms like 'largest integer'." ([394])

Unlike with Schrödinger, the negative effect of attempting to maintain a naive realist approach to the physical world has prompted a number of physicists to question the very *holding* of metaphysical views or commitment to a quantum ontology. However, ultimately, as Clifford Hooker has noted,

"Some scientists do take the empirical success of the theory, combined with its interpretational obscurity, as proof that empiricism is the correct attitude, but this has not been widespread...There are also many scientists, it must be admitted, who point to the empirical adequacy and say 'enough, get on with experiment', but these scientists have really dismissed the interpretational issue rather than decided it, and they are seen to have their pragmatic motivations. These scientists will be satisfied by crude empiricist renditions like that of the physicist Leslie Ballentine and they ought to be satisfied by the elegantly worked out, but ontologically noncommittal models van Fraassen himself has provided. The irony is, for all its elegance, van Fraassen's interpretation is being, and is bound to be, largely ignored by scientists just because it doesn't treat the theoretical problems seriously in the sense required." ([234], p. 180)

W. Michael Dickson has suggested that, in view of the empiricist alternative, realist physicists instead be committed to a specific verification criterion.

"Physicists have a loose criterion for whether some element of a theory corresponds to a real physical entity: If an experiment reveals 'somewhat directly' a physical entity corresponding to the theoretical element in question, and is more or less the way the theory says it is. Empiricists correctly point out that this vague criterion is not at all philosophically compelling. It is perhaps the correct criterion for physics, but it provides only the barest of starting point for philosophy: When the criterion is met, realists and empiricists are on even ground as regards pure physics, and must settle the dispute purely philosophically...; when it is not met then the realist should provisionally agree with the empiricist that the theoretical element in question does not refer to a real physical entity... I propose that the realist is committed to the following *verification criterion*: The existence of a supposed unobservable entity is scientifically plausible to the extent that measurable effects of the entity can (in principle) be verified, and the properties assigned to the entity by science can plausibly be said to be possessed to the extent that they can (in principle) be measured." ([136])

What is most important to recognize is that any appeal to anti-realism, empiricism, or instrumentalism to solve technical problems in quantum physics, which often revolve around behavior that is not locally causal, is suspect in itself. As Tim Maudlin has rightly pointed out,

"Realism in philosophy of science is generally contrasted with instrumentalism or empiricism, which views assert that one can have no grounds to believe that the unobservable ontology of a theory is accurate. In this sense, *theories* are neither realistic nor non-realistic, only interpretations of (or better: attitudes toward) theories. And the strongest argument in favor of instrumentalism, from Osiander onwards, is underdetermination: the existence of many incompatible theories all capable of 'saving the phenomena'. The beauty of Bell's theorem, of course, is that it is insensitive to the details of the theory suggested: *any* theory which can save the phenomena (if the phenomena include claims about the behavior of macroscopic devices located in space and time) must be non-local. Even a classical instrumentalist would be forced to accept non-locality." ([314], pp. 304-305)

Recall that the first explicitly 'localist' stance after the appearance of quantum mechanics was that of Einstein, who in this regard commented that

"on one supposition we should, in my opinion, absolutely hold fast: the real factual situation of the system S_2 is independent of what is done with the system S_1, which is spatially separated from the former." ([151], p. 85)

The not uncommon move, to jettison *realism* in light of the failure of 'local realism,' that is the failure of locality, is simply a (philosophical) mistake. Indeed, as Krips has correctly noted in accord with Dickson's points,

"Q[uantum]T[heory], or even an interpretation of [it], cannot of course prove [a realist] metaphysics. But a particular interpretation of QT can support this realist metaphysics by presenting QT as descriptive of an objective reality." ([281], p. 127)

That is, metaphysical realism gains support to the extent that quantum theory *is* well interpretable non-instrumentalistically in such a way that the reality described by quantum theory is objective, say by physical quantities taking values when unmeasured and in a mind-independent way, given that the theory makes correct predictions and is explanatory whether it is local or not.

Following the language of the EPR paper and other seminal works in the foundations of quantum mechanics, the phrase *realist interpretation* is now common in the literatures of both philosophy and physics in relation to quantum mechanics. Krips offers the following characterization.

"What does it mean to adopt a 'realist interpretation' of [quantum theory]? I shall take it to mean accepting that [quantum theory] is true, that the objects [it] refers to (electrons, protons, etc.) exist, that the properties it refers to are 'real,' and in particular that the physical quantities it refers to are 'real'; in short it also means that we can interpret [quantum theory] 'literally,' in the sense that we can take all its referential terms as genuinely referring and not just as convenient fictions or metaphors for the real." ([281], p. 126)

This roughly characterizes the class of interpretation Einstein had in mind, and one to which even Heisenberg and Bohr might have had no objection.[6] As Fine has also noted,

"While there can be no doubt that Einstein turned away from positivism to realism, or that realism was important in his thinking about the quantum theory, there is considerable room for speculation concerning exactly what Einstein's realism involves." ([174], p. 87)

Clearly, a metaphysical realist world view *does* contribute to our understanding of the physical world, however unexpected the consequences of such a world view might be, such as when it suggests the presence of objective indefiniteness. The most important point is that naive realism is not the *only* form of realism available as an attitude toward quantum theory.

For Einstein, metaphysical realism was not to be held out of absolute necessity, but was seen as the best foundation for a coherent world view.

---

[6] See also the Einstein's statement to Born quoted in Section 1.7.

"[The] 'real' in physics is to be taken as a type of program, to which
we are, however, not forced to cling *a priori*. No one is likely to be
inclined to attempt to give up this program within the realm of the
'macroscopic'... But the 'macroscopic' and the 'microscopic' are so
inter-related that it appears impracticable to give up on this program
in the 'microscopic' alone." ([152], p. 674)

Indeed, Fine has argued that Einstein's realism, which Fine calls "motivational
realism," was much weaker and more subtle than standard metaphysical realism ([174], Chapter 6). Again, a central concern for Einstein was avoiding
observer-dependence.

"The sore point lies less in the renunciation of causality than in the
renunciation of the representation of a reality thought of as independent of observation." (*cf.* [428], p. 374)

It is widely recognized that more effort has been expended interpreting
quantum theory than any other physical theory in recent times simply because physicists have clear and natural motives for seeking interpretations of
quantum mechanics, as Hooker has pointed out.

"The theory is strikingly empirically adequate, as well understood
mathematically as any other theory (*i.e.* modestly so), but poorly
understood conceptually and ontologically... In fact, no theory has
drawn more interpretational discussion in the history of science, and
not just (or even mainly) by philosophers but primarily by scientists
themselves seeking theoretical understanding... With few (any?) ex-
ceptions, all of the outstanding scientists of this century have worried
the problem, seeking theoretical insight." ([234], p. 180)

For example, Bell remarked

"When I look at quantum mechanics I see that it's a dirty theory.
The formulations of quantum mechanics you see in the books involve
dividing the world into an observer and an observed, and you are not
told where that division comes—on which side of my spectacles it
comes, for example—or at which end of my optic nerve. You're not
told about this division between the observer and the observed... So
you have a theory which is fundamentally ambiguous, but where the
ambiguity involves decimal places remote from human abilities to
test." ([119], p. 54)

Although Bell's interest was mainly a practical one, based on criteria internal
to physics, he strongly objected to "FAPP" (for all practical purposes) ratio-
nalizations, involving approximations that could not be rigorously justified.

The Copenhagen interpretation which, like the Basic interpretation, was
a widely accepted one, presents quantum mechanics as a *final* theory of mechanics as well as providing with a relatively intuitive basis on which to use

it. Popper called advocates of such finalism—Pauli, for example—"end-of-the-road people" ([361], p. 13) clearly differing from Einstein and Bell.

> "[Einstein said] 'it seems to me that those physicists who regard the ways of description of quantum mechanics as in principle final ('*definitiv*') will react to these considerations as follows: they will drop the requirement... of the independent existence of physical real things in distant parts of space; and they could rightly claim that quantum mechanics nowhere makes implicit the use of any such requirement'." ([361], p. 21)

Einstein was bothered by finalism on philosophical grounds; he continued to struggle to bring quantum theory and the broader common-sense and scientific conceptions of the world into harmony and, like Schrödinger, to provide arguments against the validity of the world picture provided by the Copenhagen interpretation, in particular, to challenge the idea of a holistic participatory universe that interpretation appeared to accommodate. In order to get to the heart of the matter, he delved directly into the philosophical issues the neglect of which enabled what was, for him, a serious threat to the integrity of the scientific world view. For example, the EPR paper contended that

> "Any serious consideration of a physical theory must also take into account the distinction between objective reality, which is independent of any theory, and the physical concepts with which the theory operates. These concepts are intended to correspond with the objective reality, an by means of these concepts we picture this reality to ourselves." ([153])

In a letter to Schrödinger, Einstein similarly spelled out his discomfort with quantum mechanics under the subjectivist approach to interpretation that was common at the time.

> "Most [contemporary physicists] simply do not see what sort of risky game they are playing with reality—reality is something independent of what is experimentally established. They somehow think that quantum theory provides a description of reality, and even a *complete* description; this interpretation is, however, refuted most elegantly by your [cat experiment]... Nobody really doubts that the presence or absence of the cat is something independent of the act of observation." ([365], p. 39)

The option to put aside philosophical considerations and simply to accept the most pragmatic interpretation of quantum theory is seductive to physicists, but risks taking current physical theory too seriously; Einstein warned that the Copenhagen interpretation can serve as "a gentle pillow for the true believer." Even Omnès in the end finds himself committed to the view that "the logic of common sense is a secondary feature relying upon a somewhat deeper logical structure having its roots in physical reality" ([329], p. 204).

Although under a different interpretation, some have even been open to the having physics dictate *logic*. The view of logic as *a posteriori* is, perhaps, epitomized by the erstwhile claim of Putnam that "logic is just as empirical as geometry" in light of the discoveries of quantum logic [366]. Stachel correctly criticized this move, as follows.

> "At any given state of scientific development what we confront is not 'reality,' but some particular theoretical structure, and its accompanying modes of experimental protocol, which enable us to understand and cope with some aspects of the world... The danger is such consolations may divert us from confronting tensions within the existing theoretical and experimental structure, or between that structure and other, unassimilated elements—tensions that could lead to a deeper comprehension of, changes to, or even the complete overthrow of, that structure." ([427], p. 233)

To put the matter in simple terms, although a conceptual shift in physics can be an extremely valuable thing, a 'corresponding' shift in *metaphysics* or *logic* is something altogether more dramatic and if an error more likely to be grave. As is again shown below in the case of the Radical Bayesian interpretation, a philosophical move motivated by a desire to finesse a paradox that does not also address a genuine and essentially philosophical problem can actually threaten to greatly hinder progress by shifting attention away from physics.

The only robust new mechanical theory to emerge from the continual probing of the foundations of quantum mechanics has been Bohmian mechanics, an alternative model which is readily seen to cohere with realism and follows the spirit of Einstein's philosophy, even if not to the satisfaction of Einstein himself or, thus far, the physics community as a whole. The basis for this sort of mechanics was set out in essence early on by de Broglie and Bohm and will not be discussed in detail here; let us simply note Bohm's own description of its origins.

> "[Blokhinzhev and Terletzky] made it clear that ... one may consistently regard the current quantum theory as an essentially statistical treatment, which would eventually be supplemented by a more detailed theory permitting a more nearly complete treatment of the behavior of the individual systems. Then in 1951, partly as a result of the stimulus of discussions with Dr. Einstein, the author began to seek such a model; and indeed shortly thereafter he found a simple causal explanation of the quantum mechanics which, as he later learned, had already been proposed by de Broglie in 1927... partly as a result of additional suggestions made by Vigier, de Broglie then returned to his original proposals, since he now felt that the decisive objections against them had been answered." ([53], p. 110)

Bohmian mechanics is a different version of quantum theory with a capacity for producing predictions differing from those of standard quantum mechanics.

By contrast, the recently introduced Radical Bayesian interpretation is an unsatisfactory approach to the interpretation of quantum mechanics because it introduces no new empirical predictions while *with the same stroke* risking greatly stunting the progress of physics, indeed, risking to do so more radically than any previous interpretation previously suggested; as Redhead has said "Setting dogmatic limitations on scientific theorizing, on the basis of obscure philosophical preconceptions, is a dangerous prejudice from the standpoint of a conjectural-fallibi[li]st approach to the nature of scientific activity" ([371], p. 51).

It should be recalled that realism as a philosophical stance toward the quantum world has also been targeted for rejection on the basis of *empirical* evidence, in addition to being considered merely metaphysically 'extravagant.'

"Recently, a new bogeyman seems to have been found: realism. Thus [Lucien] Hardy states: 'In 1965 Bell demonstrated that quantum mechanics is not a local realistic theory. He did this by deriving a set of inequalities and then showing that these inequalities are violated by quantum mechanics' [206]. The conversational implication is that Bell's theorem only applies to local *realistic* theories, so that locality (and hence perhaps also consistency with Relativity) can be recovered if one only jettisons realism. But no standard theory, even a non-realistic one (whatever that means) can be local, so we must reject the conversational implication." ([314], p. 304)

EPR's 'reality criterion,' which was presumably conflated in that instance with the form of metaphysical realism that nonetheless likely motivated it, is perhaps a legitimate target for rejection on physical grounds as an element of an interpretational program. Nonetheless, *metaphysical realism* is not. Indeed, Bell himself, who always expected his inequality to be violated by rigorous experimental testing, was deeply committed to realism, as reflected in his response to the query would you "prefer to retain the notion of objective reality and throw away one of the tenets of relativity: that signals cannot travel faster than the speed of light?" which was

"Yes. One wants to be able to take a realistic view of the world, to talk about the world as being there even when it is not observed. I certainly believe in a world that was here before me, and will be here after me, and I believe that you are part of it! And I believe that most physicists take this point of view when they are being pushed into a corner by philosophers." ([119], p. 50)

Note also that Bell did not view (at least the Many-worlds version of) the Collapse-Free approach as a solution to the difficulties presented by quantum phenomena, despite its capability of being presented realistically, because he viewed that interpretation as "radically solipsistic," despite contrary claims about it by various advocates (*cf.* [24], p. 136).

Finally, consider the following 1961 comment of Wigner.

"The problems of epistemology and ontology have an increased interest for the contemporary physicist. The reason is, in a nutshell,
that physicists find it impossible to give a satisfactory description of
atomic phenomena without explicit reference to the consciousness."
([500], p. 33)

Although this reflects Wigner's own theoretical bias toward the particular
possibility of a psychophysical approach to problems in the foundations of
physics, it is certainly the case that an exploration beyond the traditional
scope of physics is required to the interpret of quantum mechanics, relative
to the interpretation of other theories of physics, because standard quantum
theory is peculiar. At a minimum, it forces objective indefiniteness on physics.

A fair characterization of the current situation in the progression toward
a more solid foundation for quantum mechanics through the deployment of
physical and philosophical tools is the following, offered by R. I. G. Hughes.

"The theory uses the mathematical models provided by Hilbert
spaces, but it's not clear what categorical elements we can hope
to find represented within them, nor, when we find them, to what
extent the quiddities of these representations will impel us to modify
the categorical framework whose elements have their images within
it; we obtain an interpretation by the dialectical process of bringing
to the theory a conceptual scheme, and then seeing how this conceptual scheme needs to be adjusted to fit it. Because there are several
solutions to this problem, there can be competing interpretations of
the same theory." ([244], p. 176)

This situation is sometimes characterized as the "underdetermination" of *interpretation* of quantum theory, which is addressed in a later section of this
chapter. Although this situation is sometimes bemoaned, it differs from *theoretical* underdetermination. There is no *a priori* reason that a correct interpretation must be uniquely determined. That quantum mechanics is *not*
self-interpreting is a reflection of the non-triviality and importance of the
activity of theory interpretation as science continues its advance into new
realms. What one should seek are consistent interpretations of theories that
best assist us in deepening and broadening our understanding of the world
without requiring *insufficiently warranted* ontological commitments. Regarding quantum mechanics, the most prudent attitude is to allow neither physics
nor philosophy alone to dictate our interpretation, but rather a balanced combination of the two, for example, as both Bell and Einstein sought.

The process of attaining a better understanding of quantum mechanics
through the consideration of alternative interpretations of its formalism had
been an ongoing one even before von Neumann's mathematically rigorous
formulation of the theory was available. It has centered on the meanings and
roles of the the operator matrices and/or the wave-function and of quantum
probabilities. This activity continues. Let us now consider several of the most
significant such interpretations to date.

3.2 The Basic Interpretation

The interpretation given to the quantum formalism by von Neumann and Dirac is distinct from the approaches that preceded it, in particular, the Copenhagen interpretation with which it has sometimes been less than carefully identified; as Paul Feyerabend has correctly noted, "when dealing with von Neumann's investigations, we are not dealing with a refinement of Bohr—we are dealing with a completely different approach" ([167], p. 237). Ballentine, the most fervent contemporary advocate of the other preceding interpretation, which attempted to remain true to naive realism, calls the interpretation of von Neumann and his Princeton-associated colleagues that of the *Princeton school* [16]. Whitaker, likely following Ballentine, has also called it the *Princeton interpretation* ([496], p. 194). Bub has referred to it as simply the *basic approach* [87]. Here, it is called the *Basic interpretation*.

The Basic interpretation emphasizes the fundamental nature of the Born probability rule and combines it with the projection postulate for specifying the state of a quantum system upon measurement. Like the Copenhagen interpretation, it regards the quantum state as a *complete* description of quantum phenomena as opposed, for example, to Einstein's view of the quantum state as an incomplete description of the quantum world. However, the Basic interpretation does not require measuring apparatus to be describable in classical terms—unlike the Copenhagen interpretation, it rests on the comparatively explicit context-independent and mathematically internal description of the measurement process discussed in the previous chapter. Indeed, the Dirac–von Neumann approach was the first to involve a detailed mathematical investigation of measurement explicitly internal to quantum mechanics, although it should be recalled that Bohr was later driven to invoke a somewhat quantum mechanical treatment of measurement in order to argue, *contra* Einstein, that Heisenberg's uncertainty relation could not be violated by a rigged double-slit apparatus, as discussed in Section 1.2.

The Basic interpretation takes upon itself the requirement that quantum mechanics describe not only measurement outcomes but also the independently existing, although possibly indefinite physical quantities of systems about which these measurements provide data. However, unlike versions of the later Collapse-Free interpretation which also put no restrictions on the description of measurement apparatus but are often understood to describe very large numbers of universes, there is no question that the theory describes the evident universe as a *single* universe. It is consistent with the realist metaphysical stance, despite its reference of the experience of observers whose physical *bodies* are within its purview in the context of measurement.

Its realist character was spelled out later and variously by others; for example, "the classical tradition of simply located objects characterized independently of experiment, was presupposed by Born and von Neumann and imposed on the data with the help of an informal language of 'particles' and 'states'." ([497], p. 71)

Its relation to psychophysical parallelism, also upheld in the Copenhagen interpretation, differs importantly from that of, for example, Heisenberg.

Von Neumann explicated the principle of psychophysical parallelism as follows.

"[T]he measurement or the related process of the subjective perception is a new entity relative to the physical environment and is not reducible to the latter. Indeed, subjective perception leads us into the intellectual inner life of the individual which is extra-observational by its very nature... Nevertheless, it is a fundamental requirement of the scientific viewpoint—the so-called principle of psycho-physical parallelism—that it must be possible so to describe the extra-physical process of the subjective perception as if it were in reality in the physical world." ([477], pp. 418-419)

That is, although the subjective experience of the observer is not physical, the observer's experience must be describable consistently with the physics of its body. Von Neumann adds, "Indeed, experience only makes statements of this type: an observer made a certain (subjective) observation; and never any like this: a physical quantity has a certain value" ([477], p. 420). Thus, the interpretation does not consider the domain of quantum theory to be that of experience or knowledge, as von Neumann clearly states, despite invoking the Heisenberg *Schnitt*.

"[In] any case, no matter how far we calculate [up the chain of observation toward the interior of the observer's body] at some time we must say: and this is perceived by the observer. That is, we must always divide the world into two parts, the one being the observed system, the other the observer. In the former, we can follow up all physical processes (in principle at least) arbitrarily precisely. In the latter it is meaningless..." ([477], p. 419)

Thus, experience must only be consistent with events in the external world, it need not *constitute* or *create* the external world. The interpretation is clearly fully consistent with metaphysical realism, something not evident in the case of the Copenhagen interpretation. Quantum mechanics is also taken to be a fundamental theory of the physical world and is irreducibly probabilistic, the idea which prompted Einstein's famous protest, "God does not play dice with the world." [7]

---

[7] Bell commented, "I would like to qualify this 'God does not play dice' business. This is something which is often quoted, and which Einstein did say rather early in his career, but afterwards was more concerned with other aspects of quantum mechanics than with the question of indeterminism. And indeed, Aspect's par-

The following are central elements of the Basic interpretation given by von Neumann in his Hilbert-space formulation of quantum mechanics.[8]

(1) Every physical system is attributed a separable Hilbert space $\mathcal{H}$.

(2) Every physical magnitude is associated with an Hermitian self-adjoint, not necessarily bounded, linear operator O, called an *observable*, and vice versa.

(3) Every state of a physical system is assigned a *statistical operator* ρ that is a linear bounded self-adjoint positive trace-class operator, and vice versa.

(4) The Born rule provides expectation values for physical quantities.

(5) All facts about a physical system at time t are described via $\rho(t)$.

Von Neumann considered any use of the Born rule, as in (4), as rendering the associated interpretation a probabilistic one ([477], p. 210); elements (3)-(5) render the theory *irreducibly* probabilistic. The Born rule, as clarified by Pauli, is an element of this and *all* later probabilistic interpretations of the quantum state.[9] Element (5) is among the most distinctive of the Basic interpretation, stating in essence that quantum theory is *complete*. Given (5), the Born rule for assigning probabilities (4) requires each Hilbert-space ray, that is, pure quantum state to provide the expectation values for all physical magnitudes. Another assumption of the Basic interpretation is the rule regarding the wave function $|\psi\rangle$ itself, provided by Dirac, that is, the rule for connecting pure quantum states $\rho = |\psi\rangle\langle\psi|$ with physical magnitudes.

(6) (*The eigenvalue–eigenstate link.*) "The expression that an observable 'has a particular value' for a particular state is permissible in quantum mechanics in the special case when a measurement of the observable is certain to lead to the particular value, so that the state is in an eigenstate of the observable... In the general case we cannot speak of an observable having a value for a particular state, but we can speak of its having an average value for the state. We can go further and speak of the probability of its having any specified value for the state, meaning the probability of this specified value being obtained when one makes a measurement of the observable." ([142], p. 253)

This statement can be viewed as incorporating the idea of objective indefiniteness of physical magnitudes. Von Neumann similarly characterized the probabilistic character of the quantum state ρ. A further distinctive assumption of the Basic interpretation was stated by Dirac in 1927, namely,

---

ticular experiment [confirming Bell inequality violation] tests rather those other aspects, specifically the question of no action at a distance" ([119], p. 46).

[8] The postulates of quantum mechanics are given in standard form in Appendix B.

[9] Born initially favored the view of elementary quantum systems as "corpuscles" ([256], p. 39); he was awarded the Nobel prize in 1954 "for his fundamental work in quantum mechanics and especially for his statistical interpretation of the wave function."

(7) (*The projection postulate.*) "[T]he state of the world at any given moment [is described] by a wave function ψ, which normally varies according to a causal law in such a way that its initial value determines its value at any later moment. It may happen, however, that at a given moment t_1, ψ may be expanded into a series of the form $\psi = \sum_n c_n \psi_n$ in which the ψ_n are wave-functions that cannot interfere with each other at times later than t_1. In this case, the state of the world at times later than t_1 will not be described by ψ but by one of the ψ_n. One can say that nature chooses the particular ψ_n that is suitable, since the only information given by the theory is that the probability that any one of the ψ_n will be selected is $|c_n|^2$. Once made, the choice is irrevocable and will affect the entire future state of the world. The value of n chosen by nature can be determined by experiment and *the results of all experiments* are numbers that describe such choices of nature." ([303], p. 262)

The causal law to which this statement refers is the Schrödinger equation. Von Neumann and Dirac viewed both the unitary evolution and the state projection as regarding probabilities provided by the statistical operator ρ.

It is also clear that Dirac considered observation to coincide with the state projection.

"Consider an observation, consisting of the measurement of an observable α, to be made on a system in the state ψ. The state of the system after the observation must be an eigenstate of α, since the result of a measurement of α for this state must be a certainty." ([139], p. 49)

Dirac understood the instantaneous change of state resulting from the measurement involved in observation to incorporate this 'indeterministic' and *physical* 'jump,' which he viewed as that arising from the unavoidable disturbance of the system during measurement.

"When we measure a real dynamical variable ξ, belonging to the eigenvalue ξ', the disturbance involved in the act of measurement causes a jump in the state of the dynamical system. From physical continuity, if we make a second measurement of the same dynamical variable immediately after the first, the result of the second measurement must be the same as the first. Thus after the first measurement has been made, there is no indeterminacy in the result of the second. Hence after the first measurement is made, the system is in an eigenstate of the dynamical variable ξ, the eigenvalue it belongs to being equal to the result of the first measurement. This conclusion must still hold if the second measurement is not actually made. In this way, we see that a measurement always causes the system to jump into an eigenstate of the dynamical variable that is being measured, the eigenvalue this eigenstate belongs to being equal to the result of the first measurement." ([142], p. 36)

Unless the system has previously been precisely measured for an observable compatible with the corresponding projector being measured, or chance has it that the measurement made is compatible with that projector, the system state will change instantaneously upon measurement, which essentially involves decoherence of the state in the corresponding eigenbasis. The problem of consistently describing measurements, described in Section 2.4 above, remains a significant concern for physicists up to this day, particularly with a two-process (**1.** projective. **2.** unitary.) dynamics. To aid in the resolution of this problem, a number of physicists have sought a specific mechanical model of the wave-function 'jump'; for example, Penrose, who has appealed to gravity for this purpose, has expressed the following sentiment.

"Taking this formalism at face value, we have a statevector $|\psi\rangle$ which evolves for a while according to the completely deterministic Schrödinger equation.... Then, at odd times, when an 'observation' is deemed to have been made, the Schrödinger evolved statevector is discarded and replaced by another, which is selected in a random way, with specific probability weightings, from among the eigenvectors of the operator corresponding to the observation. As has been argued on innumerable occasions, this is a wholly unsatisfactory procedure for a fundamental description of the 'real world.' "([345], pp. 107-108)

However, he then adds,

"In the first place, it is often argued that $|\psi\rangle$ itself should not be regarded as giving an objective description of the world (or part of it) but as providing information merely of 'one's state of knowledge' about the world. This view I really cannot accept." ([345], p. 108)

This represents the view of most physicists that, despite the difficulty of the measurement problem, a realist world view is to be preferred to a subjectivist one, such as offered by some interpretations discussed below in later sections.

In addition to his relatively detailed theory of measurement, von Neumann also provided in the *Grundlagen* the "no-go theorem" on hidden variables, discussed in Section 1.7, which he believed established the impossibility of dispersion-free quantum states. This conclusion turned out to be premature, because this proof contained an unwarranted assumption; in 1952, Bohm constructed an explicit model showing that the implications of von Neumann's proof were other than he at the time understood. However, von Neumann's views on the question of such "hidden parameters" were more subtle than generally appreciated.[10]

Regarding the question of causality, von Neumann understood the term *causal* to pertain in two senses, one more formal than the other.[11]

---

[10] See, for example, [19], pp. 33-35.

[11] For similar comments by Dirac, see [142], pp. 46, 132-134.

"The two interventions **1.** and **2.** are fundamentally different from
one another. That both are formally unique, i.e., causal, is not important; indeed, since we are working in terms of the statistical properties of mixtures, it is not surprising that each change, even if it is
statistical, effects a causal change of the probabilities and expectation values. Indeed, it is precisely for this reason that one introduces
statistical ensembles and probabilities! On the other hand, it is important that **2.** does not increase the statistical uncertainty existing in U, but that **1.** does: **2.** transforms [pure] states into [pure]
states... while **1.** can transform [pure] states into mixtures. In this
sense, therefore, the development of a state according to **1.** is statistical, while according to **2.** it is causal. " ([477], p. 357)

Margenau later argued that the quantum state description describes reality
in accordance with causality under the quantum description whereas an attempted description of a classical sort of the same behavior would be *acausal*.

"The causally evolving ψ-states are not immediately tied to observations; they refer... to aggregates of observations... correspondence
of a hitherto unexpected type had to be introduced to restore causality... no juggling of 'hidden parameters' will wring knowledge of individual observational events from states. But states continue to evolve
in a causal fashion." ([309], p. 300)

Although it is clear that an interpretation of quantum mechanics can be
presented consistently with realism by its example, the Basic interpretation as
presented by von Neumann provides no direct explanation for the intersubjective agreement between results of measurements *between* different observers,
such as might be explained by providing some physical mechanism corresponding to the postulated projection upon observation. If two observers observe
the same object, there is no *a priori* reason that the random outcomes for
their observations would be predicted to agree; agreement must simply be
assumed. Preferably, such a mechanism would be describable by Process **2**.

After many decades of attempts, the measurement problem has still not
been explained away. This problem still appears insurmountable within standard quantum mechanics, as was recognized long ago by a number of physicists, such as London and Bauer [302] and Wigner [499] who, like Dirac,
not only viewed state projection as a physical process but also went beyond
the principle of psychophysical parallelism and the view that the observer
'makes the choice of eigenbasis' at measurement to a stronger relation that is
pscyho-physical *simpliciter*, making the *mind* of the observer the cause of the
state-projection.

Although the Basic interpretation was a great advance beyond the alternative at the time of its inception, namely, the early-stage Copenhagen interpretation, it cannot be the final word on the question of quantum phenomena;

Shimony has expressed the following view regarding this, which emphasizes the importance of retaining an epistemology founded in science (*cf.*, [369]).

"[I]n order to hold on to epistemological naturalism at the foundational level without resorting to physicalism a profound metaphysical and scientific revolution is required... In this revolution, physics— in its current formulation or in extrapolations from it that can be envisaged—would not be the basic natural science. Mentality would have a fundamental status in nature, either coordinate with physical reality or yet more fundamental. There have also been speculations that quantum mechanics points to, or at least is hospitable to, such a mentalistic revolution... None of them are convincing, but to me physicalism is even less so... Quantum mechanics is plagued with... 'the measurement problem.' There are, of course, serious proposals to resolve this problem without modifying the formalism of quantum mechanics, but there are students of the subject, among whom I am one, who believe that phenomenology is right and quantum mechanics is wrong, and that a successor to present quantum mechanics will account in a natural way for the occurrence of definite events." ([416], pp. 306-307)

This view has been a primary motivation for the appearance modified versions of quantum dynamics, such as pursued by Penrose, Philip Pearle and others.[12] Bohmian mechanics and theories such as that of Penrose including gravitational collapse involve not standard quantum mechanics but *modified* theories of quantum mechanics, even though these make use of wave-functions.

For his part, after providing the Basic interpretation, von Neumann described the probabilistic nature of quantum mechanics in his formulation as follows.

"[T]he present system of quantum mechanics would have to be objectively false in order that another description of the elementary process than the statistical one be possible." ([477], p. 325 of the 1955 edition)

It is important to recognize that von Neumann did exhibit an openness to the possibility of modifying quantum theory to make further progress; in their seminal quantum logic paper of 1936, Birkhoff and von Neumann comment that they were careful "to avoid being committed to quantum theory in its present form" [48]. Von Neumann remarked that "quantum mechanics has, in its present form, several serious lacunae, and it may even be that it is false, although this possibility is highly unlikely" ([477], p. 327 of 1955 edition) but also that no 'fundamental' physical theory has *remained* truly so ([478], p. 2).

---

[12] *Cf.* Pearle's overview of his and related attempts [341], as well as Shimony's list of desiderata for such attempts [412], p. 55.

The influential Collapse-Free approach to interpreting quantum mechanics, discussed below, includes only interpretations of quantum mechanics that do not involve modifying or adding to the standard quantum dynamics but rather reinterpreting the meaning of state projection.

Yet other relatively new approaches involve expanding or replacing the formal elements of quantum mechanics. For example, Jordan suggested that quantum mechanics needs an additional axiom in order to provide a consistent description of the world as observed.

> "Let us acknowledge that it is both possible and necessary to formu-late a physical axiom not formulated hitherto... a special axiom to express the empirical fact that... each large accumulation of micro-physical individuals always shows a well defined state in space and time—that a stone never, unlike an electron, has indeterminate coor-dinates. One often vaguely believes this to be guaranteed by Heisen-berg's $\Delta p \Delta q > h$; but in fact this relation only provides a possibility and not a necessity for the validity of our axiom... " ([264])

3.3 The Copenhagen Interpretation

The historically first well rounded interpretation of the quantum mechanics was the Copenhagen interpretation. It is based on ideas articulated by Bohr that predated the full Basic interpretation, although Dirac's early version of that interpretation appeared at roughly the same time as Bohr's first article on New quantum mechanics.

The Copenhagen interpretation, as currently understood, is actually the product of several investigators.

> "What is commonly known as the Copenhagen interpretation of quantum mechanics, regarded as representing a unitary Copenhagen point of view, differs significantly from Bohr's complementarity inter-pretation, which does not employ wave packet collapse in its account of measurement and does not accord the subjective observer any privileged role in measurement.... [T]he Copenhagen interpretation is an invention of the mid-1950s, for which Heisenberg is chiefly re-sponsible, various other physicists and philosophers, including Bohm, Feyerabend, Hanson, and Popper, having further promoted the in-vention in the service of their own philosophical agendas." ([243])

The interpretation as considered here includes Heisenberg's ideas, which were influenced by Pauli's views as much as Bohr's founding ideas.[13]

---

[13] For more on Pauli's contribution see, for example, [226].

The Copenhagen interpretation focused mainly on measurement and the limitations of its efficacy in the quantum realm, for example, as described by uncertainty relations. The widely recognized initial basis for the interpretation was laid down by Bohr in 1927 in his now famous Como lecture ([55], and [256], Section 4.1). Because Bohr's views were continually evolving, that lecture cannot be considered a definitive statement of the interpretation. It is, nonetheless, widely regarded as the *best* single statement of Bohr's version as given at any one time. Heisenberg described its seminal character as follows.

"Bohr considered the two pictures—particle picture and wave picture—as two complementary descriptions of the same reality. Any of these descriptions can only be partially true, there must be limitations to the use of the particle concept as well as of the wave concept, else one could not avoid contradictions. If one takes into account those limitations which can be expressed by the uncertainty relations, the contradictions disappear. In this way since the spring of 1927 one has had a consistent interpretation of quantum theory, which is frequently called the 'Copenhagen interpretation'." ([219], p. 43)

As history progressed, Bohr spoke progressively less of wave–particle duality and increasingly more of kinematic–dynamic complementarity. Whether there indeed exists a consistent unitary Copenhagen interpretation remains a matter of debate. Here, we primarily consider Bohr's views as given in statements at various later times together with those of Heisenberg and survey the elements of the interpretation as consistently as possible.

On the Copenhagen interpretation, the state-vector is considered as providing, without reservation, as complete a description of the individual quantum system as can be given by physics, much as it is understood to be on the Basic interpretation, although the details of the quantum formalism itself rarely appear explicitly in the writings of Bohr. Dugald Murdoch has described the relation between Bohr's interpretation and a simple statistical one as follows.

"[Although] Bohr did not think that the statistical interpretation was mistaken, he saw no point in insisting upon it, since the state vector provides as complete a description of an individual object as is possible... [He] believed that on its own it did not provide a complete basis for the interpretation of quantum mechanics." ([322], p. 119)

Also like von Neumann, Bohr required measuring apparatus to be considered genuine physical objects, rather than mere idealizations, which was a requirement that served him well in debates with Einstein. A key difference from the Basic interpretation is that Bohr believed that the wave-function description must necessarily be supplemented by classical concepts.

One of the clearer summaries by Bohr of the elements of the Copenhagen interpretation is the following.

"The unambiguous account of proper quantum phenomena must, in principle, include a description of all relevant features of the experimental arrangement. . . In the case of quantum phenomena, the unlimited divisibility of events implied in such an account is, in principle, excluded by the requirement to specify the experimental conditions. Indeed, the feature of wholeness typical of proper quantum phenomena finds its logical expression in the circumstance that any attempt at a well-defined subdivision would demand a change in the experimental arrangement incompatible with the definition of the phenomena under investigation." ([64], p. 3)

In particular,

"*the finite magnitude of the quantum of action prevents altogether a sharp distinction between a phenomenon and the agency by which it is observed*, a distinction which underlies the customary concept of observation and, therefore, forms the basis of the classical ideas of motion. With this in view, it is not surprising that the physical content of the quantum-mechanical methods is restricted to a formulation of statistical regularities in the relationships between those results of measurement which characterize the various possible courses of the phenomena." ([57], p. 11)

And, crucially,

"it is decisive to recognize that, *however far the phenomena transcend the scope of classical physical explanation, the account of all evidence must be expressed in classical terms.*" ([63], p. 39)

The last of the above comments identifies a distinctive aspect of the Copenhagen interpretation, clearly differentiating it from the approach of the Basic interpretation, namely, that all experimental arrangements are required to be *classically specifiable*; quantum phenomena are viewed as dependent on the (classical) measuring apparatus involved in measurement. The latter is reflected in the following comment by Bohr on Einstein's argument regarding the completeness of the quantum mechanical description.

"Of course there is in a case like [that of the EPR scenario] no question of a mechanical disturbance of the system under investigation during the last critical stage of the measuring procedure. But even at this stage there is essentially the question of an influence on the very conditions which define the possible types of predictions regarding the future behavior of the system. Since these conditions constitute an inherent element of the description of any phenomenon to which the term 'physical reality' can be properly attached, we see that [there is no justification for claims that the] quantum-mechanical description is essentially incomplete." ([58], pp. 60-61)

Bohr was definitive regarding the centrality of epistemology to his interpretation of quantum mechanics and believed that it neither requires nor implies a particular metaphysical position ([412], p. 310). In particular, he viewed the description of quantum phenomena as involving little realist ontological commitment although, importantly, he explicitly rejected idealism ([63], pp. 78-79). Bohr viewed his approach as, in a very particular sense, objectivist. According to him, although there are what he called "structures of pure thinking," the

> "transition to reality is made by theoretical physics, which correlates symbols to observed phenomena... these very structures are regarded by the physicist as the objective reality... This procedure leads to structures which are communicable, controllable, hence objective. It is justifiable to call these by the old term 'thing in itself'. They are pure form, void of all sensible qualities... But that they are perfectly empty does not fit the facts. Remember what practical use can be made of them." ([62], pp. 227, 232; cf. [281], p. 24)

Wheeler, who was strongly influenced by Bohr, pointed out that

> "Indeed, in Bohr's very last taped interview a few months before his death, he singled out certain philosophers for particular criticism. He said, '...they have not that instinct that is important to learn something and that we must be prepared to learn something of great importance... they did not see that it (the complementarity description of quantum theory) was an objective description and that it was the only possible objective description.' That represents the centre of his thinking on quantum theory. I think the word objective in Bohr's sense referred to the idea of dealing with what's right in front of you: the perceptions that you experience and the measurements you make, rather than Einstein's idea of a universe existing 'out there'..." ([119], pp. 58-59)

Wheeler's characterization of Bohr's views reflects the extent to which Bohr's emphasis on the role of the mind brings him very close to a form of idealism. Again, however, it is unjustified to consider Bohr's complementarity interpretation idealist or subjectivist as has, for example, Diederik Aerts in the following characterization.

> "The complementarity principle introduces the necessity of a far reaching subjective interpretation of the theory. If the nature of the behavior of a quantum entity (wave or particle) depends on the choice of the experiment that one decides to perform, then the nature of reality as a whole depends explicitly on the act of observation of this reality. As a consequence it makes no sense to speak about a reality which exists independently of the observer." ([2], Section 3.2)

Neither did Bohr regard himself as a positivist, despite being sometimes labeled so. Positivism has been defined in several ways, a common element being that under it knowledge claims must be entirely and directly grounded in experience.[14] That Bohr's philosophy was *influenced* by positivism was argued, for example, by Popper who, it should be recalled, was always on guard against denials of realism [93]. Bohr himself commented that the

> "Positivist insistence on conceptual clarity is, of course, something I fully endorse, but their prohibition on any discussion of the wider issues, simply because we lack clear-cut enough concepts in this realm, does not seem very useful to me—this same ban would prevent understanding of quantum theory." ([222], p. 208)

In the final analysis, "Bohr was, basically, a realist" ([361], p. 9). His approach contrasts with that of the early Heisenberg who, in 1927, argued that physics must be merely a "formal description of the relation between perceptions" ([233], p. 16) and who later expressed the view that, in the light of quantum mechanics, "objective reality has evaporated" [218]. Thus, while arguably in agreement with Bohr, the early Heisenberg had taken a philosophical step beyond Bohr's position, in that Bohr only claimed that the *physical content of the quantum theory* was limited. Nonetheless, late in his career Heisenberg also distanced himself from positivism.

> "The positivists have a simple solution: the world must be divided into that which we can say clearly and the rest, which we had better pass over in silence. But can any one conceive of a more pointless philosophy, seeing that what we can say clearly amounts to next to nothing. If we omitted all that is unclear, we would probably be left with completely uninteresting and trivial tautologies." ([222], p. 213)

One of the benefits of Bohr's approach is that it allowed him to account for the mutually exclusive character of non-commuting observables by pointing out that the apparatus for precisely measuring two such quantities of microsystems cannot be simultaneously used for two such measurements, as evidenced by the Bohr–Einstein debate and the thought experiments around which it turned [62, 319]. This and other debates were pivotal in the development of the Copenhagen interpretation and its legitimacy. Wheeler commented on this as follows.

> "[Bohr's view was] battle tested. Bohr argued and discussed with everyone who had point of view, so that in the end I would say that nobody has had a better picture of what quantum theory is and what it means." ([119], p. 59)

When the term *complementarity* was first introduced by Bohr, he applied it to causality itself:

---

[14] For a description of the details of the so-called 'received view,' see [13, 178, 201].

"The very nature of the quantum theory thus forces us to regard the space-time co-ordination and the claim of causality, the union of which characterizes the classical theories, as complementary but exclusive features of the description, symbolizing the idealization of observation and definition, respectively." ([56], p. 580)

Indeed, Bohr was of the view that

"the viewpoint of complementarity presents itself as a rational generalization of the ideal of causality." ([61])

On Bohr's view, quantum objects in absolute isolation are entities to which no specific properties or conceptions are applicable (*cf.* [354], p. 12). It would be most appropriate simply to say that on his view they do not exist in themselves in the way that the objects of common sense do.

Although Bohr defended his interpretation very effectively in debates, systematic descriptions of the Copenhagen interpretation are best sought in the work of others, including Pauli, Fock, and Heisenberg, which in some cases differ strongly with Bohr in one aspect or another; some of the elements of these descriptions differ from Bohr's views, which form the basic core of the interpretation, to the extent that a unified account can be given at all. It may be that—as has been argued, for example, by Mara Beller—no fully consistent version of the Copenhagen interpretation exists, and only "the appearance of consensus was achieved despite fundamental disagreements among the proponents" [26]. Stapp has similarly commented that "the diversity... in prevailing conceptions of the Copenhagen interpretation itself" is "striking."

"The cause of these divergences is not hard to find. Textbook accounts of the Copenhagen interpretation generally gloss over the subtle points... The writings of Bohr are extraordinarily elusive... Heisenberg's writings are more direct. But his way of speaking suggests a subjective interpretation that appears quite contrary to the apparent intentions of Bohr... [Furthermore, t]he writings of Bohr and Heisenberg have... not produced a clear and unambiguous picture of the basic logical structure of their position." ([434])

He believes the Copenhagen interpretation can be clarified by identifying its "logical essence," consisting of two assertions:

"(1) The quantum formalism is to be interpreted *pragmatically*.
(2) Quantum theory provides for a *complete* scientific account of atomic phenomena,"

which he goes on to spell out in some detail (*cf.* [434], pp. 1105-1107). The concise, detailed and useful resumé of the interpretation given by Hans Primas,[15] as eight theses underlying the interpretation which betray some of the

---

[15] Primas refers to it as the "Copenhagen interpretation of Pioneer quantum mechanics" ([364], p. 98).

tensions among the original Copenhagen interpreters, serves adequately to characterize the interpretation for our purposes.

(1) Quantum mechanics refers to individual objects.
(2) The probabilities of quantum mechanics are primary.
(3) The placement of the cut between observed object and the means of observation is left to the choice of the experimenter.
(4) The observational means are to be described in classical terms.
(5) The act of observation is irreversible and creates a document.
(6) The quantum jump is a transition from the potentially possible to the actual.
(7) Complementary properties cannot be revealed simultaneously.
(8) Pure quantum states are objective but not 'real.'

Bohr's key interpretational concept, the *principle of complementarity*, appears in relation to physical magnitudes in thesis 7: magnitudes corresponding to non-commuting observables are mutually exclusive, being precisely those that are 'complementary.' The clearest characterization of the principle of complementarity by Bohr himself is that, in quantum mechanics, one is forced

"to adopt a new mode of description designated as *complementary* in the sense that any given application of classical concepts precludes the simultaneous use of other classical concepts which in a different connection are equally necessary for the elucidation of the phenomena." ([57], p. 10)

It has been argued by Healey that the first of the above theses is both incompatible with Bohr's personal views and untenable because, with its inclusion, the Copenhagen interpretation inherits the measurement problem ([211], p. 11). It, therefore, can be regarded as inessential to the interpretation; it is included here because it was certainly held by *some* of those who formulated the interpretation, and as already noted, Bohr saw "no point in insisting upon it" ([322], p. 119). However, Healey himself has strongly criticized what he calls the "strong version" of the Copenhagen interpretation, wherein the first thesis is *not* included and quantum states are allowed to refer *only* to ensembles, that is, sets of identically prepared quantum systems, as follows. Although rejecting the theory's reference to individual quantum systems may help such a "strong version" avoid the quantum measurement problem, one result of the move is that the interpretation inherits instead the need to *assume* definite measurement outcomes in each measurement, because quantum mechanics is otherwise unable to show that measurements produce definite outcomes *at all*; such an assumption is untenable, because it *either* requires that all systems have definite values for all their dynamical variables at all times, which is incompatible with the complementarity, the seventh thesis, *or* is the interpretation rendered descriptively and explanatorily incomplete ([211], pp. 15-18).

It was the individual process in a suitable well specified experimental situation to which Bohr referred when using the term *phenomenon*.

"It is certainly far more in accordance with the structure and interpretation of the quantum mechanical symbolism, as well as with elementary physical principles, to reserve the word 'phenomenon' for the comprehension of the effects observed under given experimental conditions." ([60], p. 25)

A form of thesis 1, albeit with "objects" replaced by "phenomena," appears in this manner in the following remark of Bohr.

"To my mind there is no other alternative than to admit in this field of experience, we are dealing with individual phenomena..." ([63], p. 51).

Thesis 2 is reflected in Bohr's statement above that "the physical content of the quantum-mechanical methods is restricted to a formulation of statistical regularities in the relationships between those results of measurement which characterize the various possible courses of the phenomena," which some have mistakenly taken as evidence that Bohr was an empiricist, a point addressed below.

Thesis 3 imposes Heisenberg's *Schnitt*. It and thesis 4 were both addressed together by Heisenberg in the chapter of his *Physics and philosophy* dedicated by title to the Copenhagen interpretation.

"Certainly quantum theory does not contain genuine subjective features, it does not introduce the mind of the physicist as part of the atomic event. But it starts from the division of the world into the 'object' and the rest of the world, and from the fact that at least for the rest of the world we use the classical concepts in our description. This division is arbitrary and historically a direct consequence of our scientific method; the use of the classical concepts is finally a consequence of the general human way of thinking. But this is already a reference to ourselves and in so far as our description is not completely objective." ([219], pp. 55-56)

Pauli differed somewhat with Heisenberg with respect to the above characterization of the objectivity of quantum mechanics and was a strong advocate of thesis 4, arguing as early as 1923 that in quantum theory "we give up the *laws* of classical theory, but still always with the *concepts* of that theory" [334]. The following imperative statement of Bohr relates to theses 4 and 5 and makes clear that, on the Copenhagen interpretation, not only must the measurement apparatus be classically described but also the associated data must enter a classical *record*. "[I]t is decisive to recognize that, *however far the phenomena transcend the scope of classical physical explanation, the account of all evidence must be expressed in classical terms*" ([63], p. 39). This is, perhaps, the most peculiar of the Copenhagen requirements. In essence, it precludes the independence of quantum physics from classical physics; otherwise classical mechanics itself is, in at least this sense, reduced to quantum mechanics, which would then render Bohr's requirement on the description of quantum phenomena both superfluous and misleading; Heisenberg referred to "the paradox of the quantum theory, namely, the necessity of using the classical concepts" and argued that "it would be a mistake to believe that [the] application of the quantum-theoretical laws to the measuring device could help to avoid the fundamental paradox of quantum theory" ([219], p. 56).

Although it is tempting to view thesis 4 conservatively as a direct implication of the assumption that *observers* are inherently classical, another imperative statement by Bohr reflecting thesis 5 precludes this.

"[It] is also essential to remember that all unambiguous information concerning atomic objects is derived from the permanent marks... left on the bodies which define the experimental conditions... The description of atomic phenomena has in these respects a perfectly objective character, in the sense that no explicit reference is made to any individual observer." ([64], p. 3)

Thesis 6 appears in the remark of Bohr that "no elementary phenomenon is a phenomenon until it is a registered (observed) phenomenon" [56], which served more recently as the starting point for Wheeler's "it from bit" program, which is discussed in Chapter 4 below. Similarly, Heisenberg remarked that in quantum mechanics observation involves the actualization of "possibilities or better tendencies ('potentia' in Aristotelian philosophy)" ([219], p. 53). However, as Shimony has pointed out,

"This historical reference should perhaps be dismissed, since quantum mechanical potentiality is completely devoid of teleological significance. What it has in common with Aristotle's conception is the indefinite character of certain properties of the system. One does not find Aristotle saying, however, that a property becomes indefinite because of observation and that the probabilities of all possible results are well determined, whereby the quantum mechanical potentialities acquire a mathematical structure." ([412], p. 314)

As for the actualization of potentialities, a concept introduced by Heisenberg, Heisenberg commented that

"we may say that the transition from the 'possible' to the 'actual' takes place as soon as the interaction of the object with the measuring device, and thereby with the rest of the world, has come into play; it is not connected with the registration of the result in the mind of the observer." ([219], pp. 54-55)

This latter view distinguishes the process of measurement under the Copenhagen interpretation somewhat from the imposition of state projection in the

Basic interpretation. Nonetheless, Bub sees a close relationship between the two interpretations.

> "Bohr's complementarity interpretation can be regarded as a 'minimal revision' of the orthodox von Neumann–Dirac interpretation without the projection postulate, constrained by the determinateness of the measured observable R and the requirement of maximizing the set of propositions that (i) are determinate in the (unprojected) quantum state (according to the orthodox interpretation) and (ii) can be maintained as determinate, consistently with the determinateness of R." ([87], p. 203)

Heisenberg noted that the philosophical tendencies of physicists typically differ from Kant's highly influential position. In particular, Heisenberg remarked that Kant's

> " 'thing-in-itself' is for the atomic physicist, if he uses this concept at all, finally a mathematical structure; but this structure is—contrary to Kant—indirectly deduced from experience." ([87], p. 91)

Nonetheless, there are convincing reasons for viewing Bohr as a transcendentalist whose methodology relates to Kant's, as argued by John Honner who, in his study of Bohr, pointed out that

> "a transcendental argument entails not a syllogism but a blunt assertion about what cannot but be the case,"([233], p. 13)

and that

> "Bohr's framework of 'complementarity' is perhaps the most contested of recent transcendental claims." ([233], p. 14)

Thesis 8 as formulated by Primas is better stated with "real" replaced by "real in the traditional sense." For example, Bohr made the following claim.

> "Now, the quantum postulate implies that any observation of atomic phenomena will involve an interaction with an agency of observation not to be neglected. Accordingly, an independent reality in the ordinary physical sense can neither be ascribed to the phenomena nor to the agencies of observation. After all, the concept of observation is in so far arbitrary as it depends on which objects are included in the system to be observed." ([57], p. 54)

The *quantum postulate* of Bohr is the idea that, whenever an interaction involving a microscopic object takes place, there is a discontinuity; it is Bohr's way of providing an interpretational counterpart of Planck's *quantum of action* (*cf.* [354], pp. 9, 75) and is closely related to thesis 6 as Bohr held it. It can be seen as introducing a 'process' somewhat analogous to that of the projection postulate in the Basic interpretation.

Regarding the philosophical character of the interpretation as a whole, Beller has claimed that

"different scholars, with good textual evidence, have provided conflicting interpretations of [writings of proponents of the Copenhagen interpretation]: while Popper presented Bohr as a 'subjectivist,' Feyerabend found him an 'objectivist'; more recently, Murdoch concluded that Bohr was a realist, while Faye argued with equal competence that Bohr was an antirealist." ([26])

Honner has pointed out one explanation for the diversity of understandings of Bohr's position.

"Bohr wrote many cryptic essays on atomic physics and human knowledge and, despite his scrupulousness about clarity, his way of thinking has remained too enigmatic for professional philosophers. His interpretation of the implications of quantum mechanics and, in particular, his notion of 'complementarity', remain and open to divergent expositions. This may partly be due to the fact that 'complementarity' itself appears to embrace contrasting positions like idealism and pragmatism. But *pace* Wittgenstein, Bohr would claim that it is as important to stammer about the profound as to speak clearly about the obvious." ([233], p. 3)

Similarly, Carl von Weizsäcker commented

"I would prefer to call it the Copenhagen interpretation of the formalism, an interpretation by which the formalism is given a sufficiently clear meaning to become part of a physical theory. In so saying I express the opinion that the Copenhagen interpretation is correct and indispensable. But I have to add that the interpretation, in my view, has never been fully clarified. It needs an interpretation itself, and only that will be its defence." ([481], p. 25)

All considered, Heisenberg better articulated the Copenhagen interpretation and its relation to the measurement process than Bohr himself—for example,

"It should be emphasized... that the probability function does not in itself represent a course of events in the course of time. It represents a tendency for events and our knowledge of events. The probability function can be connected with reality only if one essential condition is fulfilled: if a new measurement is made to determine a certain property of the system. Only then does the probability function allow us to calculate the probable result of the new measurement. " ([219], pp. 46-47) and

> "The result of the measurement again will be stated in terms of
> classical physics.... [T]he theoretical interpretation of an experiment
> requires three distinct steps: (1) the translation of the initial exper-
> imental situation into a probability function; (2) the following up
> of this function in the course of time; (3) the statement of a new
> measurement to be made of the system, the result of which can then
> be calculated from the probability function. For the first step, the
> fulfillment of the uncertainty relations is a necessary condition. The
> second step cannot be described in terms of the classical concepts;
> there is no description of what happens to the system between the
> initial observation and the next measurement. It is only in the third
> step that we change over again from the 'possible' to the 'actual'."
> ([219], pp. 46-47)

This clear and simple prescription for avoiding the conceptual confusion likely
to be encountered by working physicists when contemplating the behavior of
quantum mechanical systems in experimental situations is likely as responsible for the powerful influence of the Copenhagen interpretation as Bohr's
pronouncements are.

Although Bohr's success against Einstein was crucial, the consistency of
the complementarity interpretation as presented in the Bohr–Einstein debate
during the Fifth Solvay Congress may be seen as questionable. In the context
of double-slit experiment, Bohr was forced to apply the uncertainty principle
to the *macroscopic* screen which, as the measuring apparatus, is required
on the interpretation to be described *entirely in classical terms*. Although
this may not actually be *inconsistent*, it does involve the application of a
quantum mechanical principle to a classically characterized system (under
the interpretation), which undermines the assumption of a clear boundary
separating the classical realm from the quantum realm, which is clearly a
background assumption of the interpretation.[16]

Another difficulty related to thesis 4 was pointed out by von Weizsäker,
namely, that there is insufficient elucidation of the meaning of the term *classical*. "Bohr's statement [that all experiments are to be described in classical
terms] implies an apparent paradox: classical physics has been superseded
by quantum theory; quantum theory is verified by experiments; experiments
must be described in terms of classical physics" ([481], p. 26). He suggested a
resolution of this problem by introducing an "interpretation of the interpretation," namely, that Bohr was stating

---

[16] See, for example, the description in [496], pp. 205-210.

"a truism, though a philosophically important one: a measuring instrument must be described by concepts appropriate to measuring
instruments. It is then not unnatural further to assume that classical
physics, in the form in which it developed historically, simply describes the approximation to quantum theory appropriate to objects
as far as they really can be fully observed... no further adaptation
of our intuitive faculty to quantum theory is needed or possible."
([481], p. 26)

This suggestion also serves somewhat to mitigate the problem of Bohr's applying a quantum mechanical rule to the measuring instruments in response
to Einstein in their discussion of the rigged double-slit experiment.[17]

Both the Basic and Copenhagen interpretations have been viewed as *the*
quantum orthodoxy contrasting with Einstein's "Interpretation I." Given the
choice, which of these 'orthodox interpretations' one prefers, and indeed within
each the emphasis one chooses to place on the role of observing agents, turns
on the question of how formally quantum a description of physics one prefers.

3.4 Orthodoxy and Explanation in Quantum Physics

The Copenhagen interpretation has most often been called *the* orthodox interpretation of quantum mechanics, despite questions about its internal consistency and the diversity of views of its advocates. This was emphatically
expressed by Peierls.

"I object to the term Copenhagen interpretation... this sounds as
if there were several interpretations of quantum mechanics. There
is only one. There is only one way in which you can understand
quantum mechanics. There are a number of people who are unhappy
about this, and they are trying to find something else. But nobody
has found anything else which is consistent yet, so when you refer
to the Copenhagen interpretation of the mechanics what you really
mean is quantum mechanics. And therefore the majority of physicists
don't use the term; it's mostly used by philosophers." ([119], p. 71)

To the extent that philosophers concern themselves with quantum mechanics
per se, it is true that questions of interpretation and explanation interest
them most. They are the philosopher's *métier* but also highly appropriate
for physicists to weigh in on. To the extent that physicists have failed to
engage foundational questions of quantum mechanics, which is clearly a lesser
extent than Peierls claimed it to be, this can be explained naturally by the
fact that physicists typically do not concern themselves explicitly with the
interpretations of the theories they use; they typically use them as best they

---

[17] See the discussion of this portion of the Bohr–Einstein debate given in Section
1.2.

can as they apply methods they find intuitive, have learned and devised, that work quite well for calculation and experimentation and that allow for the fine characterization of physical details.

Contra Peierls, Itamar Pitowsky is of the following opinion.

"I believe that von Neumann's [343] contribution to the foundations of quantum theory is exceedingly more important than that of Bohr. For it is one thing to say that the only role of quantum theory is to 'predict experimental outcomes,' and that different measurements are 'complementary.' It is quite another thing to provide an understanding of what it means for two experiments to be incompatible, and yet for their possible outcomes to be related; to show how these relations imply the uncertainty principle; and even, finally, to realize that the structure of events dictates the numerical values of the probabilities (Gleason's theorem)." ([351])

Moreover, the increasingly visible collection of the physicists who have explicitly adopted during the past few decades interpretations other than the Basic interpretation, which is that most often that used in deed if not in word, have adopted some form of the Collapse-Free interpretation not simply identifiable with the Copenhagen interpretation.

Before the emergence of quantum information science as a specific field of investigation, Bub referred to the Collapse-Free approach as the "new orthodoxy" and claimed that there is a "growing consensus" among "most physicists" that accepts it as the "modern, definitive" version of the Copenhagen interpretation[18] which he also regards as a 'minimal revision' of the Basic interpretation by the exclusion of the projection postulate. He has also claimed that for most physicists

"the measurement problem would hardly rate as even a 'small cloud' on the horizon." ([87], pp. 207-212)

Many, for example Peierls, have been in agreement with Bub on the question of the existence of a quantum orthodoxy. However, as can be seen by inspection of the broad range of views and concerns expressed by a number of the best physicists who *have* concerned themselves with the interpretation of quantum mechanics, whose views are exhibited throughout these pages, the measurement problem has remained more than a "small cloud on the horizon," except for Copenhagen true believers. However, Bub has also argued that the measurement problem is widely misunderstood [85], although he reaches this conclusion through a rather forced quantum-logical approach. This also minimizes the divergence of ontological commitments among the various advocates of the two interpretations that is clear when surveying the various versions of the Collapse-Free approach as is done in the next section.

---

[18] Note that, for Bub, the traditional "orthodoxy" is what he calls the Dirac–von Neumann interpretation, that is, the Basic interpretation ([87], Chapters 7-8).

There is no single orthodoxy among the set of interpretations into which Einstein's original I–II dichotomy has devolved. For example, the various interpretations that have emerged often differ significantly in their ontological commitments. Nonetheless, if the ontological differences among the versions of the Collapse-Free interpretation are put aside, then that rather popular approach could be seen to accord in a number of respects with at least one understanding of the Copenhagen interpretation. That option is assisted by both the vagueness of Bohr's presentation and the fact that, in Heisenberg's clearer articulation, quantum mechanics serves only to predict the results of measurements, which are required to be only classically describable, rather than to describe the microscopic world between measurements, say as a means for explaining what is observed. Again, however, only such an instrumentalist view of science—which, as Bell pointed out, is not a philosophical view to which the majority of physicists would assent if explicitly faced with the question—would enable the association of the Collapse-Free interpretation with Copenhagen interpretation much beyond the shared view of the projection postulate as non-fundamental.

Regarding the relationship of interpretation and explanation, d'Espagnat has commented that

"[T]o account for the phenomena, present-day physics seems to favor some kind of a two-level explanatory scheme. The level that may be termed 'the strictly scientific one' consists in explaining both the regularity of the observed phenomena and the observed intersubjective agreement by referring to laws. This, of course, is just the standpoint normally taken up in science, but with here the significant difference, motivating the notion of a possible 'second level,' that while in classical physics the laws were objectively interpretable—they supposedly described what exists—in quantum physics they are but observational predictive rules." ([126], p. 225)

Again, however, physicists overwhelmingly are, if only implicitly, scientific realists, that is, they are realists about the entities of current science which, in case of the microscopic ones, are not classical describable. The purpose of interpreting quantum mechanics for them is precisely to provide and understanding of the relationship between the quantum formalism and the world of quantum objects, rather than to merely explain appearances. Like Bell, Einstein, Penrose, Schrödinger, and many others have, they would rather avoid such a "second level," by resolving the measurement problem.

The questions of whether there is a quantum orthodoxy and whether there is such a "two-level explanatory scheme" at the moment are of little help for us in coming to understand the physical world; they will be of greater importance to historians of science.

3.5 The Collapse-Free Approach

The Collapse-Free interpretation of quantum mechanics is distinguished by the assumptions that fundamental quantum state change is always and only unitary and that such state change occurs at the scale of the entire universe. It assumes the burden of demonstrating that the occurrence of definite measurement outcomes can be explained without using the projection postulate. Although the interpretation was first explicated by Everett, later variants have received equal if not greater attention as a result of their utility in quantum cosmology.

Everett himself wrote only three articles on the interpretation [163–165]; the last of these was "The theory of the universal wave function," the title of which emphasizes the applicability of the quantum mechanical state description to the entire universe which, of course, includes the bodies of all observing agents. Thus, despite Bub's characterization of the Collapse-Free interpretation as emerging naturally from the Copenhagen interpretation, the former is clearly distinguished from the latter by this assumption—*cf.* the conjunction of (i) the third of the Copenhagen theses stated in the previous section, which requires the observer to be distinct from the observed object by a cut the location of which is not determined by physics alone, and (ii) the fourth thesis, which takes the observer necessarily to be classically described.

An explicitly 'many-worlds' version of the Collapse-Free interpretation of particular significance was later offered by the Bryce S. DeWitt and R. Neill Graham. This was later followed by others—most notably by Deustch's 'multiverse' version—that, like it, involve additional analyses within Everett's basic approach. De Witt and Graham collected Everett's work and their own in the volume *The many-worlds interpretation of quantum mechanics* and described his initiation of the Collapse-Free approach as follows.

"In 1957, in his Princeton doctoral thesis, Hugh Everett, III, proposed a new interpretation of quantum mechanics that denies the existence of a separate classical realm and asserts that it makes sense to talk about a state vector for the whole universe. This state vector never collapses, and hence reality as a whole is rigorously deterministic... the state vector decomposes naturally into orthogonal vectors, reflecting the continual splitting of the universe into a multitude of mutually unobservable but equally real worlds, in each of which every good measurement has yielded a definite result and in most of which the familiar statistical quantum laws hold." ([135], p. v)

An element of Everett's original version is that measurement outcomes that would ostensibly be witnessed by agents 'in' these various possible situations are representable by reference to the universal wave-function applied to all of physical reality, however circumscribed. The above rendition involves the shared universe(s) of any given collection of observing agents being among an extremely large number of equally real 'universes.' Later versions assume a more limited ontology, while retaining the idea of the universal wave-function.

The Collapse-Free interpretation has gone from something of a curiosity, as was the case immediately after Everett's own work, to an important element of theoretical physics, albeit in a special context; this approach became important to cosmologists using quantum mechanics to study the early universe to the extent that it became "rather difficult to think of any interpretation of quantum cosmology that does not invoke this view in one way or another" ([247], p, 183). More dramatically, according to Bub, this interpretation forms the core of a "new orthodoxy," which involves considerable syncretism.

> "This 'new orthodoxy' weaves together several strands: the physical phenomenon of environment-induced decoherence, elements of Everett's relative-state formulation of quantum mechanics, popularized as the 'many worlds' interpretation, and the notion of 'consistent histories' developed by Griffiths and extended in different ways by Omnès, Gell-Mann and Hartle, and others." ([87], pp. 212-213)

As already noted, however, there is considerable disagreement among advocates as to the ontological commitment of the interpretation. "The maze of many worlds and many minds interpretations of quantum mechanics is by now sufficiently serpentine to make one think twice about entering" ([137], p. 48). Nonetheless, its conceptual and historical significance warrants giving it serious consideration. Here, as unified a description and evaluation of the approach as is possible is presented. The problems shared by its various versions, arising primarily from their common denial of the existence of a dynamically fundamental 'wave-function collapse' process, are also taken up.

The preservation of psychophysical parallelism can be seen as an element the interpretation has in common with its predecessors; one goal of the approach is to demonstrate that comprehensive collapse-free state evolution is consistent with this requirement. Advocates see the Collapse-Free approach as enabling an understanding of quantum mechanics by reference to the differing experiences of subjects appearing in the universal wave-function as physical systems in superpositions of the sort in which Wigner's friend finds himself, resolving the corresponding paradox; the approach "postulates that every system that is subject to external observation can be regarded as part of a larger isolated system" ([164], p. 316). This postulate, although virtually trivial, does serve to connect all observed quantum systems to a universal system, that is, the greater 'universe.' Quantum correlation enables this.

> "In order to bring about this correspondence with experience for the pure wave mechanical theory, we shall exploit the correlation between subsystems of a composite system that is completely described by a state function." ([165], p. 9)

Everett claimed that the measurement problem in its simplest form can be avoided exactly by *embracing* the mathematical description quantum mechanics naturally provides of measurements if only the Schrödinger equation is allowed, that is, by dispensing with von Neumann's process 1. "[T]he general validity of pure wave mechanics, *without any statistical assertions*, is assumed for *all* physical systems, including observers and measuring apparata [*sic*]" ([165], p. 8). The standard statistical predictions of quantum mechanics, as given by the Born rule, are then to be *derived* rather than assumed. Quantum state projections, which play an important role in the Basic interpretation, namely, that of describing the objectification of measurement results, to the extent they are considered relevant, are merely secondary elements of an ultimately non-dynamical nature: They serve only to provide the so-called *relative states* 'within' the universal wave function associated with quantum systems capable of measurement, which are in turn taken to correspond to the experience of observers making measurements. This idea is reflected in the title of Everett's short article " 'Relative state' formulation of quantum mechanics."

The Collapse-Free approach has been touted both as the first truly 'self-interpreting' version of the theory and, as we have seen, a natural successor the Copenhagen interpretation. Similarly to others after the initial formative period, Everett saw his approach as removing the need for philosophy to play a primary role in the conception of the theory. For him,

"The wave function is taken as the basic physical entity with *no a priori interpretation*. Interpretation only comes *after* an investigation of the logical structure of the theory. Here as always the theory itself sets the framework for its interpretation." ([164], p. 316)

However, like any other, Everett's approach to quantum mechanics involves epistemic elements and ontological commitments, whether explicitly or implicitly and whether this is acknowledged by its advocates or not. For example, some adherents have claimed that within it "the symbols of quantum mechanics represent reality just as must as do those of classical mechanics" ([135], p. 167), that is, are amenable to naive realist interpretation. This assertion underlines the specious nature of the claim that the Collapse-Free interpretation is a direct extension of the Copenhagen interpretation.

The appeal now to a number of physicist advocates of the interpretation is exactly that it can be seen as taking the quantum mechanical formalism to describe reality differently from the way the Copenhagen interpretation is often taken to, namely, as a reality wherein the choices of experiments by observing agents play an central role. The greatest challenge of the approach is to provide an adequate account of the probabilities of quantum mechanics for the outcomes of all specific individual measurements in the way envisioned by Everett. Another great challenge is that of providing a consistent meaningful explication of the "branching" of systems supposed to occur during measurements, as shown below.

The following set of elements of the Collapse-Free approach was laid out by DeWitt in a version meant to improve Everett's original, which has been criticized as incomplete by both supporters and detractors.

(1) The mathematical formalism of quantum mechanics is sufficient as it stands. No metaphysics needs to be added to it.

(2) It is unnecessary to introduce external observers or to postulate the existence of a realm where the laws of classical physics hold sway

(3) It makes sense to talk about a state-vector for the whole universe.

(4) This state-vector never collapses, and hence the universe as a whole is rigorously deterministic.

(5) The ergodic properties of laboratory measuring instruments, although strong guarantors of the internal consistency of the statistical interpretation of quantum mechanics, are inessential to its foundations.

(6) The statistical interpretation need not be imposed *a priori*.

Assertion 1 involves either empiricist agnosticism or a philosophical position that might be generically assumed to underwrite all of physics if one believes specific metaphysics is implicit in the theory. Again, assertions 2 and 3 clearly distinguish the approach from the Copenhagen interpretation in that all observers are described within the theory, that is, are described within it in the same way as other physical objects are; furthermore, they are not in need of an inherently classical mechanical description as in that interpretation.

The vagueness of Bohr's language and the diversity of views identified with the Copenhagen interpretation provides one motive for instead adopting the Collapse-Free approach. DeWitt saw the Basic and Copenhagen interpretations as having introduced unjustified elements that are to be discarded because they threaten quantum mechanics with inconsistency [134]. His approach *prima facie* reduces the number of assumptions required for interpreting quantum mechanics. He attempted to show these elements to be superfluous by providing derivations of what, there, are fundamental theoretical components for which those elements are traditionally understood to account. He also viewed the relationship of this interpretation and the Basic interpretation as one not only between interpretations but as one between *theories*. Like Everett, he exploited the principle of psychophysical parallelism in this attempt to eliminate 'superfluous' interpretational elements.

> "The new theory is not based on any radical departures from the conventional one. The special postulates in the old theory which deal with observation are omitted in the new theory. The altered theory thereby acquires a new character. It has to be analyzed in and for itself before any identification becomes possible between the quantities of the theory and the properties of the world of experience. The identification, when made, leads back to the omitted postulates of the conventional theory that deal with observation, but in a manner which clarifies their role and logical position." ([164], p. 315)

The "omitted postulates" are denied the status of first principles but, importantly, are not entirely rejected. Rather, the *exceptionality* of the measurement process is rejected; observers are supposed to be shown to play no special role. *A fortiori*, consciousness is explicitly denied a physical role.

As mentioned above, the Wigner friend thought experiment is viewed as a motivational example serving to emphasize the goals of the Collapse-Free approach. For example, Everett explicitly considered the problem of making consistent the experiences of two observers (A and B, B observing A who observes the inanimate system S) but rejected Wigner's solution of giving consciousness a privileged role in quantum mechanics. He commented on the difficulty traditionally understood to be presented by the Wigner friend scenario, but ostensibly overcome by the Collapse-Free approach, as follows.

> "If we are to deny the possibility of B's use of a quantum mechanical description (wave function obeying the wave equation) for A+S, then we must be supplied with some alternative description for systems which contain observers (or measuring apparatus). Furthermore, we would have to have a criterion for telling precisely what type of systems would have the preferred positions of 'measuring apparatus' or 'observer' and be subject to alternative description. Such a criterion is probably not capable of a rigorous formulation." ([165], p. 4)

Because, from the point of view of B, there is continuous state evolution of the joint system A+S, there is an inconsistency in allowing A to induce a discontinuous collapse of the state of S. Everett's goal was to show instead that

> "this concept of a universal wave mechanics, together with the necessary correlation machinery for its interpretation, forms a logically self consistent description of a universe in which several observers are at work." ([165], p. 9)

By quantum mechanically modeling all observers, Everett also hoped to account for the quantum probabilities without the use of a physical projective state process.

> "We shall be able to introduce into [the theory] systems which represent observers. Such systems can be conceived as automatically functioning machines (servomechanisms) and which are capable of responding to their environment... we shall deduce the probabilistic assertions of [Dirac and von Neumann's] Process 1 as *subjective* appearances to such an observer, thus placing the theory in correspondence with experience. We are then led to the novel situation in which the formal theory is objectively continuous and casual, while subjectively discontinuous and probabilistic. While this point of view shall ultimately justify our use of the statistical assertions of the orthodox view, it enables us to do so in a logically consistent manner, allowing for the existence of other observers." ([165], p. 9)

Omnès has assessed Everett's prescription for using the quantum formalism differently from and more critically than DeWitt.

"Everett's contribution is not so much a theory but essentially a *representation of reality*... a way of looking at reality originating essentially from one's own philosophical inclinations. It is not science because no experiment can show it to be wrong and it is not a theoretical truth because there can be no proof of it. It is not nonsense because one cannot prove it to be inconsistent." ([329], p. 345)

Omnès considers there to be no physical difference between Everett's treatment and the standard theory. Indeed, it is not obvious that the predictions of observers in the Basic interpretation will differ from those in Everett's version. However, DeWitt was correct when he referred to the approach as an "altered theory" of quantum mechanics; the Collapse-Free approach, if successful, would reduce the number of fundamental theoretical elements needed in the theory because the projection postulate is removed with *conceivable* differences in predictive consequences. Deutsch, for example, disagrees with the claim that there are no new predictions under the Collapse-Free approach, arguing that there are specific predictive and explanatory differences between the Basic and the Collapse-Free interpretations. The criticism of the interpretation on the specific grounds that it makes no *new* predictions is without merit. The history of interpretations, in particular, which interpretation of the formalism first appeared, is irrelevant. The primary motive for interpreting quantum mechanics is to find a valid way of understanding quantum mechanics that allows one to provide consistent physical explanations. An interpretation is superior when it helps one to construct superior explanations. A reduction of the number of theoretical assumptions is preferable as well.

Omnès also objects to various versions of this interpretation on the grounds that they are insufficiently empiricist. However, his criticisms would also apply to most other interpretation of quantum mechanics, because these typically do present the theory as more than a simple predictive device. Omnès' critical characterization can be understood as an indictment of any interpretation that takes the formalism to be anything but a machine for predicting measurement outcomes. His objections do not primarily target the theoretical aspects of the Collapse-Free approach themselves; rather, that there are (sometimes implicit) metaphysical elements is objectionable to him.

Substantive criticism or rejection of an interpretation of quantum mechanics must involve a genuine *interpretational* failure, such as a logical inconsistency, conceptual confusion, or explanatory shortcoming. One such criticism of the Collapse-Free interpretation, by H. J. Groenwald, is that

"the statistical interpretation [of quantum mechanics] leaves for example no place for speaking about the quantum state of the universe." ([198], p. 45)

Another issue, at least with Everett's original version, is that it is ontologically vague. In any version of the interpretation, the components of a state superposition for the set of systems consisting of object, measuring apparatus, and environment on which it relies are in principle capable of interfering with each other even though the observed quantities are considered determinate. On this interpretation, how are the *individual entities* distinguished?

Criticism was leveled by Bell against the Collapse-Free interpretation in DeWitt's improved version on the basis of its imprecision, in that it associates an independent universe with each of Everett's branches and, therefore, involves an explicit commitment to multiple unseen universes—a tremendous increase in that commitment over that of previous interpretations, and their ill-defined branchings.

"The idea that there are all those universes which we can't see is hard to swallow. But there are also technical problems with it which people usually gloss over or don't even realize when they study it. The actual point at which a branching occurs is supposed to be the point at which a measurement is made. But the point at which a measurement is made is totally obscure. The experiments at CERN for example take months and months, and at which particular second on which particular day the measurement is made and the branching occurs is perfectly obscure. So I believe that the many-universes interpretation is a kind of heuristic, simplified theory, which people have done on the backs of envelopes but haven't really thought through. When you try to think it through it is *not* coherent." ([119], p. 55)

That is, removing the privileged role of observation in Everett's way accomplishes little overall because branching is forced to take on the burden of describing measurement with a branching process but that process is not physically well specified.

A basic difficulty associated with the branching process is simply the co-presence of 'simultaneous' actual but different measurement outcomes. Wheeler, who was an early supporter of the Everett approach, later abandoned it essentially on that account.

"I supported this to begin with, because it seemed to represent the logical follow-up of the formalism of quantum theory. I have changed my view of it today because there's too much metaphysical baggage being carried along with it, in the sense that every time you see this or that happening you have to envisage other universes in which I see something else happening. This is to make science into a kind of mysticism." ([119], p. 60)

Murray Gell-Mann and James Hartle, who proposed another successor version of the interpretation, criticized this aspect of Everett's version, viewing it as underdeveloped and, perhaps more importantly, inadequate in the cosmological context, precisely where it is best applied. Gell-Mann and Hartle were concerned that this version

> "did not adequately explain the origin of the classical domain or the meaning of 'branching' that replaced the notion of measurement. It was a theory of 'many worlds' (what we would call 'many histories'), but it did not sufficiently explain how these were defined or how they arose. Also Everett's discussion suggests that a probability formula is somehow not needed in quantum mechanics, even though a 'measure' is introduced that, in the end, amounts to the same thing." ([184], p. 430)

This criticism was later echoed by Adrian Kent, most pertinently to DeWitt's version.

> "[O]ne can perhaps *intuitively* view the corresponding components of [a wave function] as describing a pair of independent worlds. But...the axioms say nothing about the existence of multiple physical worlds corresponding to wave function components." ([269])

The failure of the interpretation to address this problem, which relates directly to its ontological aspect has, nonetheless, not discouraged enthusiastic proponents.

Deutsch, who has personally found the Collapse-Free interpretation exceptionally useful for understanding quantum computing, has offered a subtler view the branching process but in a way that embraces the 'multi-verse' to which it can be understood to give rise.

> "In my favorite way of looking at this, there is an infinite number of [universes] and this number is constant; that is, there is always the same number of universes. Before a choice or decision is made, in which more than one outcome is possible, all the universes are identical, but when the choice is made, they partition themselves into two groups, and in one group one outcome happens and in the other group another outcome happens. Normally these two groups don't affect each other thenceforward, but as I have said, they occasionally do." ([119], p. 85)

This view of the relation between 'universes' may go some way toward resolving the issue of splitting a universe, but it does nothing to address the question of the existence of unseen 'other' universes—for him, an *infinite* number—to which Wheeler and others have objected.

As mentioned above, Deutsch also claims that there are overlooked empirical differences between quantum mechanics as understood on the Collapse-Free interpretation—at least in his version—and as understood on the Basic interpretation.

"There is a widespread belief that the conventional (i.e. wave function collapse) view and Everett's are alternative *interpretations* of
the same quantum formalism, and are identical in their experimental predictions. Everett himself believed this. But he was mistaken.
The conventional 'interpretation', unlike Everett's, is actually more
than just an interpretation of the quantum formalism. It postulates a
modification of the formalism, whose nature is unspecified but whose
effect is somehow to introduce non-unitary evolution into quantum
dynamics. More specifically, it asserts that certain physical systems
(human brains and the like) violate the quantum principle of superposition. In fact, the characteristic assertion made by *all* realistic 'interpretations' other than Everett's is that '*superpositions of distinct
states of consciousness do not occur in nature*'. Such an assertion is
not just a matter of airy-fairy metaphysics. It is, in principle at least,
an experimentally testable statement." ([128], p. 215)

However, Deutsch must then either be disputing von Neumann's claim that
quantum mechanical predictions are invariant under changes of location of
the observing-system–observer-system *Schnitt* or be a physicalist about consciousness. He claims that

"[it is] necessary, in order to perform a crucial test of Everett's quantum theory against all others, to determine experimentally whether
or not the superposition principle holds for states of distinct consciousness." ([128], p. 220)

Deutsch has proposed several thought experiments, in the *literal* sense. Of
these, the significant one is predicated on his claim that alternative 'universes'
"sometimes do" affect each other. However, this is an inherently incoherent notion. Deutsch equivocates when considering alternative universes to be *actual*.
It is the nature of the universe that it is *universal*: nothing can be external
to the universe. Therefore, the universe cannot be affected from the 'outside.'
Recognizing this, Deutsch attempts to draw a distinction between the usual
notion of the universe and the sort of reality he has in mind.

"The word 'universe' has traditionally been used to mean 'the whole
of physical reality'. In that sense, there can be at most one universe. We could stick to that definition, and say that the entity we
have been accustomed to calling 'the universe'—namely, all the directly perceptible matter and energy around us, and the surrounding
space—is not the whole universe after all, but only a small portion
of it. Then we should have to invent a new name for that small, tan-
gible portion. But most physicists prefer to carry on using the word
'universe' to done the same entity that it has always denoted...A
new word, *multiverse*, has been coined to denote physical reality as
a whole." ([130], pp. 45-46)

Such a manifold is one in which similar objects in different 'universes' must be identified in order that they can superpose under the superposition principle, as he claims they should if his picture is correct. However, then each object will have multiple actual values of its physical magnitudes; the measurement problem resurfaces. If his claims are to be sustained and his version of the Collapse-Free interpretation is to involve more than under-developed "metaphysical speculation," a new metaphysics of objects allowing for this is needed, one that Deutsch has failed to supply.

Before examining yet another influential version of the Collapse-Free interpretation, that of Gell-Mann and Hartle, it is helpful first to consider a related version offered jointly by DeWitt and Graham and *its* treatment of the branching problem. That version retains DeWitt's assertions (1)-(6) above and adds further detail relating to the branching process. Like Deutsch's version, and surprisingly in light of DeWitt's own assertion 1 above, this involves an extreme degree of ontological commitment. In particular, they assert that

"reality, which is described *jointly* by the dynamical variables and the state vector, is not the reality we customarily think of, but is a reality composed of many worlds... equally real worlds," ([135], p. v)

a view they attribute to Everett—although he only refers to their being equally real 'branches'—each reflecting a unique sequence of measurement results in the experience of the observer. For them, applying quantum mechanics to cosmological situations requires an explication of the emergence of classical behavior involved in the explicit 'splitting' of distinct real worlds, for which they add an axiom. Bell viewed this formulation as rendering the Collapse-Free interpretation a "pilot-wave theory without trajectories" ([22], p. 133).

DeWitt and Graham describe the branching process as follows.

"By virtue of the temporal development of the dynamical variables the state vector decomposes naturally into a multitude of mutually orthogonal vectors, reflecting a continual splitting of the universe into a multitude of mutually unobservable but equally real worlds, in each of which every good measurement has yielded a definite result and in most of which the familiar statistical quantum laws hold." ([135])

For them, like Deutsch, the single universe is composed of many such "worlds." Splitting is said to occur when a "measurement-like" interaction takes place. Although such interactions remain imprecisely specified, splitting is required to be such that the state of each resulting world serves as a good measurement record, similarly to Copenhagen thesis (5). In their discussion, DeWitt and Graham explicitly considered such an interaction between a coherent sum of pertinent eigenstates and a system including a pointer. The determinateness of outcomes as seen by individual conscious observers was to be underwritten by the predicted correlations between those observers and pertinent outcomes in each world. The "measurement-like" interactions involved in this approach result in entangled states of the system and environment. Formally, one can view such states as providing maps defining the "relative state." For example, consider an entangled state of the subject (apparatus A) and the object (system S).

$$|\Xi\rangle = \sum_{ij} c_{ij} |\psi_i\rangle |\chi_j\rangle \ . \tag{3.1}$$

The joint state $|\Xi\rangle$ provides a map $\xi : \mathcal{H}_A \to \mathcal{H}_S$ given by

$$|\psi\rangle \mapsto \sum_j \langle \psi \chi_j | \Xi \rangle |\chi_j\rangle = \sum_j \langle \psi | \psi_i \rangle c_{ij} |\psi_j\rangle \ . \tag{3.2}$$

The conceptual grounding of the quantum probabilities remained a key goal in the DeWitt–Graham proposal. However, in this version, like Deutsch's, it seems there may be differences between what is observed and what is predicted by standard quantum mechanics, although they claim that "in most of [these worlds] the familiar statistical quantum laws hold" which, however, is also to say not in *all* worlds, although a theorem was offered that purports to show that the collection of such 'deviant' cases constitutes something like a set of measure zero. An account of the probabilities to be attributed to experimenters, which was considered sufficient proof that quantum theory is 'self-interpreting,' was provided in the form of the "EWG metatheorem" sketched in an influential 1970 Physics Today article by DeWitt, based on the above standard description of pre-measurement [134]. To be self-interpreting here appears to mean "requiring no metaphysics." The statement of the theorem is: "The mathematical formalism of the quantum theory is capable of yielding its own interpretation." Its proof involves showing: (i) how the conventional probability interpretation of quantum mechanics emerges from the formalism itself, and (ii) how correspondence with reality can be achieved even though the wave-function never "collapses." The starting point of the theorem is again "to take the mathematical formalism of quantum mechanics as it stands, to deny the existence of a separate classical realm, to assert that the state vector never 'collapses' " and specifically to assume that "the world must be sufficiently complicated that it be decomposable into systems and apparatuses."

The account involves, by this era of the universe, the simultaneous existence of 10^{100} very slightly differing 'copies' of the objects in the universe, including embodied observers who are 'unable to feel' the constant splitting of their worlds. The vector-space amplitudes of the universal wave-function are given no a *priori* interpretation; their interpretation is to follow from the consideration of sequences of measurements made by the apparatus—presumably to be unproblematically identified despite the repeated splittings into different new worlds with each additional measurement—on an ensemble of identical systems, that is, systems prepared in the same pure state, each measurement being of the type of Equation 3.2. In each case, the universe is assumed well decomposable into ensemble and apparatus. The apparatus sequentially observes all systems of the ensemble, each measured exactly once.

After such a sequence of measurements, the joint quantum system they constitute is described as a state of product form $|\Psi_0\rangle = \prod_i |\psi_i\rangle|\Phi\rangle$, where $\langle s|\psi_i\rangle = c_s$ for all values of i, with successive resulting states described in the basis $\{|s_1\rangle|s_2\rangle \cdots |A_1, A_2, \ldots\rangle\}$, where the nth measurement process is described as

$$U_n(|s_1\rangle|s_2\rangle \cdots |A_1, A_2, \ldots, A_n, \ldots\rangle)$$
$$= |s_1\rangle|s_2\rangle \cdots |A_1, A_2, \ldots, A_n + gs_n, \ldots\rangle \qquad (3.3)$$

so that after a number of measurements the joint state will be

$$|\Psi_n\rangle = \sum_{s_1, s_n, \ldots} \prod_i |\psi_i\rangle|\Phi[s_1, s_2, \ldots s_n]\rangle , \qquad (3.4)$$

where

$$|\Phi[s_1, s_2, \ldots s_n]\rangle = \int dA_1 \int dA_2 \ldots |A_1 + gs_1, A_2 + gs_2, \ldots\rangle\langle A_1, A_2, \ldots |\Phi\rangle, \qquad (3.5)$$

the results of these measurements being, in general, non-identical.

The memories of the apparatus serve as pointers corresponding to the resulting sequence of values, providing a distribution for each set of quantum mechanically allowed values for the system observable. The *relative frequency* function

$$f(s; s_1, s_2, \ldots, s_n) = \frac{1}{N} \sum_{i=1}^{N} \delta_{ss_n} \qquad (3.6)$$

and a hierarchy of functions of $f(s; s_1, s_2, \ldots, s_N)$ for measuring deviations from complete randomness is then considered. Using this construction, DeWitt showed that the state $|\Psi_n\rangle$ is negligibly different from a state differing from it by terms in the superposition in which $\sum_s [f(s; s_1, s_2, \ldots, s_N) - w_s]^2 < \epsilon$, for some arbitrarily small number ϵ, where w_s are the 'probabilities' of the measurement outcomes—identical to those dictated by the Born rule when assuming wave-function "collapse" upon measurement, in the limit $n \to \infty$ — and similarly for all the other statistical functions pertinent to differentiating a random distribution from a non-random one [134]. In this way, the quantities w_s are shown "naturally" to appear despite the assumption of a non-collapsing wave-function. Given this result, any observer presumably should, *contra* Deutsch, be incapable of experimentally differentiating between his being in a universe with collapse or a 'multi-verse' without collapse on the basis of quantum measurements.[19]

Everett's goal is ostensibly achieved without contradiction by the EWG meta-theorem by reference to the 'experience' of observers, despite the concerns of von Neumann that motivated his introduction of the projective process. However, the above-mentioned distributions appear to exist *simultaneously* along several branches, bearing in mind that space-time is not included

---

[19] This would presumably preclude the success of the experiments suggested by Deutsch, mentioned above.

as a quantum object in standard quantum theory; the random events to which
they apply therefore no longer uniquely appear, which brings into question
the status of these quantities as probabilities at all. This problem has come
to be known as the *problem of probabilities*.

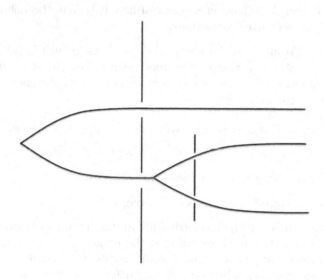

Fig. 3.1. Branching of histories with serial measurements of non-commuting idempotent observables in schematic form, with each history following one of the available
full trajectories leading from left to right [115]. On multiple-universe versions of the
Collapse-Free interpretation, this represents the 'simultaneous' existence of an exponentially increasing number of 'universes,' with the appearance of branches.

In addition to the question of the status of these probabilities as such, there
remains the equally serious problem of the presumed existence of an (at least)
exponentially growing number of worlds—on the order of 10^{100} according to
DeWitt—that have never been observed by anyone in our world or ever will
be observable. This is ontological excess with a vengeance, violating Ockham's
principle to an extraordinary degree, in that not only are unnecessary entities
postulated, but unnecessary entire *universes* of entities (*cf.* [410]).

Some have argued that versions of the Collapse-Free interpretation such
as those of Deutsch and of DeWitt and Graham are distortions of the world
picture that is natural to the approach. For example, Frank Tipler has represented the situation a mathematically simpler way and has argued that

"Many presentations of the MWI have made it appear more counter-intuitive than it really is. For example, many accounts assert that 'the entire universe is split by measurement'. This is not true. Only the observed/observer system splits; only that restricted portion of the universe acted on [by] the measurement operator M splits." ([453])

Tipler has offered the following account of how it is that "branching" occurs during measurement-like interactions.

"If we let |Cosmos⟩ be the parts of the universe which is [sic] not acted on by M—and this will be most of it—then the state of the entire Universe before the measurement is in $|\psi\rangle|n\rangle|$Cosmos⟩, and the effect of the measurement is

$$M|\psi\rangle|n\rangle|\text{Cosmos}\rangle = \alpha|\uparrow\rangle|u\rangle|\text{Cosmos}\rangle + \beta|\downarrow\rangle|d\rangle|\text{Cosmos}\rangle$$
$$= (\alpha|\uparrow\rangle|u\rangle + \beta|\downarrow\rangle|d\rangle)|\text{Cosmos}\rangle \,, \qquad (3.7)$$

which shows that |Cosmos⟩ is *not* split." ([453])

He continues by considering subjective perceptions.

"If a human being is the measuring apparatus, then any interaction with the rest of the universe will split the human being if—but only if—the interaction is such that it could result in different states of the human organism which are distinguishable by the human sensory system. The splitting is denoted by the correlations between the human and the various subsets of the Universe with which he interacts... the set of possible measurements can split a human being into only a finite number of pieces. A very rough estimate of an upper bound for the number of pieces... is 2 raised to the 10^{26} power. Admittedly a very large number, but still finite." ([453])

The above equation does not, in itself, demonstrate that the "Cosmos" does not "split" in *some* basis, in light of the superposition principle. What would be the physical property corresponding to the "Cosmos eigenstate"? If there is some such property, is it clear that the corresponding physical property is *not* correlated to the measurement outcomes? In the case of the Stern–Gerlach apparatus typically considered, for example, measurement outcomes *are* correlated with paths of systems in the resulting beams lying *outside* the "human organism" and "sensory system." Moreover, for example, quantum cosmology *does* include space-time, which is continuously parameterized, in the Cosmos. This measurement model also still fails to address the fundamental question of the in-principle inaccessibility of branches other than the one a given agent perceives. Tipler's is one more rendering of the Collapse-Free approach that is in need of a principle or mechanism for its own particular articulation of the branching process that could preserve the supposedly physically 'natural' character of the approach. It also still suffers from metaphysical excess due to its naive realist interpretation of the wave-function.

Unless an anthropomorphic view is taken, it is not clear why one should be particularly concerned with the splitting of *human beings*; but that would run contrary to the realist character of Tipler's rendering. He finally adds, "I personally think the enlarged ontology of the MWI is really a very strong argument for *accepting* rather than rejecting it," arguing that

> "the ontological enlargement required by the MWI is precisely analo-
> gous to the spatial enlargement of the universe which was an implica-
> tion of the Copernican theory...philosophers in Galileo's time used
> Ockham's principle to support the Ptolemaic and Tychonic systems
> *against* the Copernican system...Similarly, I would contend that the
> MWI should be accepted because it is the most elegant interpretation
> of the quantum mechanical formalism." ([453])

The above descriptions of measurement processes implicitly contain a pre-scription for specifying state branches: one is expected to simply "read off" branches from the given state representations. A *preferred basis problem* per-sists: how do branches objectively arise? In particular, how can they objec-tively arise given that there are an infinite number of bases in which there remain state superpositions 'after' the measurement-like interactions? Shi-mony has brought this issue directly to bear on the question of the viability of the interpretation.

> "The Everett interpretation is sometimes defended by an analogy
> to the Copernican theory: the true physical situation seems to be in
> contradiction with appearances because naive people fail to take into
> account their own physical states. I believe, however, that this anal-
> ogy is faulty, and its weakness is instructive. The reconciliation of
> the apparent turning of the celestial sphere with the correct physical
> kinematics is achieved by an analysis of visual appearances in terms
> of the actual relative motions of the bodies observed, including that
> of the observer. But the Everett interpretation lacks an ingredient
> needed to implement this analogy: namely, a principle in terms of
> which one branch of the subject is endowed with subjective immedi-
> acy." ([412], p. 160)

Therefore, the Copernican analogy fails to bolster the approach.

> "Without such a principle there is only potentiality...But with such
> a principle the interpretation would lose its purity, and indeed there
> would be a reduction of a superposition based upon an interaction
> of the physical world with the mind." ([412], p. 160)

In particular, it contradicts DeWitt's assertion 2 which so fundamental to the distinctiveness of contemporary forms of the interpretation.

Like Deutsch, DeWitt, and Tipler, Gell-Mann and Hartle have taken a specific metaphysical position, something not done by Everett in his writings, that they claim to be most appropriate to the Collapse-Free approach. They explicitly reject realism, suggesting a different rendering of Everett's interpretation.

> "The problem with the 'local realism' that Einstein would have liked is not the locality but the *realism*. Quantum mechanics describes *alternative* decohering histories and one cannot assign 'reality' simultaneously to different alternatives because they are contradictory. Everett and others have described this situation, not incorrectly, but in a way that has confused some, by saying that the histories are all 'equally real' (meaning only that quantum mechanics prefers none over another except via probabilities) and by referring to 'many worlds' instead of 'many histories." ([184])

By appealing to histories in this way, Gell-Mann and Hartle have brought the approach closer to the interpretations preceding it. This variant of the Collapse-Free approach has come to be known as the *Consistent-histories interpretation* of quantum mechanics. It is the version of the approach that most plausibly plays the role that Bub identified for it, the "New orthodoxy."

Relative to the DeWitt–Graham version, the Gell-Mann–Hartle account contains more specific physical content. It also relies heavily on the efficacy of quantum decoherence.

> "[T]here was a unique initial state of the universe, perhaps even a pure quantum state described by a gigantic wave function. The content of the universe was probably initially very erratic, when the quantization of space was supposed to dominate; it later contained practically pure thermal radiation, according to the big bang models, so that there were not many classical phenomena to be mentioned and there was practically no fact to be seen, except for very global ones. Some time after, phenomena became possible... consider the birth of a galaxy as such a phenomenon. According to pure quantum mechanics... the corresponding reduced density operator is diagonal in the collective variables of the galaxy, but that does not tell us where the galaxy is. There is only a probability for the whole galaxy to be in some place rather than another. Two states of that kind... have no possible relation and they completely ignore each other because of decoherence... Decoherence implies that, when two classical objects (e.g. two observers) belong to two different histories of the universe, they must completely ignore each other... all the histories of the universe are real." ([184], p. 455)

Note, however, that here they themselves describe alternative histories as "equally real." DeWitt's assertion 2 is also again violated if a distinct class of *classical* objects is required, as suggested here by virtue of the apparent requirement that they be present for phenomena to be possible. This rendering appears to invoke the spirit of the Copenhagen approach.

What, according to Gell-Mann and Hartle distinguishes observers?

"Both singly and collectively we are examples of the general class of complex adaptive systems. When they are considered within quantum mechanics as portions of the universe, making observations, we refer to such complex adaptive systems as information gathering and utilizing systems (IGUSes)... an IGUS uses probabilities for histories and therefore performs further coarse graining on a quasiclassical domain. Naturally, its course graining is very much coarser than that of the quasiclassical domain since it utilizes only a few of the variables in the universe. The reason such systems as IGUSes exist, functioning in such a fashion, is to be sought in their evolution within the universe. It seems likely that they evolved to make predictions because it is adaptive to do so. The reason, therefore, for their focus on decohering variables is that these are the *only* variables for which predictions can be made. The reason for their focus on the histories of a quasiclassical domain is that these present enough regularity over time to permit the generation of models (schemata) with significant predictive power... the IGUS evolves to exploit a particular quasiclassical domain or set of such domains." ([184])

Although this version presents the most detailed picture of the interpretation, it provides no genuine solution to the preferred-basis problem. It also equivocates regarding the ontological status of worlds, just as Everett's original version had done. The emphasis on 'real histories' as opposed to real worlds does little to remove the incoherence of postulating multiple actualized universes unless strong empiricism is also taken on board.

Let us now return to the preferred-basis problem and consider some of the specific ways in which it has been engaged. In general, the correlation between the measuring apparatus and measured system involves an entangled state that does not uniquely correspond to an element of the eigenbasis of *any particular* pairing of Hermitian operators, even though it correlates these two subsystems as strongly as is quantum mechanically possible; it is the same issue that arises in the quantum measurement problem. The difficulty this presents has been powerfully pointed out by Shimony.

"Part of the appeal of the Everett interpretation is the metaphor of a trunk with many branches... The metaphor tacitly presupposes, however, a preferred basis, in which the vector representing the state of the system is expressed: the vector itself is a trunk, and the projections upon the basis vectors are the branches. But objectively there is no preferred basis. There is a continuum of possible bases, all on the same footing. But when one speaks in a more accurate way about the mathematics of the quantum state, the quasi-familiarity of the original metaphor is lost, and with it the appeal." ([412], p. 159)

Bell was also highly critical of this metaphor (*cf.* Figure 3.1).

> "At the microscopic level there is no such asymmetry in time as would be indicated by the existence of branching and non-existence of debranching. Thus the structure of the wave function is not fundamentally tree-like. It does not associate a particular branch at the present time with any particular branch in the past any more than with any particular branch in the future. Moreover, it seems reasonable to regard the coalescence of previously different branches, and the resulting interference phenomena, as *the* characteristic feature of quantum mechanics. In this respect, an accurate picture, which does not have any tree-like character, is the 'sum over all possible paths' of Feynman." ([24], p. 135)

The closest to Bell's suggestion the interpretation has mustered, without becoming instead a version of the Process interpretation discussed in a following section, is Deutsch's "favorite version" of the interpretation portrays a 'multiverse,' although Deutsch is quite clear that the branches correspond to distinct co-existing worlds that all always exist, even before 'interacting.'

Considerable attention has been paid to quantum decoherence in hope of a solution to the preferred-basis problem. However, as the decoherence-theorist Erich Joos has succinctly assessed the situation,

> "It should be emphasized that [the decoherence of the joint apparatus–system state] does not solve the measurement problem. In particular, an ignorance (*i.e.* ensemble) interpretation of [the state], as suggested by some authors, would wrongly identify an improper mixture with a proper one." ([262], p. 42)

To see this, it is instructive to return to the simple 'split portion' of the universe in Tipler's spin measurement. It is clear that the claim that a branch is objectively selected by the measuring apparatus is unjustified; the way he has chosen to portray the state in Equation 3.6 involves an ambiguity, namely, that a specific spin *direction* has not been identified Let us remove the ambiguity by taking this direction to be the z–direction by writing

$$M|\psi\rangle|n\rangle = (\alpha|\uparrow_z\rangle|u_z\rangle + \beta|\downarrow_z\rangle|d_z\rangle)|\text{Cosmos}\rangle \qquad (3.8)$$

without contesting, for the moment, his claim that the "Cosmos" factors out. Recall that the 'measurement operator' M, on the Everett interpretation, is a *unitary* operator.[20] For specificity, let us also take $\alpha = \beta = \frac{1}{\sqrt{2}}$, that is, let the first factor above be the Bell state $|\Phi^+\rangle$. With some thought, one recognizes that the state of the right-hand side of Equation 3.8 can be rewritten in the same general form but with equally strong correlations for measurement of the x–spin, namely,

---

[20] This operator is therefore better labeled U than M, which in the literature typically denotes a *nonunitary*, projective operator.

$$M|\psi\rangle|n\rangle = \frac{1}{\sqrt{2}}|\uparrow_x\rangle|u_x\rangle + \frac{1}{\sqrt{2}}|\downarrow_x\rangle|d_x\rangle \, , \qquad (3.9)$$

which remains an instance of $|\Phi^+\rangle$. That is, there is a simple relationship between z–spin and x–spin eigenstates, namely, $|\uparrow_z\rangle = \frac{1}{\sqrt{2}}(|\uparrow_x\rangle + |\downarrow_x\rangle)$ and $|\downarrow_z\rangle = \frac{1}{\sqrt{2}}(|\uparrow_x\rangle - |\downarrow_x\rangle)$. Of course, this is true not only in the case of the spin eigenstates, but is so for the apparatus eigenstates $|u_z\rangle$, $|d_z\rangle$ as well, that is, $|u_z\rangle = \frac{1}{\sqrt{2}}(|u_x\rangle + |d_x\rangle)$ and $|d_z\rangle = \frac{1}{\sqrt{2}}(|u_x\rangle - |d_x\rangle)$. One sees that the required perfect measurement correlations exist simultaneously for *both* z–spin and x–spin, which are eigenstates of *incompatible observables*. In the absence of Process 1 or complete decoherence, which typically would be achieved only in the limit of infinite time, any 'branching' of worlds is ill defined. Furthermore, there is nothing special about this particular manner of choosing the spin axis: any direction will do. This is to what Shimony refers when he speaks of a *continuum* of possible bases, all on the same footing.

Again, the process of quantum decoherence is usually relied upon to solve the preferred-basis problem, as in the approach of Gell-Mann and Hartle above. However, if, as is typical, the two subsystems become spacelike-related, external noise will act locally on the state which, if it is a spin-singlet state as it readily can be, is *immune* to phase decoherence. That is, there exist *decoherence-free* states of the sort commonly understood to arise in measurement. Again, because of the purity of the overall state required by the approach, the appeal to decoherence, this will not help matters from the purely physical point of view. Unless the way that physical theory is related to reality is altered in a way depending on interpretational moves, discussed below, that are far more radical than would be seriously considered doing mere physics, are required to finesse this problem. As von Neumann realized long ago, adding the remainder of the universe to the chain of measurement changes nothing essential. The pertinent correlations between the remainder of the universe and the object system have exactly the above form; despite the extremely large number of degrees of freedom introduced, any developed correlation or chain of correlations that matters to measurement, say of a spin-1/2 system, can be reformulated as the correlation of the measured system with the pertinent remainder of the universe having two pertinent eigenstates.

A related attempt to resolve the preferred-basis problem is to examine the Schmidt decomposition to see whether it might be involved in the 'selection of basis.' The basis involved in this decomposition need not particularly be connected with any specific measurement, nor will the basis typically correspond to that of dynamically preserved measurement outcomes. Nonetheless, specific moves of this kind have been investigated to help resolve the problem. For example, Bub has considered, following Żurek [522], states not only of the Hilbert spaces of system and apparatus, but these together that of the environment; Bub and Andrew Elby showed that for states in this tripartite composite Hilbert space, a unique Schmidt-*like* decomposition of states exists even if the coefficients of the expansion coefficients involved are degenerate

[159] as, for example, in the case of the GHZ state. In particular, they produced the following *tridecompositional uniqueness theorem*.

> Suppose $|\Psi\rangle = \sum_i \alpha_i |a_i\rangle |b_i\rangle |c_i\rangle$, where $\{|a_i\rangle\}$ and $\{|c_i\rangle\}$ are linearly independent, while $\{|b_i\rangle\}$ are merely noncollinear. Then there exist no alternative linearly independent sets of vectors $\{|a_i'\rangle\}$ and $\{c_i'\}$, and no alternative noncollinear set $\{|b_i'\rangle\}$, such that
>
> $$|\Psi\rangle = \sum_i \alpha_i |a_i'\rangle |b_i'\rangle |c_i'\rangle \ . \tag{3.10}$$
>
> (Unless each alternative set of vectors differs only trivially from the set it replaces.)

This theorem generalizes in the obvious way to an arbitrary finite number of component systems, say Tipler's "Cosmos." One may consider the system of the third Hilbert space to be a compound environment with a state space of countable Hilbert-space dimension. Unfortunately, there is no guarantee of the existence of a generalized *Schmidt* decomposition [347].

The above result only proves the uniqueness of a Schmidt decomposition *should* one exist, which is the assumption the theorem takes as its starting point. As Peres pointed out in his note providing the necessary and sufficient conditions for the existence of a generalized Schmidt decomposition,

> "A natural question is whether [the Schmidt decomposition] process can be extended to more than two subspaces. Such an extension of Schmidt's theorem would be useful for modal interpretations of quantum theory, and triple sums can indeed sometimes be found in the literature on that subject. The unlikeliness of occurrence [*sic*] of multiple Schmidt decompositions can readily be seen by counting the free parameters involved: for example, if there are three particles, each of which described by a d-dimensional space, their combined (pure) state, in a d^3-dimensional space, depends on $2(d^3 - 1)$ real parameters (after discarding overall normalization and phase factors). On the other hand, the three unimodular unitary transformations which can be performed for these three particles have only $3(d^2 - 1)$ free parameters, not enough to solve the problem in general." ([347])

Moreover, this decomposition is typically unavailable for measurements that are nonideal [158]. In the article in which the tridecompositional uniqueness theorem appeared, Elby and Bub assessed the situation as follows.

> "Decoherence theorists can invoke tridecompositional uniqueness to support their claim that environmental interactions make the pointer reading become 'classical' in some sense. And modal interpreters can claim that when a quantum state vector can be a tridecomposed (with two of the three bases orthogonal), the observables picked out by the tridecomposition have definite values." ([159])

Nonetheless, they pointed out,

> "[One of us has argued] that ideal measurements of most observables
> are impossible. If this is correct, then the unique biorthogonal decom-
> position of the measuring device with the rest of the universe, if it
> exists, usually picks out some apparatus basis other than the pointer-
> reading basis. Furthermore, a nonideal measurement usually results
> in a particle-apparatus-environment state that cannot be tridecom-
> posed. Our technical results cannot help to solve these deep-seated
> technical problems associated with nonideal measurements." ([159])

Some advocates remain undaunted. For example, David Wallace has claimed
that

> "the preferred-basis problem looks eminently solvable without chang-
> ing the formalism. The main technical tool towards achieving this has
> of course been decoherence theory... there are no purely conceptual
> problems with using decoherence to solve the preferred-basis prob-
> lem,... the inexactness of the process should give us no cause to reject
> it as insufficient." ([482])

However, with this degree of laxity, the same could be said of the measurement
problem itself, that is, of the main issue motivating the search for interpretations different from the widely used Basic interpretation in which it first arose.
Furthermore, to date, no technical result exists that shows that decoherence
helps one solve the preferred-basis problem.

In addition to the pre-measurement correlation between the system of
interest and measuring apparatus in the observable–pointer basis, there is ongoing interaction between the apparatus and its environment during measurement. Interaction with the environment can to a great extent diagonalize the
density-matrix of the state of the system plus measuring apparatus in a specific basis, and rapidly do so in complex natural environments [261, 522]. The
overall interaction is entirely unitary, but the reduced joint system–apparatus
state may become increasingly diagonal as the three-component state evolves
toward environmental states that can rapidly become increasingly orthogonal, although generally not fully so. However, this is as far as decoherence
itself can assist any interpretation. Most specifically, why *one outcome versus
another* appears in a given measurement basis cannot be explained by decoherence. Żurek, in his decoherence-based 'existential interpretation' of the
wave-function, has been driven to claim, along the lines of Everett's original
expositions, that each possible measurement outcome simply *does* occur in
some universe within the 'multi-verse.'

For his part, Everett was, rather surprisingly, satisfied with the following
prescription for assigning post-measurement states.

"Subsystems do not possess states that are independent of states of the remainder of the system, so that the subsystem states are generally correlated with each other. One can arbitrarily choose a state for one subsystem, and be led to the relative state for the remainder." ([164], p. 317)

This position has recently been championed by some philosophers in a way that represents an extremely subjectivistic approach to the physical world of the sort Bell considered radically solipsistic. A far more objectivist tack has been pursued by Deutsch, who has suggested that an additional rule be appended to the quantum formalism, specifying "how the interpretation basis depends (solely) at each instant on the physical state and dynamical evolution of a quantum system" that is

"based on an idea of Everett (private communication) that at any rate *during measurements* the basis is determined by the requirement that in that basis the interaction indeed take the form of a measurement," ([127], p. 2)

in the context of a theory of measurement along the lines of that offered by DeWitt. However, as we have seen, this move is insufficient, as Deutsch has largely acknowledged.

Deutsch has also disputed the claim of DeWitt and Graham that the statistics of measurements can be obtained directly from the Everett treatment because

"there is nothing in the formalism telling us that a set of worlds of measure zero must 'occur with zero probability.' Indeed, if we had been willing to identify the Hilbert space norm measure with a physical probability, then [their] elaborate argument about sequences of measurements would be redundant, since the *norm* of the [pertinent state component] is in any case the desired value." ([127], p. 20)

In an attempt to tackle these issues, Deutsch has proposed a "slight change" in Everett's treatment by introducing the following axiom. "The world consists of a continuous infinite-measured set of universes" [127]. The interpretation of the joint system–apparatus state $|\Psi\rangle = \sum_{a_1} c_{a_1} |a_1, t''; A_2(a_1)t''\rangle$ for the measurement of an observable $\hat{\phi}_1(t')$ is then that the set of universes is composed of n_1 disjoint sets, where the a_1th subset is of measure $|a_{a_1}|^2$, each corresponding subset being a "branch" and consisting of a continuous infinity of universes playing the "same role as individual universes do in Everett's original version," but with the probabilistic interpretation now "built in" [127]. With this move, Deutsch then considers "an ensemble of universes" described by the density operator

$$\rho_\epsilon(t) = \sum_\alpha |\langle \psi | \alpha, t \rangle|^2 |\alpha, t\rangle\langle \alpha, t| , \qquad (3.11)$$

where $|\alpha, t\rangle$ is the relative state of the αth branch at time t. The operator $\rho_\epsilon(t'')$ then characterizes results of measurements considered to have ended by time t''. He then identifies a preferred "interpretation basis" $\{|\alpha, t''\rangle\} = \{|a_1, t''; a_2, t''\rangle\}$ that would allow one to consistently regard measurements as a generic physically process, unlike in most other interpretations, through a rather elaborate construction. Deutsch has thereby offered the most frank and subtle Collapse-Free interpretation that can be clearly identified as such that does not alter the theory beyond recognition or revert to instrumentalism.

As mentioned above, further attempts to come to terms with the shortcomings of the Collapse-Free approach have involved radical moves of a rather more conceptual than mathematical character. In these, subjectivity is often given a place more primary than physics. Wallace has argued that, given the effect of decoherence,

"branching effects in such a theory can be understood, literally, as replacement of one classical world with several—so that in the Schrödinger Cat experiment, for instance, after the splitting there is a part of the quantum state which should be understood as describing a world in which the cat is alive, and another which describes a world in which it is dead. This account applies to human observers as much as to cats... Each future observer is (initially) virtually a copy of the original observer, bearing just those causal and structural relations to the original that future selves bear to past selves in a non-branching theory." ([482])

Wallace views this move as entirely valid: "since the existence of such relations is all that there is to personal identity, the post-branching observers can legitimately be understood as future selves of the original observer..." [482]. Maximillian Schlosshauer has defended a similar position, suggesting that definiteness during measurement events is not a valid requirement of fundamental physics.

"We demand objective definiteness because we experience definiteness on the subjective level of observation, and it should not be viewed as an a priori requirement for a physical theory." ([390], p. 1271)

However, again, this eliminates much of the motivation for the Collapse-Free interpretation as an alternative to the Basic interpretation.

It is similarly claimed by some philosophers that progress is to be made in solving the problems faced by the Collapse-Free interpretation by moves that are likely to be seen as even more objectionable to physicists. The grounds on which the theory is to be evaluated become radically different, as Simon Saunders suggests.

"[T]he question is no longer what is the space of possibilities?—but: What is the space of possibilities *in which we are located*? That we are ourselves constituted of processes of a very special sort is in part explanatory; because our nature is in part contingent, a matter of evolutionary circumstance, the demand for explanation is blunted." ([388])

Two key aspects of the problem of reconciling the stochastic nature of observed measurement outcomes and quantum probability with the deterministic unitary dynamics of the Collapse-Free approach are noted by Saunders: the *incoherence problem* and the *quantitative problem*. The latter problem is that of finding a well defined sense in which the Collapse-Free picture can support quantum probabilities as *uncertainties* that correspond to the quantities given by the Born rule. However, this appears to differ little from what is offered by hidden-variables treatments under the pilot-wave theory of quantum mechanics. Wallace has claimed that substantial progress has been made by Saunders toward resolving the problem associated with the first aspect with his "subjective uncertainty theory," wherein "an agent awaiting branching should regard it as *subjectively* indeterministic. That is, he should expect to become one future copy or another but not both, and he should be uncertain as to which he will become" [482]. Saunders refers to his overall approach as *relativism*. On this view, value-definiteness and probability are understood as *relational*, which is to be understood in the sense that no dependence on essential characteristics is allowed; "temporal relations are tenselessly true; probabilistic relations are deterministically true" [387].

As seen from the perspective of Bell a decade earlier, this points to what can be seen as the most novel element of the Everett theory, namely, "a repudiation of the concept of the 'past', which could be considered in the same liberating tradition of Einstein's repudiation of absolute simultaneity" ([24], p. 118). Bell was highly critical of Everett's treatment, which he called "a radical solipsism" due to its reliance on the replacement of the past by memories, and about which he said that

"if such a theory were taken seriously it would hardly be possible to take anything else seriously... It is always interesting to find that solipsists and positivists, when they have children, have life insurance." ([24], p. 136)

Spelling out Saunders' proposal involves a further development of the version of the Collapse-Free interpretation of Gell-Mann and Hartle, which relies on solutions analogous to those that have been proposed in attempts to understand the nature of time. What precisely does it mean for value-definiteness to be relational? For him, this means that along a given Everett 'branch,' specified by a choice of basis and a selection of outcome, values are definite. He assumes that decoherence is sufficient for the selection of the preferred basis. What does it mean for *probability* to be relational? "Probabilistic facts are not made true by chance," rather, decoherence wherein a general formula for the relative state (relative to $|\psi_j\rangle$)

$$|\phi\rangle_{\psi_k} = \sum_i \frac{1}{\sqrt{\sum_i |c_{ik}|^2}} c_{ik} |\phi_i\rangle , \qquad (3.12)$$

given the universal wave-function

$$|\Psi\rangle = \sum_{ij} c_{ij} |\phi_i\rangle |\psi_j\rangle , \qquad (3.13)$$

is considered. An operator treatment, which has projectors describing pure states—something that, as discussed in Section 2.1, is also used to consider the quantum analogues of the characteristic functions on the phase space of classical mechanics—is typically used [387]. This allows the consideration of "decoherent histories"; the evolution of the universal state is represented using a set of "mutually exclusive histories," given by a sequence of projectors, where interference between "disjoint histories" is insignificant [197, 328].

It was Robert Griffiths who first found necessary conditions for the consistency of the additivity of quantum amplitudes and called those satisfying them *consistent histories* [197]. With the projector representation, histories $h = \{P_1(t_1), P_2(t_2), \ldots P_n(t_n)\}$, each associated with a series of events involving a sequence of (non necessarily compatible) observables $\{A_i\}$ (in the Heisenberg picture) at sequential times t_i are attributed weights

$$w_\rho(h) = \mathrm{tr}\big(P_n(t_n)\ldots P_2(t_2)P_1(t_1)\rho_0 P_1(t_1)P_2(t_2)\ldots P_n(t_n)\big) , \qquad (3.14)$$

where ρ_0 is the initial state and the weights of disjoint histories will sum to one. The spectrum of each of the A_i is written as a combination of disjoint subsets $D_i^{\alpha_i}$. One then considers time-dependent projectors $P_i^{\alpha_i}(t_i)$ onto the subspace spanned by the eigenvectors of the A_i with eigenvalues lying in the $D_i^{\alpha_i}$. In the Gell-Mann–Hartle approach, one makes use of this construction as the basis for a decohering histories based version of the interpretation. They considered a complex-valued *decoherence functional* $D(h, h') \equiv \mathrm{tr}(h\rho h'^\dagger)$ for pairs of histories h, h' and for collections of histories $[P^\alpha]$, imposing the conditions that the sequences $\{P_i^{\alpha_i}(t_i)\}$ of which they are constituted as a product (i) sum to the identity operator and (ii) are mutually orthogonal within the history. Histories are considered fully fine-grained if they are given by a set of projectors onto rays and coarse-grained if some or all the projectors project onto larger subspaces. Gell-Mann and Hartle also introduced a natural decoherence condition on collections of histories, namely, that for off-diagonal elements both the real and imaginary parts of the functional approximately vanish for decoherent collections, where the resulting *matrix of values* is

$$D([P^\alpha], [P^{\alpha'}]) \equiv \mathrm{tr}\big(P_n^{\alpha'_n}(t_n)\cdots P_1^{\alpha'_1}(t_1)\rho_0 P_1^{\alpha_1}(t_1)\cdots P_n^{\alpha_n}(t_n)\big) , \qquad (3.15)$$

which is sufficient for the inconsistency of histories but not necessary [197, 328]. Any set of disjoint exhaustive histories is referred to as a *decoherent history space*; such sets are referred to as *quasi-classical domains* whenever further fine-graining would result in a loss of decoherence [184].[21]

On Saunders' treatment, an Everettian Lüders-type rule provides the relative states associated with the eigenvalue i:

$$\rho \to \rho' = \frac{(\mathbb{I} \otimes P_i)\rho_0(\mathbb{I} \otimes P_i)}{\mathrm{tr}(\mathbb{I} \otimes P_i\rho_0)} \ . \tag{3.16}$$

Then, a state P_i of the object system will be 'actual' relative to the state P_i of the measuring system exactly when $\mathrm{tr}(P_i \otimes \mathbb{I}) = 1$. He has attempted to extend this rule without requiring a tensor space structure, writing the relative state in a new notation, namely,

$$\mu_\rho(P'/P) \equiv \frac{\mathrm{tr}(P'P\rho PP')}{\mathrm{tr}(P\rho P)} \ , \tag{3.17}$$

and saying that the pure state P is 'actual' relative to pure state P' when $\mu_\rho(P'/P) = 1$, where the projectors P, P' do not necessarily commute. The projectors for histories are again constructed using resolutions of the identity $\{P_i\}$ at each time t, but with elements that do not necessarily commute, and considers their time-ordered products, denoted $C = P_k^{\gamma_k} P_{k_1}^{\gamma_{k_1}} \cdots P_1^{\gamma_1}$. Saunders then requires that their real part be zero if and only if $C \neq C'$. Needless to say, this amounts to the consideration of a theory differing in its formal elements from standard quantum mechanics, because the tensor product structure of the theory, from which entanglement arises, is abandoned.

A resolution of the 'quantitative problem' mentioned by Wallace has been proposed by Deutsch, who invokes decision theory. Recall that this is the problem of supporting quantum probabilities as uncertainties that correspond to the quantities given by the Born rule. Unfortunately, an adequate presentation of the Deutsch treatment is beyond the scope of this book—let us simply note that Huw Price has pointed out a significant difficulty associated with Deutsch's move [363]. Wallace has also noted this, but is not bothered by it. He provides the following contextual characterization of this theory.

"The underlying principle is essentially that of symmetry: if there is a physical symmetry between two possible outcomes there can be no reason to prefer one to another. Such arguments have frequently been advanced in non-quantum contexts but ultimately fall foul of the problem that the symmetry is broken by one outcome rather than another actually happening (leading to a requirement for probability to be introduced explicitly at the level either of the initial conditions or of the dynamics to select which one happens). They find their natural home—and succeed!—in Everettian quantum mechanics, where all outcomes occur and there is no breaking of the symmetry." ([483])

---

[21] When considered from the point of view of Żurek's environmentally induced decoherence, coarse graining is the averaging over environmental degrees of freedom.

Such subjectivist quantum cosmological approaches to constructing a new foundation may turn out to be valid but, again, they involve moving significantly away from the original quantum theory the interpretation of which is the true problem. For our purposes, it is more productive to return to the consideration of standard quantum mechanics than to pursue these further.

3.6 The Naive Interpretation

The Naive interpretation of quantum mechanics takes the formalism to refer only to ensembles of quantum systems and never to individual systems or, necessarily, to beams of particles. This is the most fundamental principle of the interpretation, which we can call principle 0. The adjective *naive* as used here serves, among other things, to more precisely distinguish this interpretation by reference to its association with naive realism. Another characteristic of the interpretation is that it, like the Collapse-Free approach, attempts to minimize the number of assumptions used to interpret quantum mechanics.

Among contemporary physicists, the best known advocate of this interpretation has been Ballentine. "Ballentine's paper has proved to be one of the most stimulating sources of discussion and dialog" regarding statistical interpretations [232]. He has referred to his interpretation as simply "the statistical interpretation" [16], although this phrase has been consistently used throughout the years in describe *any* interpretation of quantum mechanics that takes quantum probabilities *not* to refer to individual processes. Some advocates claim that the approach does not require all dynamical variables to have definite values at all times, although without this assumption the interpretation is neither fully defined nor distinct from others. For that reason, this is not considered here to be an element of the interpretation.[22] The interpretation has also been (incorrectly) attributed to John Taylor, who refers to his version as the "ensemble interpretation"; importantly, Taylor rejects the precise value principle ([119], Chapter 7); if this rejection is consistently maintained, then there is little to distinguish his version from a generic statistical interpretation that rejects applying quantum probabilities to individual systems. Nonetheless, his views will be discussed here because they illuminate important aspects of and motives behind the approach.[23]

Some or all of the distinguishing components of the Naive interpretation were explicated at various times, either early on in the history of the theory or in later textbooks, in the decades following the introduction of the theory. Those contributing some, although not all, of these elements include John Clarke Slater [425] and Einstein ([152], p. 665), whose position lies closer to Taylor's than to Ballentine's. Three principles underlying the interpretation

---

[22] *Cf.* Principle 1 below.

[23] For a comprehensive review of the related cluster of interpretations, one can consult [232].

as considered here are the following, which have been identified by Healey who refers to it as the *naive realist interpretation* ([211], Section 3.1).

(1) The *precise value principle*. Every observable has a precise value for every state of the ensemble.

(2) The *relative frequency principle*. The quantum probabilities provide the relative frequencies of measurement outcomes for the ensemble.

(3) The *faithful measurement principle*. Every good measurement reveals a pre-existing value of the observable.

This interpretation is distinguished from others by principle 3 (in conjunction with principle 0). Principles 1 and 3, were the interpretation consistent, would provide a solution to the quantum measurement problem, in that they explain the appearance of definite outcomes in individual experiments no matter how the experimenter chooses to measure a quantum system. Nonetheless, a solution to the measurement problem is not forthcoming under it, as shown below, essentially as a result of the principle 2, which relates ensemble states to measurement statistics in accordance with the Born rule.

On principal 2, quantum probabilities are defined as relative frequencies by considering ensembles as ordered sets and taking the probability as the relative frequency of occurrence of outcomes in the set of events in the large-number limit.[24] Other interpretations of quantum mechanics have made use of this interpretation of probability, which is relatively uncontroversial. In particular, Ballentine views quantum ensembles as infinite sets of individual systems, with measurements constituting selections of putative subensembles from the quantum ensemble.

Perhaps the strongest motive behind the Naive interpretation is the desire to avoid paradoxes presented by the well known thought experiments, which on it are seen as arising from the assumption that the state-vector is a description of individual systems, as in the Copenhagen interpretation, rather than as one of ensembles only. Einstein had considered such a move on the basis that, in the context of his other views, the alternative would imply that quantum mechanics is incomplete.

> "The statistical character of the theory would... have to be a necessary consequence of the incompleteness of the description of the systems in quantum mechanics, and there would no longer exist any ground for the supposition that a future basis of physics must be based on upon statistics..." ([151], p. 87)

Indeed, he was

---

[24] The term *relative frequency principle* for (2) was introduced by Hughes ([244], p. 163).

"convinced that everyone who will take the trouble to carry through such reflections conscientiously will find himself finally driven to this interpretation of quantum-theoretical description (the ψ-function is to be understood as the description not of a single system but of an ensemble of systems)." ([152], p. 671)

Specific examples of the claim that paradox can be resolved by a strict adherence to a statistical interpretation without the introduction of subjectivity are the following of Taylor, involving two crucial thought experiments already discussed. He first considers the EPR experiment.

"[I]t's clear that a paradox arises there, because we're assuming that when a measurement is made, say, of the spin of a particular particle, we can also measure the spin of a far away particle whose properties are correlated [with it]... This would be paradoxical if you believe that you are indeed measuring individual systems because it would seem that you're actually able to influence that far away particle, and in some ways determine its spin simply by making a measurement on the nearby particle. The ensemble interpretation says, however, that we're looking at a whole ensemble of such systems... we only know about ensembles of such situations." ([119], p. 107)

Ballentine has similarly pointed out that his interpretation would avoid the EPR quandary in the same way, that is, by virtue of principle 0. Taylor later considers the Schrödinger cat thought experiment.

"[A]ccording to any interpretation of quantum mechanics which attempts to describe individual systems... the quantum mechanical state is composed of the cat being alive for half the time, and the cat being dead for the other half. In other words, the cat doesn't know whether it's dead or alive, which is absolutely absurd! Now if you take the ensemble interpretation, then in 50% of the cases the cat is alive and 50% it's dead. That's quite reasonable... there is no way of saying whether it is alive or dead in any particular case. It's a meaningless question." ([119], pp. 110-111)

However, there is nothing specifically *quantum mechanical* about the ensemble approach used to resolve the above quandaries in the minimalist context of the naive interpretation, relative to any other theory of statistical mechanics; one must explain the peculiarities of the statistics of quantum mechanics, given by the Born rule, for this to represent an advance in the understanding of the quantum world. But such an explanation, given the remaining assumptions of the interpretation, ultimately must be by reference to some some hidden-variables theory of individual systems (*cf.* [371], pp. 45-48). Perhaps recognizing this, Ballentine is quite open to such an explanation of quantum mechanical predictions. The underlying theory of individual systems would need to explain the inability of experimenters to prepare dispersion-free statistical states, that is, states in which ensembles simultaneously have well defined values of properties corresponding to non-commuting observables in accordance with the precise-value principle. However, one then faces the fatal difficulty that the Kochen–Specker theorem contradicts that principle.

In the Bohr–Einstein debate, Einstein unsuccessfully attempted to provide a thought-experiment that would show that such ensembles could be prepared by the appropriate use of a double-slit apparatus [303]. This situation is faced together with the fact of the failure of local hidden-variable theories as evidenced by the empirical violation of Bell-type inequalities as well as the lack of any clear line delimiting the boundary between the quantum and classical realms, as seen above in our discussion of the Copenhagen interpretation which also (at least implicitly) assumes such a strict distinction. A detailed theory of the dynamics of the hidden variables involved would also be highly desirable, although it may not be necessary for establishing the consistency of the interpretation.

As noted above, aspects of the measurement problem also remain under the interpretation. If one begins, as on this view, with a system in a pure quantum state to be interpreted as that of an ensemble, it must evolve to a quantum *mixed state*, given by Equation 2.2, after measurement. The entangled state of the measurement apparatus and the measured system resulting from pre-measurement and given in Equation 2.1 must, like all quantum states, represent an ensemble of 'fundamental states' on this interpretation. A subensemble of this ensemble will be picked out by measurement, which increases one's knowledge of the state in which the system has been prepared; the behavior of the subensemble picked out during measurement, which is described by $|p_n\rangle|o_n\rangle$, can be traced backward in time in accordance with the Schrödinger equation. This would allow the determination of the initial fundamental ensemble state, which must also be the ensemble before measurement began, because the measured value pre-exists, namely, that described by $|\Psi_0\rangle|o_n\rangle$ where $|\Psi_0\rangle$ is the initial apparatus state. The backward-evolved state must also describe an ensemble of 'fundamental states' all of which must be different from those when the system is originally prepared in a superposition with the joint system $|\Psi_0\rangle(\sum_j |o_j\rangle)$, a contradiction. Thus, the measurement problem remains for this interpretation.[25]

Von Neumann's 'no-go' theorem, which was later shown to be of limited scope, has also led many to disregard the Naive interpretation because it depends on the existence of hidden variables ([256], p. 448). The theorem aside, the Naive interpretation is fundamentally incoherent for that very reason. Shimony has succinctly pointed this out as follows.

---

[25] See also the analysis in [190].

"Once you say that the quantum state applies to ensembles and the ensembles are not necessarily homogeneous you cannot help asking what differentiates the members of the ensembles from each other. And whatever are the differentiating characteristics these are the hidden variables. So I fail to see how one can have Ballentine's interpretation consistently...", ([413])

a point also made by Fock [3]. Schrödinger also dismissed such an interpretation in the 1935 cat paper, in the section entitled "Can one base the theory on ideal ensembles?"

"The essence of this line of thought is precisely this, that one practically never knows all the determining parts of the system...To describe an actual body at a given moment one relies therefore not on one state of the model but on a so-called Gibbs ensemble...an ideal, that is, merely imagined ensemble of states, that accurately reflects our limited knowledge of the actual body. The body is then considered to behave as though in a single state arbitrarily chosen from this ensemble. This interpretation had the most extensive results. Its highest triumphs were in those cases for which *not* all states appearing in the ensemble led to the *same* observable behavior... At first thought one might well attempt likewise to refer back the always uncertain statements of Q.M. to an ideal ensemble of states, of which a quite specific one applies in any concrete instance—but one does not know which one. That this won't work is shown by... [careful consideration of the behavior of] the oscillator energy..." ([394])

Various attempts at finessing these arguments by bringing into question broader fundamental concepts of physics have been made but have been, at best, of extremely limited benefit. Much like the other alternatives to the well established Basic and Copenhagen interpretations, the approach fails to rival or surpass them.

"[The interpretation] rapidly loses its beguiling simplicity. One might say that its *chief* attraction is that measurement may merely record, and not, in any sense, create. But as soon as basic conservation laws, certainly obeyed in experimental results, are not respected in the premeasurement values, this feature must be lost. And with the talk on the 'experimental situation', again the main selling point of the [ensemble in which dynamical variables all have determinate values] appears to have disappeared." ([496], p. 216)

3.7 The Radical Bayesian Interpretation

The Radical Bayesian interpretation is currently only a proto-interpretation of quantum mechanics. Its distinctiveness lies in the use of a thoroughgoing Bayesian subjectivist interpretation of quantum probabilities, and hence fully subjective interpretation of the quantum state. Because this interpretation is currently more of a research program in a relatively early stage of development, no list of basic principles is given here. Instead, various claims and approaches of its advocates will be considered. It is sufficiently distinct and interesting to warrant separate consideration here because of its close connection to quantum information science.

Advocates of this approach have argued that the Copenhagen interpretation, to which it is most closely related, differs markedly from the interpretation to emerge from its program because, it is claimed, under the Copenhagen interpretation

> "a system's quantum state is determined by a sufficiently detailed classical description of the preparation device, [something impossible in the Radical Bayesian approach]." ([103])

The Radical Bayesian program is the most explicitly subjectivist approach to interpreting quantum mechanics yet to be pursued. Without dismissing the above consideration, it can be said that the program is closely related to the most subjectivistic readings of the Copenhagen interpretation.

Several elements of the program are similar to those of the Copenhagen approach, particularly in Stapp's version [434] wherein the ideas of William James are emphasized (*cf.* [434], Section IV and Appendices A and B). Heisenberg also early on made the following somewhat conflicted statement regarding the possibility of an irreducible subjective element in quantum mechanics.

> "The probability function does—unlike the common procedure in Newtonian mechanics—not describe a certain event but, at least during the process of observation, a whole ensemble of possible events. The observation itself changes [it] discontinuously; it selects of all possible events the actual one that has taken place. Certainly quantum theory does not contain genuine subjective features, it does not introduce the mind of the physicist as part of the atomic event. But... our description is not entirely objective." ([219], pp. 54-55)

Peierls, who believed there was only one legitimate way of understanding quantum theory, a way he identified with the Copenhagen interpretation, took a more avowedly subjectivist view of the quantum state than the young Heisenberg.

> "[T]he most fundamental statement of quantum mechanics is that the wavefunction, or more generally the density matrix, represents our knowledge of the system we are trying to describe." ([342], p. 11)

Furthermore, unlike in his youth, when he viewed quantum theory as almost entirely objective in character, later Heisenberg viewed it (c. 1958), in agreement with Peierls, as *almost entirely subjective*, rejecting the objectivist ideal of previous physics cited at various stages by Einstein, Schrödinger, and many others. For example, in *The physicist's conception of nature*, he claimed that the mathematics of physics

"no longer describes the behavior of elementary particles, but only of our knowledge of their behavior." ([220])

This view is echoed by the Radical Bayesian approach and lies at its core.

Recall also Bell's summary of the relationships of the founders of the Copenhagen interpretation to subjectivism.

"In the beginning Heisenberg and Pauli felt very close to Bohr. Those three were the Copenhagen trio, the Three Musketeers of the Copenhagen interpretation. In later years, Pauli seems to have decided that Bohr himself was not a complete supporter of the Copenhagen interpretation. He reproached Bohr along the following lines: Bohr insisted that there was this division between the quantum-mechanical system and the classical apparatus. He explicitly repudiated the idea that the human mind was somehow an important element in quantum mechanics—that is, that the division was between the interior world [mind] and the outer world [matter]. But Pauli was attracted to that idea.... He felt that the *real* Copenhagen interpretation did insist that the mind was something that you could not avoid referring to in formulating quantum mechanics." ([45], p. 53)

Pauli had his own particular view of the "undetached observer," which has influenced the Radical Bayesians through the idea of a "participatory universe" [494], stopping just short of Wheeler's "it from bit" thesis (*cf.* Section 4.7). His Copehagenist view differs in this regard from that of Radical Bayesians only as to precisely where and how subjectivity enters quantum mechanics.

The general idea of a Bayesian approach to quantum probability actually goes back at least as far as 1984, when Lane P. Hughston suggested that one "take a Bayesian attitude towards the wave function in quantum mechanics." However, Hughston proceeded to derive a Schrödinger-like state evolution with an additional non-linear term [245]. By contrast, the Radical Bayesians have not altered the quantum *formalism* itself which, as seen in the introduction to this chapter, they view as "self-interpreting." They advocate the revision, indeed, the virtual rejection of interpretative principles, including those of the Copenhagen interpretation. As their analytical starting point, the Radical Bayesians hold, "following de Finetti, that in the last analysis, probability assignments are always subjective..." [103]. What makes the Radical Bayesian program *radical* is that, unlike the interpretations discussed above, it *throws out of all the standard axioms* of quantum mechanics, or more precisely, suggests their replacement with other unknown principles.

What is positively asserted by its prime advocate, Fuchs, who described this form of Bayesianism as "radical," is its relationship to the subjectivity already present in the Copenhagen approach. "Quantum mechanics has always been about information. It is just that the physics community has somehow forgotten this" [181]. The meaning of the quantum state and the sort of information intended here are clear:

> "a quantum state is specifically and only a mathematical symbol for
> capturing a set of beliefs or gambling commitments." ([181])

This suggests that it is *human knowledge*, rather than specifically *information* in the technical sense, that is involved in this interpretation.[26] The distinction is important here because the Radical Bayesians point to the technical results in quantum information science as evidence of the validity of their approach. It is one thing to argue that information and knowledge are intimately connected; it is another to say that the description of the *signals* used to communicate information is only descriptive of human knowledge.

In the sense of *interpretation* considered here, the Radical Bayesian program presents quantum mechanics as less a physical theory than an epistemic one; not only is quantum mechanics fundamentally probabilistic, but its probabilities are explicitly taken to describe *only* the degrees of belief of observers and *not* the physical world. The position flies in the face of von Neumann's crucial insights regarding the place of consciousness in relation to physics. In particular, as von Neumann pointed out

> "measurement or the related process of the subjective perception is a
> new entity relative to the physical environment and is not reducible
> to the latter. Indeed, subjective perception leads us into the intel-
> lectual inner life of the individual which is extra-observational by its
> very nature. . ." ([477], p. 418)

and that

> "experience only makes statements of this type: an observer made a
> certain (subjective) observation; and never any like this: a physical
> quantity has a certain value." ([536], p. 420)

The Radical Bayesian program comes close to identifying the relationship between quantum mechanics and the world with that of *probability theory* and the world, which is something entirely different. Accordingly, in response to the question "How could a theory that does not describe physical reality give such accurate results. . . ?", the response of Fuchs and Peres was,

---

[26] The conflation of human knowledge with information, as seen in more detail in
the following chapter, is not uncommon in contemporary thought about quantum
mechanics.

"The point is that a theory need make no direct reference to reality in order to be successful or to be accurate in some of its predictions. Probability theory is a prime example of that because it is a theory of how to reason best in light of the information we have, regardless of the origin of that information. Quantum theory shares more of this flavor than any other physical theory. Significant pieces of its structure could just as well be called 'laws of thought' as 'laws of physics.' However, this does not preclude quantum theory from making *some* predictions with absolute certainty. Among these are the quantitative relationships between physical constants such as energy levels." ([183])

Nonetheless, Fuchs at the very outset of his programmatic article "Quantum mechanics as quantum information (and only a little more)," looks to Einstein for support.

"Albert Einstein was the master of clear thought.... he was the first person to say in absolutely unambiguous terms why the quantum state should be viewed as information (or, to say the same thing, as a representation of one's beliefs and gambling commitments, credible or otherwise)...Whatever these things called quantum states *be*, they cannot be 'real states of affairs' for [a single EPR system] alone. His argument was simply that a quantum-state assignment for a system can be forced to go one way or the other by interacting with a part of the world that should have no causal connection with the system of interest."

Beyond Einstein's concern over what Schrödinger called "state steering," the full paragraph in praise of Einstein of which the above is an excerpt, is rather odd, given Einstein's belief that physical theories must serve exactly to describe reality; it is exactly the presence of beables, in Bell's sense of theoretical elements that are to correspond to elements of reality in the EPR sense (*cf.* [24], p. 175), that is rejected when embracing subjectivism. Radical Bayesianism is certainly not Einstein's position regarding the nature of physical theory or quantum theory specifically. Although it is true that the position against which Einstein and his collaborators argued was the Copenhagenist position that quantum mechanics is as *complete* description of the physical world as is possible, their goal was one of establishing a complete *physical description* of an objective and external reality, one to which existing quantum mechanics, being a good physical theory, must by their lights refer [150].

One might ask how Einstein could have 'missed' the importance of subjectivism, since it would certainly be incorrect to attribute it to him, even if there may be some basis on which to argue that he committed himself to an epistemic interpretation of quantum uncertainties. Moreover, for reasons offered in the next chapter, the above picture also involves a fundamental misconstrual of information itself that also cannot be attributed to him. Given his general assumptions about complete theoretical descriptions, what Einstein likely expected that quantum mechanics would ultimately be shown to provide is a statistical state of systems of the same general sort as classical statistical mechanics gives of them. The corresponding thermodynamical state of a gas is still a description of a physical system existing independently of our beliefs about it, even though that description is, at least in the case of equilibrium, reducible to a statistical one. Einstein, along with many other physicists after him, believed that quantum mechanics was a theory essentially describing physical systems that have objectively real properties, the *same* properties that are associated with quantum mechanical observables, not that quantum mechanics is instead theory of states of belief or the knowledge of individuals.

One might imagine the difference to be simply a matter of one's understanding of the nature of probabilities, rather than of physics. However, in that case, the difference between the Bayesian interpretation of quantum mechanics and other statistical interpretations would not be a matter of understanding quantum physics as opposed to *any other* statistical physical theory, which is precisely the same difficulty encountered by the advocates of the strongly objectivist Naive interpretation, given that the theory has been shown to be irreducibly probabilistic.

Einstein, like von Neumann, presumably expected any apparent subjectivity in quantum mechanics to dissolve in light of a deeper future fundamental theory of physics, which presumably would still include several of its current postulates. Recall, however, that the method for further exploration of quantum phenomena advocated is

> "to go to each and every axiom of quantum theory and give it an information theoretic justification if we can... The raw distillate left behind—minuscule though it may be with respect to the full-blown theory—will be our first glimpse of what quantum mechanics is trying to tell us about nature itself." ([183])

There is little in the way of physical principles offered by the program. What is offered instead is the following vision.

> "... [the distillate] may be little more than a miniscule part of quantum theory. But being a clear window into nature, we may start to see sights through it we could hardly imagine before." ([181])

As shown in the following chapter, a similar attitude toward the quantum postulates has recently been used by others with different views of the interpretation of probability, ultimately with little success. Like those others, Fuchs argues that due to recent successes, "quantum information" holds the key to the "what quantum mechanics is telling us."

"In the 75 years that have passed since the founding of quantum mechanics, only the last 10 have turned to a view and an attitude that may finally reveal the essence of the theory. Quantum information—with its three specializations of quantum information theory, quantum cryptography, and quantum computing—leads the way in telling us how to quantify that idea... Quantum algorithms... Secret keys... the list of triumphs is growing." ([183])

But those triumphs are merely technological; they don't in themselves provide direct insight into the physics on which they are based. The claim is that

"Far from a strained application of the latest fad to a time-honored problem, this method holds promise precisely because a large part—but not all—of the structure of quantum theory has always concerned information." ([183])

But this merely begs the question.

Ultimately, Fuchs does hold that physics requires the removal of subjectivity, which is likely why he views the somewhat awkward invocation of Einstein as fitting. Nonetheless, if the real physics lies in the "distillate," then the attempt to understand quantum theory as regarding information in the absence of any specific new underlying mechanics is a distracting exercise. Rather than looking to quantum protocols and probability theory for a window onto deeper physics, someone seeking such a successor theory is far better off turning directly to physical phenomena in search of *anomalies* to be explained, in view, for example, of the insightful analysis of the history of science of Thomas Kuhn [284]. A thoroughgoing subjectivist interpretation is simply the wrong way to go about a revolution within physics. The Radical Bayesians, in Putnam's phrase, "seek to have a revolution and minimize it too." The progress made by quantum information science in relation to the foundations of quantum mechanics such as in entanglement theory and the study of causality has not depended on the subjective interpretation of quantum probability. Indeed, this is one reason why the ongoing disputes regarding the interpretation of probability itself have had essentially no impact on quantum theory.

Fuchs and Peres claimed to have had no *a priori* positivist incentive when laying out the essentials of this program.

"Some people may deplore [the operationalist character of this interpretation of quantum mechanics], but we were not led to reject a free-standing reality in the quantum world out of a predilection for positivism." ([183])

As apparently unknown to Radical Bayesians as it is to the overwhelming majority of physicists and philosophers today, Rothstein, who as shown above received correspondence from Einstein in which the latter emphasized his commitment to realism, had outlined a similar operationalistic program fifty years ago, in which "paradoxes and questions of interpretation in quantum mechanics, as well as reality, causality, and the completeness of quantum mechanics are examined from an informational viewpoint" ([381]).

He was also of the view that quantum theory fundamentally concerns information, although he did not invoke the particular subjectivist interpretation of *probability* that distinguishes the Radical Bayesian approach and hadn't modern quantum protocols to consider. According to Rothstein,

> "the concepts and terminology of information theory not only correspond precisely to measurement, but that a closely related concept, that of organization, corresponds precisely to laws, and to operations as used in physics.... By measurement we obtain information about the world, and by means of laws we bring some order into the otherwise chaotic observations. Measurement provides the raw data of experience; theory or laws organize it into a coherent whole... Bad choices of quantities to analyze, even when they are operationally defined, can hide whatever laws may be operating very effectively. Operational definitions are thus by no means a universal panacea. They are necessary rather than sufficient conditions for progress." (*e.g.* [381])

The similarities to the contemporary subjectivist approach are not insignificant, although there is no hint in Rothstein's writings of the view that the laws of quantum mechanics are anything like "laws of thought." His operationalist ideas, which represent a serious information-based approach to quantum mechanics a half-century previously, have made little long-term impact in the study of the foundations of quantum mechanics despite their relatively high profile at the time they were initially put forward.[27]

The differences between Rothstein's ideas and Radical Bayesianism point out even more strongly that what distinguishes the latter approach is almost entirely its adoption of de Finetti's conception of probability. It is, therefore, also noteworthy that a paper on the de Finetti conception of probability was given before, by Richard Bevan Braithwaite in a 1956 Colston symposium entitled *Observation and interpretation*, with Bohm, Fierz, Landsberg, Rosenfeld, and Vigier in attendance (*cf.* [278], p. 3). This presentation apparently had little effect. The core mathematical results from the physicists associated with the approach, Fuchs has argued, show that "all probabilities in quantum mechanics can be interpreted as Bayesian degrees of belief and that the Bayesian approach leads to a simple and consistent picture..." [103]. Even granting this claim, it remains for advocates of the approach to show

---

[27] It should, however, be noted that one of the set of pertinent articles that variously appeared in *Philosophy of Science*, *Physical Review*, and *Science* [380–383] was reprinted in the influential collection, *Quantum theory and measurement*, of Wheeler and Żurek [495].

that the physical results they obtain derive their import *primarily* from that conception of probability.

Given the zeal with which the program has been advocated, one would expect that it would have by now been shown to allow quantum theory to provide physical explanations that are superior to those provided under other interpretations. Moreover, as Shimony has commented in a more general context,

> "[if] the definite occurrence of a measurement result is interpreted epistemically rather than as a physical process... or a psychophysical one... I do not know how [this] could be proved definitively, because it is unclear what analytic and experimental discoveries would imply the impossibility of an account of actualization of a potentiality as a physical or psychophysical process." ([411], p. 57)

That is, the character of quantum mechanics, most profoundly exhibited in measurement outcomes, cannot be better attributed to a subjective probability than an objective one given, for example, of the option of invoking the actualization of potentialities as understood in the context of the Basic interpretation.

The subjectivism of the Radical Bayesian interpretation and its all-but-anti-realist character would also appear to preclude quantum mechanics from being a theory of physics at all, that is, from being at all capable of genuinely explaining physical phenomena. The Radical Bayesians have answered this concern quite directly but in a way that could only convince the converted.

> "[I]t appears that the absence of a mechanistic explanation is just the message that quantum mechanics is trying to send us. Accepting the Bell/EPR analysis at face value means accepting what might be the most important lesson about the world, or what we believe about the world... [that one should] accept the lesson of no instruction sets [behind measurement outcomes] if you wish to interpret quantum mechanics [at all]... It might still be argued that an agent could not be certain about [outcomes] without an objectively real state of affairs guaranteeing [them]. This outcome, it seems to us, is based on a prejudice. What would the existence of an instruction set add to the agent's beliefs about the outcome?" ([103])

The "prejudice" referred to here is any presumably realism stronger than what Devitt calls "Fig-leaf Realism" such as that of the Kantian noumenal world, an undifferentiated objective background reality; anything more would provide *some sort* of "instruction set" for measurement outcomes. This misses the point, of course, being of concern only to one who is already a Radical Bayesian. Why must physics concern *beliefs*?

Little or nothing seems to remain in quantum theory, so interpreted, regarding the physical world itself.

> "The wedge that drives a distinction between Bayesian probability theory in general and quantum mechanics in particular is perhaps nothing more than this 'Zing!' of a quantum system that is manifested when an agent interacts with it... Take all possible information-disturbance curves for a quantum system, tie them in a bundle, and *that* is the long-awaited property, the input we have been looking for from nature... Hilbert-space dimension will survive to be the stand-alone concept with no need of an agent for its definition."
> ([181])

Presumably, "this Zing!" is related to the quantum of action as it relates to measurement, that is, Bohr's quantum postulate. Fuchs argues that quantum mechanics is "a theory not about observables, not about beables, but about 'dingables' " [181]. To anyone who would regard quantum probabilities as objective rather than subjective on the basis that they are determined by physical law, in that "quantum states $|\psi\rangle$ are independent theoretical constructs from the probabilities they give rise to, $p(d) = \langle\psi|E_d|\psi\rangle$, through the Born rule", the Radical Bayesians reply that

> "these expressions are not independent at all... quantum states are every bit as subjective as any other Bayesian probability [sic]. What then is the role of the Born rule? Can it be dispensed with completely? It seems no... But from our perspective, its significance is indeed different than in other developments of quantum foundations. The Born rule is not a rule for *setting* probabilities, but rather a rule for *transforming* or *interconnecting* them." ([181])

Again, the resemblance here to Rothstein's operationalist view of physical laws as "organizers of experience" is striking.

One might rightly ask why the radical Bayesians believe the "Zing!" is necessary at all; why not simply drop the external world altogether from the interpretation as genuine revolutionaries might? One answer is that when one has obtained all the knowledge one can about a quantum system, that system can still do things that the physicist is required to explain, as Popper pointed out. A mechanism is still necessary to understand quantum phenomenon, but according to the Radical Bayesian view quantum mechanics simply does not provide it. (Bub, following a similar methodology of reducing mechanics to information theory, instead—and more reasonably given the lack of mechanism resulting from such a move—insists on information also being *physical*.) As for the measurement process that provides one with new knowledge about a quantum system, "the levels of subjectivity for the state and the state-change rule must be precisely the same for consistency's sake" [181]. This is a good point, which Copenhagen subjectivists (other than Pauli?) seem to have overlooked.

One result produced under the guidance of this interpretation is a derivation that shows that

"Up to an overall unitary 'readjustment' of one's final probabilistic beliefs—the readjustment takes into account one's initial state for the system as well as one's description of the measurement interaction—quantum collapse is *precisely* Bayesian conditionalization." ([181])

But this is simply a repackaging of the well known result of Stairs regarding the Lüders rule [432], which predated quantum information theory. Radical Bayesians have also claimed that

"a quantum state... explicitly someone's information—must always be known by *someone*... On the other hand, for many an application in quantum information, it would be quite contrived to imagine that there is always someone in the background describing the system being measured or manipulated, and that what we are doing is grounding the phenomenon with respect to *his* state of belief. The solution, at least in the case of quantum state tomography, is found through a quantum mechanical version of de Finetti's classic theorem on 'unknown probabilities.' " ([181])

Quantum state tomography is the determination of the unknown state of an ensemble of identically prepared quantum systems through the measurement of the full set of observables, therefore including non-commuting observables, for the state of a statistically significant number of members of the ensemble. However, no-one would dispute that quantum state tomography is about learning an unknown quantum state preparation or that Bayes' theorem is valid. An analogous situation arises in classical tomography, such as in classical optics when one desires to know, say an unknown polarization state of a light beam, one performs a full set of measurements of polarization along orthogonal axes; the procedure for finding a spin-1/2 quantum state is just the same.

The question is what *physical* insight into the quantum world is gained from this interpretation other than, to borrow a phrase from Einstein, a "gentle pillow for the true believer in the information age"?

3.8 The Process Interpretation

Yet another interpretation of quantum mechanics emphasizes quantum processes and probabilities for transitions between quantum states over individual quantum states. It was put forward by Feynman and follows from initial insights of Dirac. On this *Process interpretation*, a distinction is made between distinguishable and indistinguishable processes. A process is considered to be distinguishable if and only if a record results based on which its occurrence or non-occurrence can be determined. If there is only one possible process from initial to final conditions, classical probabilities are appropriate for describing the situation; otherwise, probabilities calculated via quantum amplitudes for the possible intermediary processes consistent with those conditions are used.

Feynman provided the basis for the Process interpretation, in the context of a Lagrangian rather than Hamiltonian formulation of the theory, as follows.

"We can take the view point, then, that the wave function is just a mathematical construction... In the more complicated mechanical systems... the state of motion of a system at a particular time is not enough to determine in a simple manner the way that the system will change in time. It is also necessary to know the behaviour of the system at other times; information which a wave function is not designed to furnish... Quantum mechanics can be worked [out] entirely without a wave function, by speaking of matrices and expectation values only. In practice, however, wave function is a great convenience, and dominates most of our thought in quantum mechanics. For this reason, we shall find it especially convenient, in interpreting the physical meaning of the theory, to assume our mechanical system is such that, no matter how complex between T_1 and T_2, outside of this range the action is the integral of a Lagrangian. In this way we may speak of the state of the system at times T_1 and T_2, at least, and represent it by a wave function. This will enable us to describe the meaning of the new generalization in terms with which we are already familiar." ([168], pp. 45-46)

Although, in broad terms, the Process interpretation can be seen to be related to the Copenhagen interpretation, it is distinct because (i) it does not require most of the theses of that interpretation, and (ii) it places less emphasis on quantum mechanics as a description of quantum states as of quantum processes in which the quantum of action cannot be neglected as Stachel, in particular, has done in fleshing out Feynman's approach.

In order to make the connection of process with wave-function language, in his Ph.D. thesis, *The principle of least action in quantum mechanics*, where the approach was first well formulated, Feynman noted, following Dirac, that

"... we may speak of the state of the system at time T_1 as being given by a wave function ψ, and of the state of the system at time T_2, by a wave function χ. We can then make the physical assumption that *the probability that the system in state ψ at time T_1 will be found, at the time T_2, in the state χ is the square of the absolute value of the quantity* $\langle\chi|1|\psi\rangle$.... We can define other physical quantities in terms of this, by determining the changes in this probability, or rather in the quantity $\langle\chi|1|\psi\rangle$, produced by perturbations of the motion." ([168], p. 46)

Feynman's reformulation was motivated mainly by Dirac's article "The Lagrangian in quantum mechanics," in which the latter argued that a Lagrangian formulation of quantum mechanics would be superior because (i) it allows a principle of least action to be used, and (ii) the action is relativistically invariant [140]. In particular, Feynman took notice of Dirac's comment that this approach bears a strong relation to the theory of contact transformations of which the transformation matrix is the quantum analogue.

One can take as a primary interpretative thesis of the approach that

(0) Quantum mechanical processes are the fundamental basis of quantum theory; conditional probabilities are fundamental, rather than states.

Further assumptions are the following [431].

(1) Quantally measurable quantities are a subset of classically measurable quantities, forming (sub)sets that can be measured in a single quantum process (of *registration*).

(1.1) Quantum mechanics deals only with open systems, in interaction with some device that *prepares* the system in a clearly demarcated way before it undergoes some sequence of interactions, during which no registration occurs, and with another device that *registers* something about the system afterwards. This entire cycle of preparation–interactions–registration constitutes a quantum *process*.

(2) Quantum mechanics describes processes undergone by (real or virtual) ensembles of physical systems, either predictively or retrodictively, whether they are pure or mixed.

(2.1) In computing probabilities in quantum mechanics, the following rules are to be used (*cf.* [169] and [427], pp. 314-315).

(2.1.1) There is a complex probability amplitude for each distinguishable process leading from an initial preparation to a final (registered) result. The probability for that process is equal to the modulus squared of its amplitude, which must be a complex number of modulus ≤ 1.

(2.1.2) If several alternative subprocesses, indistinguishable within the given physical arrangement, lead from the initial preparation to the final (registered) result, then the amplitudes for all the indistinguishable subprocesses must be added to get the total amplitude for the entire (distinguishable) process (*quantum law of superposition of amplitudes*).

(2.1.3) If several distinguishable alternative processes lead from the initial preparation to the same final result, then the probabilities for all these processes must be added to get the total probability for the final result (*classical law of addition of probabilities*).

(2.1.4) If an indistinguishable process consists of a sequence of steps, the amplitudes for all the steps must be multiplied to get the total amplitude for that process (*quantum law of multiplication of amplitudes*).

(2.1.5) If a distinguishable process consists of a sequence of steps, the probabilities for all the steps must be multiplied to get the total probability for that process (*classical law of multiplication of probabilities*).

Feynman viewed the following expression, which combines rules 2.1.2 and 2.1.4, as fundamental to his reformulation of quantum mechanics, now outlined.

$$\phi_{ac} = \sum_b \phi_{ab}\phi_{bc} , \qquad (3.18)$$

where ϕ_{ij} designates the probability amplitude for obtaining outcome j in a measurement of physical magnitude J conditionally upon performing measurement of physical magnitude I and obtaining outcome i, for a sequence of three measurement events in which magnitudes A, B, C are measured in that order; the corresponding probabilities are $P_{ij} = |\phi_{ij}|^2$, *cf.* rule 2.1.1. He noted that

"[this equation] is a typical representation of the wave nature of matter. Here, the chance of finding a particle going from a to c through several different routes (the values of b) may, if no attempt is made to determine the route, be represented as the square of a sum of several complex quantities—one for each available route." ([169])

He then proceeds to show how the above rules

"may be readily extended to define a probability amplitude for a particular completely specified space-time path." ([169])

The mathematical starting point for this construction is the following.

"Assume that we have a particle which can take up various values of a coordinate x. Imagine that we make an enormous number of successive position measurements, let us say separated by a small time interval ϵ. Then a succession of measurements such as $A, B, C, \ldots$, might be the succession of measurements of the coordinate x at successive times $t_1, t_2, t_3, \ldots$, where $t_{i+1} = t_i + \epsilon$. Let the value, which might result from measurement of the coordinate at time t_i, be x_i. Thus, if A is a measurement of x at t_1 then x_1 is what we previously denoted by a. From a classical point of view, the successive values $x_1, x_2, x_3, \ldots$, of the coordinate practically define a path $x(t)$. Eventually, we expect to go [to] the limit $\epsilon \to 0$.

The probability of such a path is a function of $x_1, x_2, \ldots x_i, \ldots$, say $P(\ldots, x_i, x_{i+1}, \ldots)$. The probability that the path lies in a particular region R of space-time is obtained classically by integrating P over that region." ([168], p. 77)

For this purpose, Feynman later introduced the postulate

"I. If an ideal measurement is performed to determine whether a particle has a path lying in a region of space-time, then the probability that the result will be affirmative is the absolute square of a sum of complex contributions, one from each path in the region." ([169])

He then noted that, so stated, the postulate is incomplete because "the meaning of a sum of terms one for 'each' path is ambiguous," something mathematically resolvable by taking appropriate limits, and that, although the pertinent integrals involve spatial coordinates, measurements of momentum can be subsumed under an appropriate analysis of pointer measurements pertaining to momentum [169].

A second postulate specifies the method for finding the probability amplitude, $\Phi[x(t)]$, for each path $x(t)$.

"II. The paths contribute equally in magnitude, but the phase of their contribution is the classical action (in units of $\hbar$); $i.e.$, the time integral of the Lagrangian taken along the path," ([169])

which is proportional to $\exp(1/\hbar)S[x(t)]$, where $S[x(t)] = \int L(\dot{x}(t), x(t))dt$ is the time integral of the classical Langrangian $L(\dot{x}(t), x(t))$ taken along the path in question. The assumption that the Langrangian is a quadratic function of velocities (and the assumption that finite although arbitrarily long time intervals are involved) suffices for the mathematical equivalence of this formulation of the theory to the usual one.

Thesis (0) accords with Einstein's view that the quantum state does not directly refer to individual systems, but rather to the behavior of ensembles of systems

"In what relation does the 'state' ("*Quantum state*") described by a ψ function stand to a definite real situation? [*Sachverhalt*] (let us call it 'real state')? Does the quantum state characterize a real state (1) completely or (2) incompletely?
The question cannot be answered at once [*ohne Weiteres*] because indeed every measurement signifies an uncontrollable real intervention [*Eingriff*] in the system (Heisenberg). The real state is thus not immediately accessible to experience and its judgement always remains hypothetical (comparable with the concept of force in classical mechanics, if we do not set up the laws of motion a priori). Assumptions (1) and (2) are therefore both possible in principle....

I reject (1)..." ([428], pp. 374-375)

Stachel has argued that violating Einstein's (2) by interpreting the quantum formalism as describing individual physical systems without reference to an ensemble of which they are members leads to error, pointing out that "different state functions must be attributed to the same individual system depending on whether it is treated as part of a predictive or retrodictive ensemble" [431], something that flies in the face of the time-symmetry of the Schrödinger evolution.

Thesis (1) of the Feynman formulation is similar to thesis (7) of the Copenhagen interpretation. Accordingly, Stachel has attributed the process approach Bohr as well as Feynman, pointing to the similarity of what we have here called "processes" to Bohr's "phenomena" [431]. An important feature of this approach is that the Feynman rules on which it is based can be used to derive the Lüders rule for updating the quantum state function (*cf.* [427], Appendix).

> "In wave function language, this rule is equivalent to the projection postulate, so my demonstration shows that the Feynman rules are bound to result in the quantum-mechanical prediction for any experiment obtained from the wave function by use of the projection postulate at some intermediate stage." ([429], p 251)

This applies both to situations involving single-part and compound systems, such as that considered by EPR. Stachel has articulated this approach within a broader philosophical perspective, which he calls *dynamic structural realism* [430]. On this view, there are entities of different natural kinds that may fall into structural hierarchies, which are potentially subject to change.

> "It seems that, as deeper and deeper levels of these structural hierarchies are probed, the property of inherent individuality that characterizes more complex, higher-level entities—such as a particular crystal in physics, or a particular cell in biology—is lost... [M]odern physics has reached a point at which we are led to postulate entities that have [essential natures] but not [unique individuality]." ([430], p. 56)

This picture is in contrast to realist world views that are both more radical, such as Putnam's quantum-logical realism, discussed in Chapter 2, and more conservative, such as the Naive interpretation, discussed in Section 3.6.

Sukanya Sinha and Rafael Sorkin have also proposed a version of this approach to quantum mechanics in this vein, in which

> "reality is (just as classically) a single 'history', e.g. a definite collection of particles undergoing definite motions; and quantum dynamics appears as a kind of stochastic law of motion for that history, a law formulated in terms of non-classical probability amplitudes." ([424])

Sinha and Sorkin have provided an analysis of the probabilities occurring in an EPR-type experiment with photons by summing over possible histories of the photons involved, without the use of state-vectors in the course of their calculations, where events are characterized entirely in terms of the possible photon trajectories. They argue that such an approach "goes a long way toward taking the 'mystery' out of quantum mechanics," although this comes at the price of "the presence of non-positive [probability] amplitudes and references to two-way paths" [424].

Needless to say, however, the introduction of negative probability values involves a considerable departure from traditional probability theory, one which Feynman has argued is consistent, nonetheless [173]. Indeed, he compared their introduction to that of the negative numbers and noted that they are similarly useful in simplifying calculations.

> "[C]onditional probabilities and probabilities of imagined intermediary states may be negative in a calculation of probabilities of physical events or states. If a physical theory for calculating probabilities yields a negative probability for a given situation under certain assumed conditions, we need not conclude the theory is incorrect. Two other possibilities of interpretation exist. One is that the conditions (for example initial conditions) may not be capable of being realized in the physical world. The other possibility is that the situation for which the probability appears to be negative is not one that can be verified directly." ([173], p. 238)

The latter option appears the most attractive one. However, it is unclear whether that does much to alleviate the concern about the conflict between negative probabilities and probability theory such as that of Kolmogorov. This approach to interpreting quantum mechanics appears primarily as a modern, minimal version of the Copenhagen approach.

3.9 Interpretational Underdetermination

Before moving on to the consideration of recent proto-interpretations that have arisen since the emergence of quantum information science, it is appropriate to consider another way that its results have influenced thinking regarding the activity of quantum theory interpretation. Some physicists and philosophers have engaged the foundations of quantum theory at a higher level of mathematical abstraction than previously, with a view toward grounding the theory in the technical notion of information as opposed to in human knowledge or subjectivity as has been done in the past, and also differently from Rothstein's pioneering approach.

The most striking example is based on the Clifton–Bub–Halvorson (CBH) theorem, which relates quantum theory directly to information-theoretic constraints [111, 204]. In light of the CBH theorem, Bub has taken a position similar to that of the Radical Bayesians, in particular, that "the appropriate aim of physics at the fundamental level is the representation and manipulation of information," although with one extremely important difference: quantum mechanics nonetheless *remains* physics, rather than being something like a set of "laws of thought." This is accomplished *by taking information itself to be physical.*

Bub has specifically argued that:

(1) a quantum theory is best understood as a theory about the possibilities and impossibilities of *information transfer*, as opposed to a theory about the mechanics of physical entities,

(2) given the information-theoretic constraints, any mechanical theory of quantum phenomena that includes an account of the measuring instruments that reveal these phenomena must be empirically equivalent to a quantum theory, and

(3) assuming the information-theoretic constraints are in fact satisfied in our world, no mechanical theory of quantum phenomena that includes an account of measurement interactions can be acceptable, and *the appropriate aim of physics at the fundamental level then becomes the representation and manipulation of information* ([88], emphasis added).

In arguing this, Bub introduces a specific notion of what constitutes "a quantum theory" that transcends standard non-relativistic quantum mechanics. In particular, for him, a *quantum theory* is one in which observables and states have a specific sort of algebraic structure, that of a C^*-algebra.[28] In von Neumann's standard formulation of quantum mechanics, the bounded operators have such a structure. However, this definition has within its scope not only the standard non-relativistic quantum theory but also the standard quantum field theories. The CBH theorem concerns theories satisfying the following requirements.

(i) the observables of the theory are represented by the self-adjoint operators in a non-commutative C^*- algebra (with individual algebras commuting).

(ii) the states of the theory are represented by C^*-algebraic states (as positive normalized linear functionals on the algebra), and spacelike separated systems can be prepared in entangled states that allow remote steering.

(iii) dynamical changes are represented by completely positive linear maps. The theorem shows that theories with observables and states of the above type can be characterized in terms of just three information-theoretic constraints.

Bub's argument is structured as follows. It is claimed that thesis (1) follows from the CBH theorem and an understanding of the problems that arise when one attempts to interpret quantum theory in the traditional manner, that is, as a theory of mechanics. Second, it is argued specifically that information-theoretic constraints preclude the possibility of such a mechanical theory including an account of the behavior measuring instruments that reveal quantum phenomena, which was exactly the interpretational focus for both Bohr and von Neumann in their approaches to quantum mechanics. This accords with the view, advocated by Bub, that quantum theory *interpretation* is nec-

---

[28] A brief summary of the mathematics of C^*-algebras is given in section 6 of the Appendix.

essarily underdetermined by any empirical evidence. Thesis (3) is then argued to follow, once one assumes that the world has built-in constraints on the acquisition, representation, and communication of information—the strongest of his assumptions.

The information-theoretic constraints considered are the following. First is the impossibility of superluminal information transfer between two physical systems by performing measurements on one of them. This condition means that when two agents perform non-selective local measurements, the measurements of one can have no influence on the statistics of outcomes obtained by the other, and vice-versa.[29] The subsystems possessed by the two agents are kinematically independent if every element of the C*-algebra of one commutes pairwise with every element of the C*-algebra of the other, that is, the algebras are mutually commuting. Second is the impossibility of perfectly broadcasting information possibly communicated using an unknown physical state, which for pure states amounts to requirement that states cannot be perfectly cloned.[30] This constraint requires each of the two algebras to be non-commutative. Third is the impossibility of communicating information so as to implement a quantum bit commitment protocol with unconditional security. The quantum bit commitment protocol is a primitive cryptographic protocol involving two agents, which can be described as follows. One agent supplies an encoded bit to a second agent as a warrant for commitment to a binary value, that is, 0 or 1. Although it must be impossible for the second agent to infer that value at this initial stage, the information provided, together with additional information supplied by the sender later, should allow one to infer that value during the revelation stage of the protocol. The receiver should also be certain that the protocol will not allow for 'cheating' by the sending of the initial information in a way that would allow the value to be changed after the initial, 'commitment' stage. The "no bit commitment" condition protects one against any a priori theoretical preclusion of entangled states violating local causality.

Bub views the CBH theorem as grounds for taking the structure of quantum mechanical states and observables to be representable as a non-commutative C*-algebra. Interpretative problems in the Hilbert-space formulation of quantum mechanics are then to be viewed as arising from its incompatibility with a "proper" phase-space representation and a corresponding problem of adequately describing state measurement [89]. As seen in previous chapters, these difficulties with the standard Hilbert-space formulation are clear and widely acknowledged. Bub then goes further, arguing from this that the rational epistemological stance is to remain *uncommitted* to any

---

[29] Selective measurements could alter the statistics of measurements performed at a distance simply because there is a change of the ensemble with regard to which statistics are taken.

[30] No-cloning for pure states is an immediate consequence in standard quantum mechanics of the superposition principle.

constructive interpretation of mechanical theory that purports to solve the measurement problem; such interpretations are understood to be empirically underdetermined. Thus, mechanistic interpretations of quantum mechanics are invalid.

Crucially, however, in order to go through, with quantum theories remaining physical theories at all, Bub's argument requires an additional and crucial background assumption, namely, "taking the notion of quantum information as a new physical primitive" [89]. It requires information to be itself physical in nature, an assumption considered in detail in the following chapter. Therefore, he believes the correct stance is to view any 'quantum theory' as "a theory about the representation and manipulation of information, which then becomes the appropriate aim of physics," rather than to provide an account of the behavior of objects *themselves* because, as he sees it, satisfying information-theoretic constraints requires whatever constructive account that might be given to keep its extra machinery (hidden variables) forever hidden in principle ([89], p. 555). However, there is little reason that information *should* be considered physical and, as shown in the next chapter, good reasons for it not to be considered so.

In light of Bub's argument, one could take a different lesson away from the BCH theorem: Because, on the basis of the assumptions of his argument, the necessary activity of understanding quantum mechanics as a theory of mechanics is precluded, there are grounds for rejecting the least justified of those assumptions, namely, the assumption that information is a physical primitive. Moreover, for reasons discussed in the following chapter, the assumption itself is incoherent.

4

Information and Quantum Mechanics

Information processing and communication by physical means, for example, electronic circuitry, antennae, and the electromagnetic field are subject to constraints prescribed by the principles of the theories governing those means. The physical principles governing quantum and classical computers and communication lines differently constrain the behavior of quantum information, distinguishing it from classical information. The development of the theory of quantum information has mainly involved the extension of methods first developed for investigating information in relation to classical systems to methods for investigating analogous situations involving quantum systems. In this way, information theory has been broadened to include situations such as communication using the polarization states of individual or entangled photons directly and, more significantly, with the latter as a resource (*cf.*, *e.g.*, [251]).

A central question when considering information in relation to the foundations of quantum mechanics is whether quantum information and classical information significantly differ, and if so, how fundamental their differences are (*cf.* [149, 449, 450]). Classical and quantum information are similar in that both can be used as resources for communication and computation. That there are, nonetheless, fundamental differences between them can be seen by recognizing that the quantity of information—as opposed to what the information might be *about*—is defined in terms of the entropies of the distributions of probabilities of the occurrence of corresponding events, more specifically, on the probabilities of proper subsets of events among sets of possible events involved in these activities. As seen in previous chapters, the respective sorts of system used to encode information in these two contexts have quite different behaviors. Two differences between quantum information and classical information can immediately be identified. (1) The entropy functions used to quantify information in these two contexts are derived from different sorts of probability distributions, specifically, distributions of different *mathematical forms*. (2) The characterizations of classical and quantum systems are *conceptually* quite different, leading to qualitatively different constraints on the signs used for coding.

G. Jaeger, *Entanglement, Information, and the Interpretation*
of Quantum Mechanics, The Frontiers Collection,
DOI 10.1007/978-3-540-92128-8_4, © Springer-Verlag Berlin Heidelberg 2009

Whether these differences are fundamental depends on the understanding of the probability distributions of which the pertinent entropies are functionals and on the interpretation of quantum theory assumed. For example, differences of this kind presumably do not exist on the problematic Naive interpretation, where all probabilities arise due to ignorance, whereas on the standard Basic interpretation they are fundamental, as is also so on other, newer interpretations considered below. The Basic interpretation of quantum mechanics will typically but not exclusively be considered here. Before analyzing in detail the resulting distinction between quantum and classical information, let us take note of some common misconceptions about the relationship of information to the physical world related to the increasingly popular idea that information is physical in nature. The greatest and most common error of this sort is naively to identify information with the physical systems that may be used in communicating it. This mistake is not unique to quantum information science; it has often been made in relation to classical information as well. The error has repeatedly been noted in the traditional information theory literature—having been pointed out in standard works such as Warren Weaver's gloss on Claude Shannon's fundamental work in their joint book, *The mathematical theory of communication*, and in Norman Abramson's *Information theory and coding*—but is far less often, if ever, explicitly noted in the quantum information literature. Abramson, for example, when commenting that one should use "the contraction *binit* for binary digit," notes that, "It is important to make a distinction between the binit (binary digit)," a physical entity, "and the *bit*," a unit of information ([1], p. 7). By contrast, the quantum *two-level system* and the *qubit* were conflated in the course of the very introduction the term into the physics literature [396]. As a result, the conflation of (information-theoretic) symbols with (physical) signs is easily made in the context of communicating by physical means.

The methodology of quantifying information aside, it is clear that information is not essentially physical because it can be manifested in *non-physical* contexts, such as in conscious experience.[1] Although physicalists may assert that consciousness is ultimately physical, this is far from having been shown; even assuming it could be in some instances, considerable additional work would be required to show a clear correspondence between the pertinent experiential states and physical states in every pertinent case. In any event, in order for information to be quantified, a specific encoding must be considered that is chosen not by physics but by the *conscious agents communicating it* via the signal systems in question. Although the physics involved clearly constrains the amount of information that *can be* encoded in these systems by some maximum value, it does not, in itself, determine the amount of information encoded. Moreover, the same information can be communicated by radically different physical systems. This confusion has not given rise to a mis-

---

[1] Recall also von Neumann's careful comments on the difference between physical and mental events in relation to quantum measurement.

understanding of the nature of classical physical world primarily because the interpretation of classical mechanics is relatively uncontroversial. However, it has done so in relation to understanding the quantum world when it has been assumed in the course of interpreting quantum mechanics.

The term *qubit* was introduced into the physics literature as follows.

"[The interpretation of] the quantum entropy of some macrostate of a thermodynamic system as a measure of resources necessary to represent information about the system's quantum microstate... is accomplished by replacing the classical idea of a binary digit [that is, binit] with a quantum two-state system, such as the spin of an electron. These quantum bits, or 'qubits,' are the fundamental units of quantum information." ([396])

Presumably, at least in part, due to this conflationary early statement, it is now common in the physics literature to use *qubit* to refer both to the most basic physical system, in the sense of that with the smallest corresponding Hilbert space, that is, a two-level system, *and* to the unit of quantum information. The equivocal usage of the term *qubit* as referring to both the unit of information and to the simplest quantum system was, early on, a source and, later, a symptom of that error and related confusions.

One explanation for the ease with which the conflation of the qubit and the quantum two-level system occurred and has been sustained is that there is a specific mean number of quantum two-level systems that are *necessary to be able physically to communicate* a given amount of information under ideal circumstances. This allows one to make a highly qualified correspondence between the number of information units and the number of physical units, that is, two-level systems by reference to physical entropy. However, it is an insufficient basis on which to make a direct correspondence between information units and physical units, and even less so to an *identity* between the two-level system and the qubit.

The phrase *two-level system* is used here to denote the system with the simplest non-trivial quantum Hilbert space, which is preferable, for example, to *two-state system* for the following reason. Any quantum system possessing more than one state, that is, any non-trivial quantum system, possesses an infinite number of states due to the superposition principle. This is particularly significant in relation to quantum signaling: it is the Hilbert-space *dimensionality*—and hence number of *orthogonal states*—that is genuinely pertinent to quantum communication, not the number of quantum states. For example, encoding n bits in states of two-level systems is inefficient when the those states are not fully distinguishable; a greater number of two-level systems than the minimum number n is then needed to communicate the information. A maximum of two states, in particular, two orthogonal eigenstates at a time can be fully distinguished in that simplest of quantum systems, for example, the energy eigenstates in the case of a spin-$\frac{1}{2}$ system in a magnetic field.[2]

A physical system may be capable of realizing a symbol of a chosen code as a sign, and so of being an element of a memory or a signal, but it cannot serve as a unit of information unless the symbol is reducible to the physical sign. This is impossible, despite the fact that physics clearly *does* pertain to signaling, as in (at least some formulations of) relativity theory in which the rate at which signals can travel is constrained from above by the speed of light. Again, the analogous point can be made with regard to classical physical systems. Traditionally, the relation between a sign and what it signifies is that the former causes the latter to 'come to mind,' either typically or necessarily.[3] Signs are entities designating other entities and, as we have seen, need not be physical, although those pertinent to quantum mechanics clearly are. Signs are only established, explicitly or implicitly, in communication by the mutual agreement of agents that they are to serve to encode symbols according to a freely chosen scheme and are distinct, in particular, from *natural signs* which are entities that occur in Nature, which are understood as such, for example, on the basis of causal relationships to other entities.

Once information units are (inappropriately) thought of as physical entities, it becomes natural to attribute spatial location to specific transmissions of information (*cf.* [452]); statements suggesting that information is contained *in* an entity, such as a computer, or otherwise localized are deeply misleading. Such usage is at best of a *façon de parler*, regardless of how convenient it may be in practice to indulge in. Strictly speaking, information, being non-physical, cannot be *contained in* a physical system, despite being conveniently described as "stored in" or "borne by" it. What *can* properly be said along these lines is that information can be obtained by a localized cognitive agent as a result of a physical process that may be *constrained* by the physics of the local entities involved in the process; although a communication signal consists of physical signs that enable the communication of information, it does not contain information in any literal sense. Moreover, even assuming that all information *is* physically processed, the processes involved may be far different from that involving some simple bit-to-flip-flop correspondence.

Consider a traditional book, for example. Such a book is a physically transportable collection of print-signs encoding symbols. However, fundamentally, books relate to information because they encode *conventional symbols*. Their specific physical form, vis-à-vis the information they may communicate, is secondary. Physics merely constrains their efficacy: without at least *some* cognitive context, a putative signal communicates no information. Once the

---

[2] This may be the motivation for Fuchs' comment that the only genuinely physical aspect of a quantum system under the Radical Bayesian interpretation is its dimensionality, *cf.* Section 3.7.

[3] This brings into play notions such as reference and truth, about which see, for example [368], Chapter 4.

chain of convention is entirely lost, no information is communicated through them. As Weaver noted, "If one is confronted with a very elementary situation where he has to choose one of two alternative messages, then it is arbitrarily said that the information associated with the situation, is unity. Note that it is misleading (although often convenient) to say that one or the other message conveys unit information" [406]. As Weaver continues,

> "The concept of information applies not to the individual messages (as the concept of the meaning would), but rather to the situation as a whole... the two messages between which one must choose, in such a selection, can be anything one likes. One might be the text of the King James Version of the Bible, and the other might be 'Yes.' " ([406], p. 9)

With the symbol–sign relation clarified, let us move on consider more carefully the relationship between information and physics. Physics constrains the behavior of information whenever physical entities and processes are used as signs in communication and data processing. The behavior of information will then depend on the theories used to describe the physical systems functioning as the signs chosen for encoding, and is then constrained by physics. This is why quantum information differs from classical information rather than, say, simply because the signs involved typically differ in scale. It is necessary, in particular, to examine correlations of chosen signs between subsystems of a composite physical system. One then recognizes that information is communicated and processed differently using quantum signs rather than classical ones, not least of all because the former can be in entangled states, wherein the states of component systems are extraordinarily well correlated. It is in this sense that quantum information is a newly discovered sort of information: It is information that is, in general, differently physically constrained from that previously considered by information theory. Understanding the behavior of information under quantum constraints illuminates the foundations of quantum mechanics, although, in light of the lack of identity between symbols and the physical signs in which they are encoded, it is unlikely to *determine* it.

Basic thermodynamical results have been considered in this regard, in particular, the constraint that the erasure of a quantity of physically encoded information comes with an associated *minimal* thermodynamical cost. This fact has also been used to argue that information is physical. Although more subtle than the simple bit–flip-flop or qubit–two-level-system correspondences just considered, such an approach is also misguided, as shown below. Before considering the issue in detail, however, the basic elements of classical information theory and quantum information theory are reviewed in the following four sections. Those already familiar with these subjects may wish to proceed directly to Section 4.5.

4.1 The Theory of Information

The traditional theory of information is founded on the characterization of information in terms of distributions of probabilities of the sort that occur in classical physics. It includes early ideas introduced by Szilard [443] and Alan Turing [457] and was firmly established in the late 1940s by Shannon in his theory of communication [405]. Information is defined in this theory as the improbability of the occurrence of events in which pre-designated sequences of symbols appear—events in which observations are made, by the agent gaining information, of what are typically, but not necessarily, physical systems.

Recall that in classical systems all physical magnitudes can in principle be simultaneously specified and that, unlike for quantum systems, all indefiniteness associated with physical states must be subjective, arising only from ignorance of a definite and actual, as opposed to a possibly indefinite and potential, state of affairs in the external world. The fundamental unit measuring information in this theory, the *bit*, is associated with sequences of symbols from among a well defined and agreed set of possible sequences.[4] Any physical system having two states, acting as signs, that are stable over a pertinent time scale is capable of encoding two symbols, for example, '0' and '1,' and so of being used to communicate one bit under appropriate circumstances. The choice of the binary unit of information, which is the smallest that can be used, results in the number 2 being the standard logarithmic base for information theory, logarithmic functions being introduced as a matter of mathematical convenience because the number of available states can be very large; it was introduced by Ralph V. L. Hartley already in 1928 [208], much before Shannon's comprehensive theory appeared.[5] Because there are 2^n states available to n identical two-state systems for use in encoding, such a collection has an associated *capacity* of $\log_2 2^n = n$ bits of information. A concrete physical example of such a set of entities is a computer memory register. A pair of symbols, '0' and '1,' is mathematically represented by a random variable x taking two values and can be considered an element of the Galois field $GF(2)$. Similarly, strings of symbols can be represented by $\mathbf{x} \in GF(2)^n$. A mathematical *field* is a set of elements, with two operations (called *multiplication* and *addition*) under which it is closed, that satisfies the axioms of associativity, commutativity, and distributivity and within which there exist additive and multiplicative identity elements and inverses. The Shannon information associated with a string composed of symbols, such as '0's and '1's, in a classical communication system is understood in terms of how improbable the string is to appear as one of a previously designated set of possible strings.

---

[4] The name *bit* is a contraction of *binary unit*. It was first formally introduced for this purpose by John W. Tukey [8].

[5] If the base e is used instead, the information unit is the *nat*. Turing used base 10 and called the associated unit the *ban*, named after Banbury, England. It is more typically called the *hartley* after Hartley.

It is significant that information is defined in Shannon's theory independently of any meanings that might be associated with it in practice, despite the fundamental requirement that there be *a priori* agreement, implicit or explicit in nature, as to the *set* of symbols that are available to agents and to which probability distributions are assigned. Information theory is designed to accommodate *any* code that could be used in communication, rather than only those involved in language which do involve meaning as an essential component. Thus, although communication and language are strongly related, information theory possesses a character fundamentally different from linguistics, which concerns itself with semantics, that is, word meanings, and syntax, that is, the ways in which words can be combined to form structured sentences, as well as various ways words and sentences are related to sounds, that is, signals sent as modulated gas pressure zones, such as in air. The central role of information in communication theory was made clear in the following further comment of Weaver.

> "[T]he real reason that [Shannon-type] analysis deals with a concept of information which characterizes the whole statistical nature of the information source, and is not concerned with the individual messages (and not at all directly concerned with the meaning of the individual messages,) is that... a communication system must face the problem of handling any message that the source can produce... the system should be designed to handle well the jobs it is most likely to be asked to do, and should resign itself to be less efficient at rare tasks. This sort of consideration leads at once to the necessity of characterizing the statistical nature of the whole ensemble of messages which a given kind of source can and will produce. And *information*, as used in communication theory, does just this." ([406], p. 14)

The information associated with an event contextualized in this way is given by reference to the *statistical characteristics* of the information source: The information content associated with a discrete event $\mathbf{x}$ is $\log_2 \left(p(\mathbf{x}) \right)^{-1} = -\log_2 p(\mathbf{x})$, where $p(\mathbf{x})$ is the probability of occurrence of $\mathbf{x}$, for example, in an associated signal. The simplest events involve the appearance of individual symbols. In the case of a binary pair of symbols, for example '0' and '1', these probabilities are $p(0)$ and $p(1)$. The information associated with a specific event is considered obtained when the event is seen to occur. This mathematical expression captures the intuition that no information is gained by learning of an event that occurs with certainty $(-\log_2 1 = 0)$ whereas learning, for example, the outcome of a fair coin toss $(-\log_2 \frac{1}{2} = 1)$ provides all the information essential to the toss and eliminates the previous uncertainty as to the outcome. More generally, an informative signaling event consists of the occurrence of a sequence of specific symbols that are elements of a large set of sequences any element of which *could* have appeared in its place.

A number of mutually exclusive such events together with their probabilities is said to constitute an information-theoretic scheme of the classical type. There is an associated, entirely epistemic uncertainty in this case. By design, only the probabilities of occurrence of these events is known to the receiver beforehand. Indeed, were there *no* uncertainty, communication would be a pointless exercise. Such a scheme differs from the analogous quantum scheme because the uncertainties appearing in the quantum mechanical context can be due in part to the objective indefiniteness of the physical magnitudes of quantum signal states discussed in previous chapters. This is the fundamental basis for the difference between classical and quantum information, which is reflected in quantum probability.

The standard measure of classical information is a functional of the discrete probability distribution $\{p_i\}_{i=1}^n$, $p_i \equiv p_X(x_i) = P(X = x_i)$, where $P(X)$ is the probability mass function for the random variable X over the n possible values in a countable sample space $S = \{x_i\}$ of events x_i, namely,

$$H(X) \equiv H[p_1, p_2, \ldots, p_n] = -\sum_i^n p_i \log_2 p_i \, , \qquad (4.1)$$

known as the *Shannon entropy*.[6] The Shannon entropy is the average information, in units of bits, which is given by the expected value of the information content over all possible events available under a given scheme, that is, the information gained, on average, by an agent witnessing its elements. When there are only two alternatives available, the Shannon entropy is just $H_{\text{binary}}(p) = -p \log_2 p - (1 - p) \log_2(1 - p)$, where p is, without loss of generality, the probability of the symbol '0' and $1 - p$ is the probability of the alternative symbol '1'. When $p = \frac{1}{2}$, so that the event is *a priori* entirely unknown, one finds that $H(p) = 1$, again as intuitively expected from the example of a fair coin toss.

The Shannon entropy applies primarily to sets of symbol strings, and by design coincides with the entropy of the appropriately formed ensembles of physical signals. In the context of coding information using classical systems, the associated entropy is related to the physical entropy appearing in classical statistical mechanics that quantifies the disorganization in such systems, that is, the incompleteness of resulting specification of the classical microstate of a member of the ensemble of available signs. The similarity of this information entropy and classical physical entropy follows from the use of classical systems for signaling. As seen below, a different measure of information is, in general, needed to described communication when quantum signals are used for communication.

In addition to the basic mathematical requirement that it be a continuous function of probabilities, the Shannon entropy measure satisfies the conditions of invariance under permutations of the probabilities, p_i, and of additivity,

---

[6] These various mathematical concepts were introduced in Section 1.3.

$H[p_1, p_2, \ldots, p_n] = H[q, p_3, \ldots, p_n] + qH[(p_1/q), (p_2/q)]$, where $q = p_1 + p_2$. Invariance under permutations of the probabilities is a natural condition to impose because the set of possible specific sets of symbols is not affected by their ordering. Additivity serves to ensure consistency between all possible ways of making choices between subsets of events constituting a message. Simple coding schemes which change the symbols used or their ordering in a word provide illustrations of the sensibility of these conditions. Generally speaking, the greater the uncertainty regarding the precise message being transmitted from among a set of choices, the greater the resources required to eliminate that uncertainty.

The Shannon joint entropy of a pair of (discrete) random variables, A and B, is

$$H(A, B) = -\sum_{a,b} p(a, b) \log_2 p(a, b) , \qquad (4.2)$$

where $p(a, b) \equiv P_{AB}(A = a, B = b)$ is the probability that both $A = a$ and $B = b$; sums are taken over the two sample spaces associated with A and B. Because $H(A) \leq H(A, B)$, the uncertainty associated with A than is less than or equal to that associated with the pair A, B. This constraint does not hold for the von Neumann quantum entropy, as discussed below, and distinguishes classical from quantum information. This is one of the grounds for the introduction of the qubit concept.

The distinction between quantum information and classical information on the basis of their respective entropy functions has incorrectly been seen as trivial by some, such as Armond Duwell, [149], and Christopher Timpson who has argued *against* the claim that the Shannon information is inadequate for characterizing quantum information [449]. This is argued prejudicially with respect to the interpretation of quantum formalism to the extent that it is assumed that any associated signal source and receiver in communication are essentially classical in description; this point is taken up again below. It also inherits the difficulty of the Naive interpretation that local hidden variables models are inadequate to account for the above difference.

In order to compare two classical discrete probability distributions, one can use the relative entropy function, known as the *Kullback–Leibler distance* [283]. Given two probability distributions, $p(a) = \{p(a_1), \ldots, p(a_n)\}$ and $p(b) = \{p(b_1), \ldots, p(b_n)\}$, this is

$$H[p(a)\|p(b)] \equiv \sum_i p(a_i) \log_2 \frac{p(a_i)}{p(b_i)} , \qquad (4.3)$$

which is simply the ratio of the actual entropy to the maximum entropy of the information source. Its complement relative to unity is the *redundancy*. The conditional entropy of a classical random variable A relative to another random variable B is $H(A|B) \equiv H(A, B) - H(B)$. The classical mutual information between two random variables, A and B, described by the joint probability distribution $p(a, b) = \{p(a_i, b_j)\}$ and marginal distributions

$p(a) = \{p(a_i)\} = \sum_j p(a_i, b_j)$ and $p(b) = \{p(b_j)\} = \sum_i p(a_i, b_j)$, respectively, is

$$H(A:B) \equiv H[p(a)] + H[p(b)] - H[p(a,b)] , \qquad (4.4)$$

which captures the degree of *correlation* between the two variables, in that it is the amount of information about A that is acquired by determining the value of B. It also captures the degree of distinguishability of a given correlated situation from a fully uncorrelated situation: $H(A:B) = H[p(a,b)\|p(a)p(b)]$. This function serves to characterize the information each distribution can provide about the other and serves as a measure of the communication resource provided by "shared randomness." In the quantum context, entanglement is associated with the extraordinarily strong correlations between subsystem signal states that are naturally associated with quantum analogue of this quantity, which is used in quantum information theory because classical correlations are demonstrably inadequate for describing the behavior of entangled quantum states, *cf.* Section 4.6.

With these basic classical information-theoretical quantities specified, one can consider with precision the behavior of classical communication systems, which involve both signal systems and information. An information source functions by producing sets of sequences of symbols characterized probabilistically. Communication channels, whether classical or quantum, have specific capacities for communicating information. Channels are also capable of further enhancement through the use of quantum entanglement. Quantum communication channels are distinguished from classical communication channels by their ability to communicate *quantum* information—however inefficient it may prove to be in traditional communication applications—which the latter are incapable of transmitting, as well as classical information. In either case, by comparing the entropy characteristic of a source with the capacity of a communication channel with which it can be associated, one can determine whether the information the source is capable of encoding can be successfully communicated using the specified channel.

A very simple but useful description of a classical communication channel involves introducing a source of potential noise. A communication system with additive noise, shown in Figure 4.1, is one wherein a transmitted signal $s(t)$ can be influenced by additive random noise, $n(t)$. In the presence of such noise, the resulting classical signal, $r(t)$, is given by $r(t) = s(t) + n(t)$. The basic memoryless classical noisy channel is the *binary symmetric channel* (BSC). In such a channel, there is a probability p that noise can introduce a symbol error, '0'$\leftrightarrow$'1', through a change of sign state during transmission. This channel is readily generalized to one capable of transmitting any finite number of symbols '0',...'N', namely, the uniform symmetric channel (USC) with each being transmitted with the same probability, $1-p$, or being changed to another symbol with probability, $q = p/(N-1)$. One is then typically interested in considering the process of error correction, that is, the elimination of such errors, so that the signal is faithfully transmitted. Because the outputs

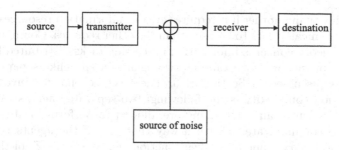

Fig. 4.1. Communication system with additive noise. The transmitter and receiver typically function so as to produce and receive physical signals of specific types. This signal may encounter noise due to some physical source, causing the signal received to differ from the signal transmitted, as shown.

of noisy communication channels depend probabilistically on their inputs, a communication channel is characterized by the distribution possible channel outputs conditional on possible inputs, that is, the mutual information $H(A : B)$.

The classical information channel capacity, defined as the maximum mutual information over all possible inputs described by probabilities $\{p_A(a_i)\}$, is the primary quantity of interest when studying classical communication systems. The classical operational channel capacity can be defined as the greatest bit-rate at which input information can be transmitted with arbitrarily low error, that is, it appears in the fundamental *noisy channel coding theorem* for classical channels: The capacity of a discrete, memoryless communication channel is simply

$$C = \max_{\{p_A(a_i)\}} H(A : B) , \qquad (4.5)$$

where A describes the channel input and B characterizes the output. In the case of a channel carrying binary signals, the capacity then lies in the range $[0, 1]$. For the binary symmetric channel described above, the capacity is $1 - H(p)$: In the case of a noiseless such channel, any transmitted symbol is received at the destination without error and the channel capacity is 1 bit-output-per-symbol. If the transmission rate is less than the channel capacity, then for any $\epsilon > 0$ there is a code having a block length large enough that the error probability is less than ϵ. Therefore, error-free classical communication is possible for rates below this (Shannon) channel capacity. Some errors are guaranteed to exist at rates above this capacity, according to the Shannon–Hartley theorem [324].

In addition to being used for the communication of information with a given efficiency, communication channels are also involved in other, more *cooperative* activities. One of the most important elements of the foundation of information theory is the study of communication complexity, which involves such activity. Roughly speaking, one information-theoretic task is considered more complex than another if it requires greater information resources, for example, communicated bits to be used than the other requires. Communication complexity theory concerns information processing tasks distributed between two distinct processors, for example, two agents in spacelike separated laboratories, as discussed in Section 1.10. The basic schema for investigation communication complexity is the following: two separated agents, A and B, each in possession of an n-bit string are allowed to perform local computations and to communicate in such a way that one of the agents is able to announce the binary value of a given function, $f : X \times Y \rightarrow Z$, of these two strings to the other agent. It is readily generalized to the case of any finite number of agents. In particular, consider the first of two parties to possess the n-component string $x \in X = \{0,1\}^{\times n}$ and the other party $y \in Y = \{0,1\}^{\times n}$, and let $Z = \{0,1\}$. It is of course possible for the second party to determine $f(x,y)$ if the first one simply communicates x to it. However, the quantity of interest is the *minimum* number of bits of information that need be communicated between A and B in order for this task to be accomplished, that is, the (classical) communication complexity, $K(f)$, of the task.

The computations required for determining the function $f(x,y)$ can be carried out either deterministically or probabilistically, giving rise to another level of classification. In the deterministic case, a communicated symbol is a function only of previously communicated symbol-values of the input from the sender, and one considers the number of bits communicated in the worst case scenario in the best possible correct deterministic protocol for computing the function $f(x,y)$. In the non-deterministic case, the symbols communicated may depend on probabilistic choices as well; a non-deterministic protocol for z is considered *correct* if it always returns $1 - z$ for $f(x,y) = 1 - z$ and for any x, y with $f(x,y) = z$ it returns z for at least one sequence of probabilistic choices made. The worst-case number of bits communicated, in the best possible correct non-deterministic protocol for z, is designated $N^z(f)$. Therefore, $K(f) \geq N^z(f)$. The case in which A and B may also share random variables is similar to the situations in which quantum states are shared and symbol values are gained from measurements in the computational basis, and so is somewhat similar to the quantum case.

4.2 The Quantum Theory of Information

Quantum information theory describes the communication and processing of information with symbols encoded in quantum mechanical systems, that is, as quantum signs, which by their nature are subject to physical constraints differing from those on classical signs. The development of quantum information theory has involved the replacement or generalization of traditional information-theoretic concepts so as to describe situations involving such signs, something that is necessary because quantum mechanical systems are described by non-standard probability distributions.

The distinction between quantum information and classical information has been drawn in several ways [101, 149, 396]. The most direct way is via the entropy function used to measure information content, that is, by noting that the (quantum) von Neumann entropy and the Shannon entropy involve different sorts of probability, only one of the two being appropriate in all cases of its type. Thus, when presenting his proof that the von Neumann entropy is exactly the mean number of qubits needed for ideal encoding using states of a quantum ensemble, Benjamin Schumacher noted that

"instead of simply applying classical information theory to probabilities derived from quantum rules, we can adopt notions of coding and measures of information that are themselves distinctly quantum mechanical." ([396])

A more formal approach is to note the differences in the abilities of agents to transmit and process information using quantum mechanical systems versus using classical systems, considering quantum procedures that are analogous to classical procedures. In the latter case, quantum and classical information are seen to differ in at least two specific respects. (1) Most of the information that can be communicated using a generic quantum-mechanical system involves of correlations between subsystems. (2) Quantum correlations can be extraordinarily strong. The differences are most clear when entangled quantum systems are involved. The first follows from the exponential growth of the number of parameters characterizing entangled states of quantum information-bearing systems with the number of subsystems, rather than the linear growth of classical systems and the subclass of fully factorable quantum states. The second is characteristic of the violations of Bell locality and of entanglement that appear in the same context. It is exhibited by the specific relatively simple example of the Bell states. In those states, which are pure, the *subsystem* (reduced) states are entirely indefinite, being fully mixed, while the states of the two subsystems are fully correlated with each other; the state of an individual subsystem is, therefore, entirely indefinite. This was clear to Schrödinger already in 1935 ([394], Section 10), a year postdating Hartley's work but predating Shannon's formal communication theory.

Entanglement between just two systems enables information processing methods that are superior to those involving merely classically correlated systems, as evidenced by the Deutsch–Jozsa algorithm, which accomplishes a specific task in just one computational step what requires several steps when performed using any classical system [132]. The potential improvement in information processing power grows more pronounced as the number of two-level subsystems used is increased. The Deutsch-Jozsa algorithm classifies binary functions, for example, those with a domain of a two-place string and a range of one symbol, $f : \{0,1\} \times \{0,1\} \to \{0,1\}$, by distinguishing members the class of constant functions, which take all input values to a single output value, from the balanced functions, in which half of the input pairs of values are taken to each element of the range. With this set of functions, the domain and range together allow the a total of 8 classifiable functions. The algorithm allows one to classifying in only *one* evaluation of a given function, $f(i, j)$, based on the value of its output, rather than several [132].

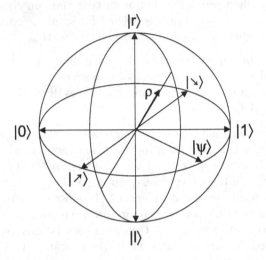

Fig. 4.2. a: Top figure. The *Bloch ball* of states of the quantum two-level system, for example, photon polarization. It is a real-valued representation of the state space based on the expectation values of the Pauli operators σ_i ($i = 1, 2, 3$) [251]. The *conjugate* bases lie along orthogonal axes. The pure states, $P(|\psi\rangle)$, lie on the periphery, the *Poincaré–Bloch sphere* [355]. The mixed states, ρ, lie in the interior; the fully mixed state $\frac{1}{2}\mathbb{I}$ lies at the origin.

To better understand how quantum systems can be used for communication and computation, first recall that the pure states of the two-level quantum system can be represented by vectors in the two-dimensional complex Hilbert space $\mathcal{H} = \mathbb{C}^2$ and note that any orthonormal basis for the Hilbert space of this system can be associated with two symbols, '0' and '1', and act as the computational basis, $\{|0\rangle, |1\rangle\}$. When encoding symbols using this system, the computational basis is typically taken either to be the vertical axis or the horizontal axis. The computational basis for the qubit Hilbert space in the latter case is shown in Figure 4.2; in the former case, these states correspond instead to the poles of the Poincaré–Bloch sphere. This basis can be put in direct correspondence with the finite Galois field. In the context of quantum information processing, for example, the pure state $|\psi\rangle = a_1|\uparrow\rangle + a_2|\downarrow\rangle$ of Equation 1.1 would be written $|\psi\rangle = a_1|0\rangle + a_2|1\rangle$, with $a_i \in \mathbb{C}$ and $|a_1|^2 = |a_2|^2 = 1$. Mixed ensembles are weighted sums of these, lying in the interior.

Now, consider from the informational point of view our very first physical example, the double-slit apparatus illustrated in Figure 1.1. The Shannon information associated with using *classical* particles plus the diaphragm as an information source—so that the uncertainty about a given signal and the encoded symbol, which could be either '0' or '1', regards the path taken by the classical particle in reaching the smallest resolvable interval about the point x—is

$$H(\{p_i\}) = -p_0 \log_2 p_0 - p_1 \log_2 p_1 = -p_0 \log_2 p_0 - (1 - p_0) \log_2(1 - p_0) \ .$$

This is an appropriate measure in that case because the two classical signal states, "from the top slit" and "from the bottom slit," are *perfectly distinguishable*. If $p_0 = p_1 = \frac{1}{2}$, that is, both symbols occur with equal frequency, one finds $H(\{p_i\}) = \log_2 2 = 1$, indicating that one bit of information per use of this apparatus is communicated, just as when a classical particle signal and a path-detection system consisting of a pair of area detectors, each assumed to perfectly detecting the system path from one of the two slits, are used.

When using an elementary *quantum* system for signaling, again two paths are available from the slits of the double-slitted diaphragm to a fixed point on the detection screen, each possible with an associated quantum probability p_i ($i = 1, 2$), namely, $p_0 = |a_1|^2$ and $p_1 = |a_2|^2$. However, as discussed in Section 1.1, the classical entropy measure is valid in the quantum case *only if only one of the two slits can be taken at a time*, that is, the two signal states, corresponding to symbols '0' and '1', are chosen to be $|\downarrow\rangle$ and $|\uparrow\rangle$, respectively, assuming a detector configuration resolving the two peaks at the right of Figure 1.1, just as in the case for classical particles. Thus, in the case of a typical quantum signal source the situation is different, due to the imperfect distinguishability of quantum signal states in general; most state pairs are imperfectly distinguishable, even in the case of noiseless channels.

In the quantum case, information is quantified instead by the von Neumann entropy, which is bounded from above by the Shannon entropy. If the sender is free to prepare any pair of quantum signal states, its upper bound of 1 can be reached, say by equi-probably sending particles into only one slit at a time, that is, sending two symbols by the encoding '0'→ $|\uparrow\rangle$ and '1'→ $|\downarrow\rangle$ involving perfectly distinguishable quantum states; if instead the set of signs produced by the transmitter consists of quantum states that are *not* distinguishable by the receiver because they are not mutually orthogonal, quantum interference will prevent that upper bound from being reached. The information transmitted per encoded symbol will almost always be *less* than 1 bit, even when symbols are sent equi-probably; for example, the two symbols could be sent equi-probably by encoding '0'→ $|\uparrow\rangle$ and '1'→ $\frac{1}{\sqrt{2}}(|\uparrow\rangle + |\downarrow\rangle)$. The standard measure of the quantum information associated with the diagonal elements of the density matrix form of ρ is the von Neumann entropy

$$S(\rho) = -\text{tr}(\rho \log_2 \rho) = - \sum_i \lambda_i \log_2 \lambda_i \ , \tag{4.6}$$

where $\{\lambda_i\}$ is the set of quantum probabilities that constitute the set of eigenvalues of ρ.[7] In the first instance, the signal ensemble is described by $\rho = \frac{1}{2}|\uparrow\rangle\langle\uparrow| + \frac{1}{2}|\downarrow\rangle\langle\downarrow| = \frac{1}{2}\mathbb{I}$, so that $\lambda_1 = \lambda_2 = \frac{1}{2}$ and $S(\rho) = 1$, whereas in the second $\rho = \frac{1}{2}[|\uparrow\rangle\langle\uparrow|] + \frac{1}{2}[\frac{1}{\sqrt{2}}(|\uparrow\rangle + |\downarrow\rangle)\frac{1}{\sqrt{2}}(\langle\uparrow| + \langle\downarrow|)]$, so that $\lambda_1 = \frac{1}{2}(1 + \frac{1}{\sqrt{2}})$, $\lambda_2 = \frac{1}{2}(1 - \frac{1}{\sqrt{2}})$ and $S(\rho) = 0.39 < 1$. If one were to simply use the Shannon entropy with the symbol-state probabilities $p_0 = p_1 = \frac{1}{2}$ as arguments, for example, rather than the above statistical operator eigenvalues, one would (incorrectly) judge that the same amount of information is communicated in the two cases.[8] Again, that procedure works *only* in the first instance, where the quantum signal states used are perfectly distinguishable, as classical states are.

The von Neumann and Shannon entropies *are* analogues, nonetheless; like the Shannon entropy, the von Neumann entropy is a measure of uncertainty associated with a state and achieves its maximum value for the maximum uncertainty. $S(\rho)$, like $H(\{p_i\})$, is non-negative and has a higher value the lower the state purity and does reaches 1 for the maximally mixed state. For the two-level system, $\rho_{\text{mix}} = \frac{1}{2}\mathbb{I}$, for which $\lambda_i = \frac{1}{2}$, $i = 1, 2$, $S(\rho) = 1$. $S(\rho) = 0$ if and only if the ensemble state is pure, $\rho_{\text{pure}} = |\psi\rangle\langle\psi|$ for any state-vector $|\psi\rangle$, so that $\lambda_1 = 1$ and $\lambda_2 = 0$ without loss of generality; noting that $\log_2 1 = 0$, one sees that $S(\rho_{\text{pure}}) = 0$. In the general case of systems with d-dimensional Hilbert spaces, one has $0 \leq S(\rho) \leq \log_2 d$, with the limits again being reached in the mixed and pure cases, respectively. However, the uncertainty is that associated with the quantum probability distribution which, as such, is not interpretable, in general, as a simple measure of ignorance of the values of corresponding magnitudes.

The difference between the classical and quantum cases is particularly clear in the case of a model of communication using the discrete version of the double-slit experiment, as in a Mach–Zehnder interferometer configuration shown in Figure 4.3, where the beam-path state of a photon more obviously involves only two possibilities because there are two non-contiguous detection areas. This involves two-level systems formed by spatial-path occupation eigenstates of systems emerging from exit ports of an initial beam-splitter, that could be coherently recombined later. In this apparatus, one considers a particular preparation $\bar{\mathcal{P}}$ of an ensemble ρ of signal systems emerging in the two beams leading from a beam-splitter. Particles enter it from the left and/or from below and leave in two exit paths. Each path then encounters a mirror, a phase shifter, a second beam-splitter, and finally a particle detector

---

[7] Here, $0 \log 0 \equiv 0$. We have assumed the set of eigenvalues of ρ to be countable; see [484].

[8] One could, of course, attempt to 'reverse-engineer' the quantum probabilities into classical probabilities, and perhaps justify this under an interpretation of quantum mechanics wherein their systems are *a priori* classical in nature, but this would, quite unnaturally, inelegantly, and problematically depend on the existence a hidden-variables theory of signal states that is generally valid.

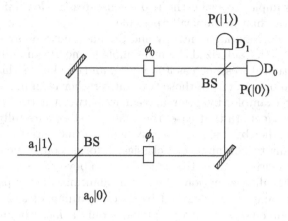

Fig. 4.3. The Mach–Zehnder interferometer providing a range of two-level system states corresponding to a range of input amplitudes a_i and phase shifts. Detectors D_i find count rates proportional to the probability of lying in the output states described by state-projectors $P(|0\rangle)$ and $P(|1\rangle)$ [251, 253]. The probabilities of detection for input amplitudes $a_0 = 0$, $a_1 = 1$ are $p_0 = \sin^2[(\phi_0 - \phi_1)/2]$ and $p_1 = \cos^2[(\phi_0 - \phi_1)/2]$.

D_i. Intervening variable phase-shifters introduce a relative phase shift between arms similarly to the way that different path-lengths from diaphragm slits to a given point on the detection screen do in the double-slit experiment. As in the double-slit experiment, changing the phases of the input amplitudes of the apparatus modulates detection rates in a way that an interference pattern appears at the photodetectors, although in this case temporally rather than spatially. In this interferometer, it is not necessary to change the location of the detection area to observe possible interferometric modulation as it is in the standard two-slit experiment. The states emerging from the beamsplitter can be proportional only to either $|0\rangle + i|1\rangle$ or to $i|0\rangle + |1\rangle$, the former if the photon were input from the left, the latter if the photon were input from below. The resulting detection rates at this receiver are proportional to the corresponding probabilities, which are given in the caption below Figure 4.3.

Recall that the extent of the entropy reduction upon a recipient's detection event, for example resulting in a pure state, can be viewed as the information gained by the associated identification of the signal state from among the set of possible states the source can provide. For a general ensemble of quantum two-level systems produced by the initial beams and beamsplitter, the eigenvalues λ_i ($i = 1, 2$) of ρ do not in general correspond to the probabilities $p_j = \mathrm{tr}(|\bar{j}\rangle\langle\bar{j}|\rho)$ of a system encoding the symbol $j =$'0','1' because the possible signals are typically indistinguishable, as shown in the second example above [253]. When the individual messages $j =$'0','1' are encoded in superposition signal states $|\bar{j}\rangle = \left(a_0^{(j)}|0\rangle + a_1^{(j)}|1\rangle\right)$ that are not mutually orthogonal, as opposed, for example, to using the two eigenstates $|i\rangle$, *less* information will be communicable than is classically possible.

Again, the von Neumann entropy and Shannon entropy agree *only* when the source can be characterized as an ensemble formed from mutually orthogonal pure quantum states, such as $|0\rangle$ and $|1\rangle$, *and* one knows that both states will be prepared only in one of those two states rather than in a superposition of the two, for example, by prior agreement between a transmitting agent and a receiving agent. In that case, the quantum states are fully distinguishable from each other by a single measurement, so that one is, in effect, using quantum systems to communicate classical information, which is described by the Shannon entropy, as in the first example above. In general, in contrast to a classical signaling situation, because a quantum state ρ prepared in an unknown basis cannot be identified based on learning of a single quantum measurement outcome, the entropy provides only a *loose* bound on this information. In practice, the signaling basis could be unknown, for example, because the receiver is an unintended one, such as an eavesdropper on a secret communication intentionally kept 'in the dark' as to the coding method in use, by virtue of the intentional use of non-orthogonal quantum states as signal states.[9]

Again, in the general case, $S(\rho)$ cannot be understood as a measure of *ignorance of the state* ρ of the transmitted system. However, in the special case that the receiver *is* given information about the signal states corresponding to $\rho = \sum_i p_i P(|i\rangle)$, where $\{|i\rangle\}$ is a specified *orthonormal* signaling basis, $S(\rho) = H(\{p_i\})$, which is the entropy of the (classical) probability distribution corresponding to $\{p_i\}$, the quantum entropy expresses classical mixing and can be understood as the average ignorance as to the *outcome of a measurement* of a system in the state ρ, although the system is non-classical. Nonetheless, even then, if the decoding process is performed inefficiently by measuring states in a different basis or if the channel is noisy, in which case the quantum states used for encoding of this classical information are altered during transmission, then the information, described by the classical mutual information between source and destination is such that $H(A:B) \leq S(\rho) - \sum_i p_i S(\rho_i)$, where ρ_i is the (possibly mixed) state in which the original transmitted pure state ends up at the receiver.

Another way of viewing the situation in quantum communication is by reference to the relationship between quantum measurement and quantum state cloning. Quantum state cloning is defined as the process by which an unknown quantum state of one system becomes also that of another quantum system without, importantly, a change in the state of the first system being induced in the process [138, 508]. A set of states from a known orthonormal basis can be perfectly cloned because quantum states from a known basis are perfectly distinguishable. This can be done in the obvious way: use the

---

[9] This relates to quantum cryptography. The quantum entropy in relation to quantum communication in more general situations is taken up in Section 4.6.

outcome, which is one of a set of mutually exclusive possible outcomes of the measurement of the observable corresponding to the known basis on a system in the "original" state, to identify the state in which to place a second system in a "blank" state. However, the no-cloning theorem shows that this procedure generally fails. The theorem can be simply and precisely stated as follows. It is impossible to make perfect copies of an unknown state of a quantum system by a unitary operation.

That entirely unknown state cloning is impossible can be easily seen. Were entirely unknown quantum states able to be perfectly so cloned, as many perfect copies as desired could be made—because the state is unknown the operation must be the same one operation regardless of any set of states considered—and used to distinguish two quantum states to whatever precision desired, in contradiction to the Holevo bound, described below, which limits the information transmission capacity of a single-qubit channel to *one bit* of information per use. A straightforward mathematical proof of the no-cloning theorem is the following. Assume that a unitary operator U could perform both of the following transformations on two *different non-orthogonal vectors* $|\psi\rangle, |\phi\rangle$: $|a\rangle|\psi\rangle \rightarrow |\psi\rangle|\psi\rangle$ and $|a\rangle|\phi\rangle \rightarrow |\phi\rangle|\phi\rangle$, resulting in perfect copies of two such two unknown vectors $|\psi\rangle$ and $|\phi\rangle$ made from a given quantum state $|a\rangle$. Such a unitary transformation would then give

$$c = \langle\psi|\phi\rangle = \langle\psi|\langle a|a\rangle|\phi\rangle \rightarrow \langle\psi|\langle\psi|\phi\rangle|\phi\rangle = \langle\psi|\phi\rangle\langle\phi|\psi\rangle = c^2 , \qquad (4.7)$$

which is possible only if $c = 0$ or $c = 1$, in which case $|\psi\rangle$ and $|\phi\rangle$ are either identical or orthogonal, contradicting the assumption (*cf., e.g.,* [512]). Note that the related condition $\langle\psi|\phi\rangle = \langle\phi|\psi\rangle\langle\psi|\phi\rangle$ is the condition that the quantum state $|\phi\rangle$ be dispersion-free for $P(|\psi\rangle) \equiv |\psi\rangle\langle\psi|$, upon squaring both sides and recalling, in considering the resulting right-hand side, that projectors are idempotent, that is, that $P(|\zeta\rangle)^2 = P(|\zeta\rangle)$ for all $|\zeta\rangle$, since $\mathrm{Disp}_\rho A = \langle A^2\rangle_\rho - \langle A\rangle_\rho^2$. Similarly to the situation just considered, this equation is only satisfied if $c = 0$ or $c = 1$, which is the case only when $|\psi\rangle$ and $|\phi\rangle$ are orthogonal or identical, respectively. Requiring this condition to hold for *all* states $|\psi\rangle$—equivalently observables $P(|\psi\rangle)$—one sees that there are no pure quantum states that are dispersion-free with respect to all observables, but only ones that are so with respect to the pertinent set of commuting observables, for which one does have $c = 0$ or $c = 1$. Therefore, for the case of pure states, perfect cloning is impossible.

Extending this result from the pure states $|\phi\rangle$ to *any* quantum state ρ, the corresponding condition $\mathrm{tr}(\rho P(|\psi\rangle))^2 = \mathrm{tr}((\rho P(|\psi\rangle)))$ for all $|\psi\rangle$ similarly requires $\mathrm{tr}(\rho P(|\psi\rangle)) = 0$ or 1 for all $|\psi\rangle$, which holds only when ρ is the zero operator in the first instance and the (non-normalized) identity operator in the second, neither of which are well defined density matrices, because neither has unit trace; the remaining possibility, that $\langle\psi|\rho|\psi\rangle$ is 0 for some $|\psi\rangle$ and 1 for others, is excluded by the fact that this quantity must vary continuously as the state $|\psi\rangle$ is changed, *cf.* the discussion of Gleason's theorem in Section 2.2 ([371], p. 62). Therefore, no unitary process can make identical copies of a general, unknown quantum state via such a process. Another argument that unknown quantum state cloning cannot be performed by a unitary operation is that the operation would then enable *simultaneous measurement* of two properties with non-commuting operators, which is precluded by the basic principles of quantum mechanics: it would enable the measurement of two such properties via the measurement of a different one of them in *each* of the identical copies.

Let us turn now to the question of *conditional* information in the quantum context, something important for the consideration of composite quantum signaling systems. As in the classical case, one can consider the quantum conditional entropy, namely, the entropy of the state of a subsystem A given the state of the other subsystem B of a bipartite system,

$$S(A|B) \equiv S(A, B) - S(B) \tag{4.8}$$
$$= S(\rho_{AB}) - S(\rho_B) , \tag{4.9}$$

which is analogous to the classical case. However, because the quantum conditional entropy can take *negative* values, a quantum system can be more definite in the joint state of two component systems than in the states of its individual components, something which is not possible in classical physics, as again can be seen by inspecting the entropies for the singlet state $|\Psi^-\rangle$. This divergence from classicality was recognized in general terms by Schrödinger in 1935.

> "Whenever one has a complete expectation catalog—a maximum total knowledge—a ψ-function—for two completely separated bodies, or, in better terms, for each of them singly, then one obviously has it also for the two bodies together... But the converse is not true. *Maximal knowledge of a total system does not necessarily include total knowledge of all its parts, not even when these are fully separated from each other and at the moment are not influencing each other at all*... [if] conditional statements occur in the combined catalog, *then it can not possibly be maximal in regard to the individual systems*. For the content of two maximal individual catalogs would by itself suffice for a maximal combined catalog; the conditional statements could not be added on." ([394], Section 10)

One, therefore, does not have additivity in the quantum case of entangled states. Such behavior goes strongly against classical intuition and allows entanglement to serve as a novel information processing resource.

The *quantum mutual information* between two subsystems described by states ρ_A and ρ_B of a composite system described by a joint state ρ_{AB} is also formally similar to but importantly different from its classical namesake:

$$I(A : B) \equiv S(A) + S(B) - S(A, B) \tag{4.10}$$
$$= S(\rho_A) + S(\rho_B) - S(\rho_{AB}) . \tag{4.11}$$

This quantity differs from the analogous classical quantity in that it can exceed the bound for the classical mutual information; the quantum mutual information reaches *twice* the maximum value obtained in the corresponding classical mechanical situation: $I(A : B) \leq 2 \min\{S(A), S(B)\}$. For a bipartite quantum pure state, $I(A : B) = 2S(A) = 2S(B)$. When the quantum mutual information exceeds the classical bound, it is *supercorrelated*. Use of this measure allows a fully quantum mechanical characterization of mutual information.

The quantum mutual information has been shown to have two different but related operational meanings [199]. The first is as the total amount of correlation, as measured by the minimal rate of randomness that is required to completely erase all the correlations in a state ρ_{AB} in a many-copy setting [199]. The second is as a relative entropy, because $I(A : B) = S(\rho_{AB} \| \rho_A \otimes \rho_B)$, where $S(\rho' \| \rho) \equiv \mathrm{tr}(\rho \log_2 \rho) - \mathrm{tr}(\rho \log_2 \rho')$, which is the *quantum relative entropy* [398]. It can, therefore, be used as an entanglement measure, as discussed in Section 1.4.

4.3 Entropy in Quantum Measurement Theory

The fact that the entropy of a quantum system is typically influenced by interactions of the environment with the system, including those involved in measurement, provides an avenue for the investigation of the foundations of physics through the behavior of quantum information. In the study of decoherence, for example in relation to the Collapse-Free interpretation, one views quantum measurement as essentially the result of the quantum mechanical interaction of an object system with its environment, as discussed in Section 3.5 above.[10]

A quantum system that interacts with other systems in the environment is an open system and is described by a statistical operator ρ having an evolution that, in many instances, can be described via a CPTP map $\mathcal{E}(\rho, t)$. In such cases, the time-dependent state can be found using the operator-sum representation, given by Equation 1.55, through the operation elements $\{K_j(t)\}$ that characterize its evolution. In particular, with the Schrödinger evolution $U(t)$ describing the composite system comprising the object-system S and its environment E, the operation elements describing the evolution of the object-system are $K_j(t) = \langle e_j | U(t) | e_0 \rangle$, where $\{|e_j\rangle\}$ is a basis for the Hilbert space of the environment and $|e_0\rangle$ is the initial state of the environment before interaction.

Consider the measurement process as taking the object system and the measurement environment to be initially uncorrelated, that is, let the initial state of this composite system be $\rho_C = \rho_S \otimes \rho_E$ and consider a set of projectors $\{P_j\}$. When the system S is involved in a measurement-type interaction

---

[10] For a recent comprehensive review, see [391].

it becomes entangled in the pre-measurement stage with the environment E including, in particular, the measuring apparatus. When a precise measurement of an observable with n distinct eigenvalues and characterized by the general projectors $\{P_i\}$ is made, the evolution of S *before selection* is

$$\rho_S \rightarrow \rho'_S = \sum_{i=0}^{n} P_i \rho_S P_i \ . \tag{4.12}$$

The entropy of the resulting state ρ'_S is always greater than or equal to that of the state ρ_S of the system before measurement. Von Neumann's proof of this was based directly on the second law of thermodynamics ([477], Section V.2), although this was later shown also to follow from Klein's inequality $S(\rho)$ (*cf.* [324], p. 515). It is the case, of course, that the mixing of states never decreases the quantum entropy.

Recall that, in his formulation of quantum mechanics, von Neumann considered measurements to be fundamental, motivating the introduction of two different processes of state evolution, namely, the 'automatic' Schrödinger evolution (Process **2.**) and that for measurement (Process **1.**), discussed in Section 2.5. Von Neumann remarked that

> "it is important that **2.** does not increase the statistical uncertainty existing in $[\rho]$, but that **1.** does: **2.** transforms [pure] states into pure states while **1.** can transform [pure] states into mixtures. In this sense, therefore, the development of a state according to **1.** is statistical, while according to **2.** it is causal... Just as in statistical mechanics, therefore, **2.** does not reproduce one of the most important and striking properties of the real world, namely its irreversibility... **1.** behaves in a fundamentally different fashion... Therefore, we have reached a point at which it is desirable to utilize the thermodynamical method of analysis, because it alone makes it possible for us to understand correctly the differences between **1.** and **2.** into which reversibility questions obviously enter." ([477], Section V.2)

The form of the quantum entropy function $S(\rho)$ was then derived by von Neumann from thermodynamical principles. Indeed, his basic goal was a consistent formalism of quantum measurement according with fundamental thermodynamical principles. Von Neumann showed, in particular, that measurements are irreversible when they induce any change in system state from that corresponding to the Schrödinger evolution. In the case of selective measurements, the evolution $\mathcal{E}(\rho, t)$ will not be trace-preserving. Any information encoded in the preparation of the object–system can be acquired through the measurement.

In some more general system–environment correlation processes that must be described by positive-operator-valued measures rather than projectors, the entropy of the state can instead decrease. The possible change of entropy in a measured system is taken up again below, in Section 4.7.

4.4 Quantum Communication and Its Limitations

Although the sending of simple messages using quantum signs is generally less efficient than that using classical means, quantum communication protocols can assist the communication of information between agents in different laboratories allowing one to reach efficiencies surpassing those achievable using only previously shared classical random data by using previously shared entangled states. Entangled states, therefore, serve as an important and novel communication resource. Nonetheless, communication cannot be carried out either using previously shared entanglement or previously shared random data alone. Were measurement of shared entangled states alone, for example, capable of signaling, the speed-of-light constraint on communication would appear to be violated; this limitation appears necessary for the consistency of special relativity and quantum mechanics.[11] Before considering the implications of quantum communication and computation protocols for the foundations of quantum mechanics in the following sections, let us first review the basic theory of quantum communication.

Any means for transmitting quantum information is formally considered a *quantum channel*. In the physical context, each quantum channel involves an equivalence class of means of signal transmission involving an ensemble of quantum systems, for example, a collection of two-level systems such as photon polarizations viewed on the basis of dichotomous magnitudes alone, which must be prepared by the sender in states ρ_i $(i = 1, 2, \ldots, n)$ encoding n symbols with corresponding probabilities p_i, in which the receiver makes appropriate measurements, as in the examples considered above in this chapter.

Quantum communication channels differ from classical channels in that, in addition to the differences between quantum and classical systems involved, there is an unavoidable interaction of any transmitted quantum signal state and the environment that manifests itself in the irreversible evolution of the local state, as briefly summarized in the previous section. Therefore, input quantum states are, in general, irretrievable by local unitary transformations from the states received as channel outputs. As in the case of classical channels, ideal quantum channels that do not introduce errors are called *noiseless*, whereas realistic channels susceptible to errors are referred to as *noisy* channels. Quantum channels are typically assumed to be stationary and memoryless; the CPTP map describing a channel is analogous to the Markov matrix describing the probabilities of outputs in terms of inputs in the description of stochastic classical channels. Therefore, the final state of an input pure statistical operator after the effect of a channel is typically assumed to be of the form $\rho' = \sum_i K_i P(|\psi\rangle) K_i^\dagger$.

The fidelity for situations in which a pure-state input $|\psi\rangle$ result final states ρ_i with probabilities p_i, that is, the statistical state $\rho = \sum_i p_i \rho_i$ is given

---

[11] For more on this question and the putative "peaceful coexistence" of quantum mechanics and relativity, see [313], pp. 148-158, [412], pp. 40, 309, and [371], pp. 106-107.

by $F\big(P(|\psi\rangle),\rho\big) = \langle\psi|\rho|\psi\rangle$. For mixed-state input ω, the fidelity is given by $F(\rho,\omega) = \big[\mathrm{tr}(\sqrt{\sqrt{\omega}\rho\sqrt{\omega}})\big]^2$, the maximum value being attained by the pure-state expression over the set of pure states in a larger Hilbert space yielding ρ_i by partial tracing [265]. A noiseless channel transmits signals with unit fidelity. A channel that completely decoheres a state ρ destroys all off-diagonal elements of the statistical operators input under the process $\rho \to \rho' = \sum \rho_{ii} P(|\psi_i\rangle)$; such a channel transmits only classical information perfectly, because it entirely destroys the coherence of input quantum states, rendering it incapable of transmitting quantum information and, in that sense, is not a truly quantum channel.

One can prepare the subsystems required for two-party communication within a larger quantum system by placing them in a shared entangled joint state ρ_{AB} having the appropriate partial traces ρ_A and ρ_B. When one desires to broadcast quantum information, a chosen state output from a quantum source is sent using a collection of copies of the state. Communication involves encoding n symbols by a process $\mathcal{E}(\rho)$ to m inputs symbols for a quantum transmission channel and then decoding them back to n symbols from m outputs of the channel by the process $\mathcal{D}(\rho)$, because this can assist in limiting the effect of noise sources acting on the communication channel. Any quantum channel is attributed several types of transmission capacity. The most basic of these are the classical capacity, the unassisted quantum capacity, and the entanglement-assisted classical capacity [39, 40].

The classical capacity, C, of a quantum channel is simply the capacity of a channel to transmit classical information using quantum systems, that is, the maximum asymptotic rate at which *bits* can be transmitted with arbitrarily good reliability using elements of the computational basis. It is the optimal asymptotic (classical) mutual information per channel use, where possibly entangled input quantum signal states are mapped back to classical data by possibly collective measurement during decoding.

The unassisted quantum channel capacity, Q, is bounded from below by C, because if a quantum channel can faithfully transmit a generic state of a two-level system then it can always at *least* transmit a computational basis state, $|0\rangle$ or $|1\rangle$. Q is known to be non-additive and may surpass the maximum value of the coherent information that can be sent by a single channel use. Prior classical communication cannot increase the quantum capacity of a channel [37]. However, if a classical back-channel is also available, allowing for two-way communication, an increase in quantum channel capacity is possible [36].

The unassisted quantum channel capacity can be described in terms of entropies as $Q(\rho) = \max_{p_i}\big(S\big(\sum_i p_i\rho_i\big) - \sum_i p_i S(\rho_i)\big)$, where $S\big(\sum_i p_i\rho_i\big) - \sum_i p_i S(\rho_i) \equiv \chi(\rho)$ is the *Holevo information*, which has a form that shows that the information communicable via the ensemble is reduced from that given by the von Neumann entropy as its components become increasingly impure. The Holevo information can also be written in terms of quantum relative entropy, $\chi(\rho) = \sum_i p_i S(\rho_i\|\rho_{\mathrm{avg}})$, where $\rho_{\mathrm{avg}} = \sum_i p_i\rho_i$, which can be understood in terms of the information accessible to the receiver by examining the quantum mutual information.

The entanglement-assisted classical capacity, C_E, is the capacity of a quantum channel to transmit classical information by making use of prior-shared quantum entanglement. It is defined similarly to Q but applies in the case where there is an interactive protocol that makes use of the quantum channel, prior-shared entanglement, and unlimited classical communication between source and destination laboratories, instead of a quantum encoding–decoding scheme. Because entangled states alone do not allow for the transmit information but are capable of improving the capacity of a quantum channel as, for example, in the quantum dense-coding protocol, the value of Q bounds C_E from below. For a channel described by the identity map, the entanglement-assisted classical capacity is twice that of the unassisted classical capacity when this protocol is used, as seen below.

Specific quantum channels are typically characterized by their effects on states of individual two-level systems and, in addition to being considered stationary and memoryless, are also often considered noisy, in order to better describe situations encountered in nature. From the point of view of foundational questions of quantum mechanics, those related to decoherence are particularly significant. For example, the (not necessarily completely) *depolarizing channel* takes a system to a fully mixed state with a probability, p, known as the strength of depolarization, or leaving it unchanged with a probability $q = 1 - p$, and so is described by $\mathcal{E}(\rho) = p\frac{1}{2}\mathbb{I} + (1 - p)\rho$, with corresponding decomposition operators $E_0 = \sqrt{1 - (3/4)p}\sigma_0$, $E_i = (1/2)\sqrt{p}\sigma_i$, where $i = 1, 2, 3$.

An important set of quantum channels for studying errors in quantum communication is that of the *Pauli channels*, including the bit-flip, phase-flip and bit+phase-flip channels. The *phase-flip channel* is described by the map $\mathcal{E}(\rho) = p(\sigma_3\rho\sigma_3) + (1 - p)\rho$, which can be specified by two decomposition operators $E_0 = \sqrt{1 - p}\sigma_0$ and $E_1 = \sqrt{p}\sigma_3$, where σ_3 is the Pauli operator corresponding to the single two-level system gate describing phase-flipping. In this channel, quantum information can be lost without energy being lost. Like the depolarizing channel, it is useful in the study of the effect of quantum decoherence. The descriptions and effects of the *bit-flip* and *bit+phase-flip channels* are similar to that of the phase-flip channel, with the Pauli operators σ_1 and σ_2, respectively, taking the place of the σ_3 in the above. Decomposition operators for these are thus $E_0' = \sqrt{1 - p}\sigma_0, E_1' = \sqrt{p}\sigma_1$, and $E_0'' = \sqrt{1 - p}\sigma_0, E_1'' = \sqrt{p}\sigma_2$, respectively.

The above channels involve no change the number of field quanta. For that reason, they are sometimes considered to involve "classical" noise, although their effects are extremely important for understanding the behavior of quantum information. A two-level system pure-state $|\psi\rangle$ that undergoes an arbitrary 'error' of the above type by coupling to an environment, taken to be in some initial state $|\bar{e}\rangle$, evolves unitarily together with the environment into the entangled state. In particular,

$$|\psi\rangle|\bar{e}\rangle = (a_0|0\rangle + a_1|1\rangle)|\bar{e}\rangle \rightarrow \quad (a_0|0\rangle + a_1|1\rangle)|\bar{e}_0\rangle$$
$$+ (a_1|0\rangle + a_0|1\rangle)|\bar{e}_1\rangle$$
$$+ (a_0|0\rangle - a_1|1\rangle)|\bar{e}_2\rangle$$
$$+ (a_1|0\rangle - a_0|1\rangle)|\bar{e}_3\rangle , \qquad (4.13)$$

where the $|\bar{e}_\mu\rangle$ ($\mu = 0, 1, 2, 3$) are states of the environment that need not be orthogonal and where each of the summands is the result of a distinct one of the possible individual Pauli 'errors.'

By contrast, the asymmetrical 'decay' of one computational-basis state to the other, for example $|1\rangle$ to $|0\rangle$, with probability p as occurs in the *amplitude-damping* channel, clearly does involve a change in field quanta. This channel is significant in the characterization of quantum decoherence exactly because it captures uniquely quantum noise effects. It can be described by the two decomposition operators, described by the 2×2 matrices E_0, which has one non-zero entry $[E_0]_{12} = \sqrt{p}$, and E_1, which is diagonal with $[E_1]_{11} = 1$ and $[E_1]_{22} = \sqrt{1-p}$.

Noisy channels are clearly the most important ones for the realistic description of quantum communication. However, even when using a *noiseless* quantum channel and quantum data compression techniques, there are specific limitations on optimal communication. Most significantly, *Holevo's theorem* provides the so-called *Holevo bound* characterizing the fundamental limit on the amount of *classical* accessible information from a source to a destination in terms of the entropy of an ensemble of quantum systems decomposable as signal states $\{p_i, \rho_i\}$; the optimal value of the mutual information $I(A : B)$ between sender's input, A, and receiver's measurement result, B, is bounded by the Holevo information, with the receiver making measurements providing outcomes m with probabilities q_m, resulting in a post-measurement ensemble $\{p_{i|m}, \rho_{i|m}\}$ and mutual information $I(i : m) = H(\{p_i\}) - \sum_m q_m H(\{p_{i|m}\})$, which the receiver desires to *maximize* over all possible measurement strategies. In this way, the destination can achieve the maximum accessible information $I(A : B) = \max I(i : m)$. In the case of a single two-level system, one sees that the bound is $\log_2 2 = 1$ bit per channel use. If the signal states are *pure states*, then $I(A : B) \le S(\rho)$, with the bound achieved if and only if the encoding states ρ_i commute and Bob measures in the basis where they are represented by diagonal matrices. This implies that $I(A : B) \le S(\rho) \le \log_2 d$, where d is the dimensionality of the Hilbert space of the encoding system, indicating that the amount of information encodable in a system is also bounded by d, which corresponds to the number of orthogonal states available for this purpose.

The general study of computational tasks and their efficiency in a context where quantum channels as well as classical communication channels are available to agents is the subject of quantum communication complexity theory [280, 510]. One can consider communication that remains largely classical but where an unlimited supply of quantum entanglement is available to the parties that have the ability to perform local operations of the sort discussed in Section 1.6. For example, rather than communicating using bits alone, parties communicate making use of entangled states as in the dense-coding protocol. The relation between entanglement and information in the quantum communication context is considered in greater detail below, in Section 4.9.

4.5 Quantum Information Processing and Speedup

Quantum information processing differs from classical information processing, as seen above in example of the Deutsch–Jozsa algorithm, much as quantum communication differs in important ways from its classical counterpart. Fundamentally, this is a consequence of the quantum state superposition principle and the corresponding objective indefiniteness of quantum physical magnitudes.

Due to the superposition principle, the size of the space of signs available to systems composed of two-level systems grows far faster than that available to analogous classical systems: The number of parameters corresponding the accessible quantum computational states grows *exponentially* in the number of two-level systems, whereas in the analogous classical system it grows only linearly in the number of two-state systems. In addition, different computational-basis states are typically simultaneously available and processed at any given time during the operation of a quantum computational algorithm. Thus, quantum computers are, in essence, complex interferometric devices, that is, generalizations of the Mach–Zehnder interferometer of Figure 4.3 involving states lying in larger Hilbert spaces, relying on quantum coherence for their operation. They are designed to exploit the computational parallelism inherent in the typically enormous computational-basis-state superpositions available within the Hilbert space of composite quantum systems.

Quantum computing is of particular interest because this extraordinary degree of parallelism makes tractable some computational tasks that have been viewed as intractable in principle in traditional computing theory. Although the operation of any quantum algorithm in principle *can* be simulated by a classical algorithm, the classical versions of key algorithms such as those for searching and factoring operate less efficiently as the size of the associated input data increases, in the latter case qualitatively so. The increase in computational efficiency that can appear in a quantum computational algorithm is known as *quantum speedup*, and is at the center of debates as to the ultimate nature of quantum computing and its relationship to classical computing.

Rolf Landauer summarized the conceptual history of the theory of quantum computing and its most significant achievement, *exponential* quantum speedup, as follows.

"Paul Benioff first understood that a purely quantum mechanical time evolution can cause interacting bits (or spins) to change with time just as we would want them to do in a computer. David Deutsch later realized that such a computer does not have to be confined to executing a single program, but can be following a quantum mechanically coherent superposition of different computational trajectories. At the end we can gain some kinds of information that depend on all of these parallel trajectories, much as the diffraction [sic] pattern in a two-slit experiment depends on both trajectories. Eventually Shor showed that this form of parallelism provides tremendous gains for the factoring problem, finding the prime factors of a large number." ([291])

Whether the quantum speedup *is* a true computational speedup has been debated. It has been argued that the appearance of a computational speedup in quantum algorithms is, in fact, an illusion, that there is no such parallelism and that there is an improper accounting of the rate of computation in the quantum case. Even some who agree that quantum speedup *is* genuine still question the usual explanation of the origin of the speedup or whether computational parallelism in any strong sense is present in quantum computing.

As seen in previous chapters in relation to many questions in the foundations of quantum mechanics, one finds here that interpretational elements of the theory—whether they originate in the Basic, Quantum logical, Naive or Collapse-Free interpretation—can play a role in where one comes down in relation to the nature and ultimate importance of quantum computing. Advocates of the Many-worlds version of the Collapse-Free approach to quantum mechanics most clearly take definite positive positions regarding both speedup and quantum computational parallelism. The thinking of Deutsch, who is one of the fathers of quantum computing, clearly exhibits this.

"When a quantum factorization engine is factorizing a 250-digit number, the number of interfering universes will be of the order of 10^{500}. This staggeringly large number is the reason why Shor's algorithm makes factorization tractable. I said that the algorithm requires only a few thousand arithmetic operations. I meant, of course, a few thousand operations *in each universe* that contributes to the answer. All those computations are performed in parallel, in different universes, and share their results..." ([130], p. 216)

In the standard approach to quantum computing, which unlike Deutsch's accords with the Basic interpretation, what classically would be multiple streams of data are seen represented together in a single quantum data set acted on by a single quantum circuit in the unique Universe so long as a measurement is not performed. On this interpretation, because of the size of the Hilbert space of a collection of two-level systems, a single quantum circuit is thought of as operating on an exponentially large data set and, therefore, operates with superior efficiency, without involving additional universes.

From another alternative perspective, Bub has argued that the speedup is genuine but that it arises from the nature of *quantum logical connectives* rather than computational parallelism, through what is a combination of quantum logical and Copenhagenist perspectives [90]. Finally, physicist Andrew Steane has opined in the philosophical literature that the speedup is only apparent but has not denied the existence or pertinence of quantum parallelism [436].

Steane has argued that the basis on which the accounting for the needed computational resources is done is improper and that Deutsch's claim that the multiple-universe view is necessary to understand quantum computing is simply incorrect. In particular, Steane claims that "in terms of the amount of information manipulated in a given time," quantum and classical computation are *equally* efficient.

> "Quantum superposition does not permit quantum computers to perform many computations simultaneously except in a highly qualified and to some extent misleading sense. Quantum computation is therefore not well described by interpretations of quantum mechanics which invoke the concept of vast numbers of parallel universes. Rather, entanglement makes available types of computation processes which, while not exponentially larger than classical ones, are unavailable to classical systems. The essence of quantum computation is that it uses entanglement to generate and manipulate a physical representation of the correlations between logical entities, without the need to completely represent the logical entities themselves." ([436], p. 469)

Although Steane is correct in noting that quantum entanglement can play an important role in quantum computation, which is certainly so in the case of Shor's factoring algorithm, it currently remains quite unclear whether entanglement must necessarily be present for every one of the unique properties of quantum computers to be exhibited.

Before considering these various positions in greater detail, let us note that there is a clear and important distinction between quantum gating operations and the operations of *quantum logic* in the sense considered in the study of the foundations of quantum mechanics following from the work of Birkhoff and von Neumann. The transformations involved in quantum information processing, the quantum logic gates, act on the states of the computational basis in a way that realizes, in this basis, the truth tables of the corresponding *Boolean* logical operations. Thus, in particular, they are *not* identical to the operations of quantum logic considered in Chapter 2 to which reference is made above.

Quantum gates acting on single two-level systems are transformations on the vector spaces of such systems appropriately mapping the computational basis $\{|0\rangle, |1\rangle\}$ to itself, as in the case of the quantum 'NOT' gate, which takes the computational-basis vectors $|0\rangle$ to $|1\rangle$ and $|1\rangle$ to $|0\rangle$, similarly in information processing effect as the classical NOT gate takes the state-value 0 to 1 and the state-value 1 to 0. Recall that this NOT gate is simply the Pauli operation σ_1; in the Poincaré–Bloch sphere representation of Figure 4.2, it reflects a state about the plane of the equator. It is often called the quantum bit-flip transformation. Indeed, if the states under consideration *were* restricted to those of the computational basis only, which amounts to considering classical information, this would precisely be a classical bit-flip operation. The Pauli matrices corresponding to state inversions along the remaining two Poincaré–Bloch sphere, σ_3 and σ_2, are also quantum logic gates, namely, phase and bit+phase flips, respectively. The Hadamard gate, H, is more obviously quantum in character, in that it takes an eigenstate of a two-level system to a balanced *superposition* of eigenstates in the initial basis. It effectively interchanges the computational and the diagonal bases: $|0\rangle \leftrightarrow |\nearrow\rangle$ and $|1\rangle \leftrightarrow |\searrow\rangle$. In matrix form, it can be written $H = \frac{1}{\sqrt{2}}(\sigma_1 + \sigma_3)$. This gate provides optimal interference between computational basis eigenstates, and has no classical analog.

In order to achieve speedup over classical computational algorithms, a number of two-level-system states are transformed in parallel, such as can be provided by tensor products of Hadamard gate operations, $H^{\otimes n}$, acting on a product of individual two-level-system states $|i\rangle|j\rangle \cdots |n\rangle$. Consider again the Mach–Zehnder interferometer of Figure 4.3. If one arranges that there be no phaseshift induced by the phaseshifter, the second beam-splitter has an effect identical to the first. The effect is that a particle exits the interferometer with the opposite eigenvalue value from the input value; there will be destructive interference in one final exit path and constructive interference in the other, the 'bright' and 'dark' ports depending on the beam-splitter port initially entered. Hence, the two beam-splitters together acting as a NOT gate operation (up to a phase factor) in the quantum computational basis, as the particle will exit in the opposite path from which it entered. The beam-splitters in this apparatus are each said to realize a "$\sqrt{\text{NOT}}$" gate. When instead phaseshifters are also placed in the paths $|i\rangle$ ($i = 0, 1$) before and after each beam-splitter and set to introduce a phaseshift of $-\pi/2$, the resulting two beam-splitter complexes perform a Hadamard transformation on the input state.

As Steane has pointed out and as shown below, quantum information processing often involves operations giving rise to *entangled* states. Significant quantum speedup in quantum computing appears to require the maintenance of highly coherent entangled superpositions of computational basis states in order to be *exponentially* more efficient, although there is still some question as to whether it is essential to quantum computation. Recall that the Hilbert space of a compound quantum system is the tensor product of the Hilbert spaces of the subsystems. The pure-state space for a system of N two-level systems is the Hilbert space $\mathcal{H}^{(N)} = \mathbb{C}^2 \otimes \mathbb{C}^2 \otimes \cdots \otimes \mathbb{C}^2$. N classical two-state systems give rise to 2^N possible classical computational states parameterized by N-bit strings $\mathbf{x}_i \in GF(2)^N$. In the quantum case, there are 2^N complex components of a vector $|\Psi\rangle \in \mathcal{H}^{(N)}$ written as a superposition state in the computational basis, which are then reduced by one by fixing the value of its unphysical global phase and normalizing it. The pure state of such a quantum system is thus parameterized by $2^N - 1$ complex numbers. The computational basis $\{|\mathbf{x}_i\rangle\}$ for the Hilbert space of many two-level systems does, nonetheless, consist of vectors which are labeled by the 2^N possible N-bit strings $\mathbf{x}_i$, which can be viewed as corresponding eigenvalues. This correspondence allows for "classical readout" of the result of quantum computation by measurement corresponding to the computational basis, although not a classical treatment of the quantum computation process itself. Thus, in this sense, the quantum case allows for parallel processing of these classical states in one processor, something that is not possible classically without multiplying the number of processors by a correspondingly large factor.

A generic state of a system composed of N two-level systems, written in the computational basis, is

$$|\Psi\rangle = \sum_{i=0}^{2^N-1} a_i |\mathbf{x}_i\rangle , \qquad (4.14)$$

where the sum is taken over all 2^N N-element strings, $a_i \in \mathbb{C}$, and the global phase angle is set to zero. A generic quantum computation is described by the evolution of this state, typically from the initial fiducial state $|\mathbf{x}_0\rangle \equiv |00\ldots0\rangle$ and implemented by a series of unitary operations and oracle evaluations, which are specific conditional gates involving an oracle function, followed by a measurement readout projecting the unitarily transformed state onto the computational basis, according to a fixed algorithm.[12]

An example of a fundamental "two-qubit" conditional logic operation is the controlled-NOT, or c-NOT, which changes or leaves unchanged the computational basis state of a second subsystem conditionally on the state of that of the first, changing it only when the first subsystem state is $|1\rangle$, acting as such component-wise on superposition states so that, for example, $(a_0|0\rangle + a_1|1\rangle)|0\rangle \rightarrow a_0|00\rangle + a_1|11\rangle$. The result of this controlled operation on the input is to take a separable state to an entangled one. A quantum oracle call is similarly a controlled operation, but one that involves an evaluation of an appropriate oracle function $f(\mathbf{x})$.

Quantum computations are ultimately irreversible and probabilistic, although the unitary logical portions of the evolution are deterministic and reversible. The operation of a standard quantum algorithm is typically viewed as involving the evaluation of one function at a time, mathematically for many

---

[12] There also exist non-standard "one-way" quantum computational methods that operate somewhat differently, in that they drop the unitary steps [370].

values at once using a specific set of quantum gates. This is typically understood to involve a form of parallelism distinct from classical computational
parallelism, at least as viewed from our single, shared universe. Nonetheless,
Bub has argued that there is no genuine parallelism involved in efficient quantum algorithms, despite his assertion that the speedup *is* genuine.

> "[E]ven before the application of the final transformation, the sub-
> space representing the global property of the function already con-
> tains the state, i.e., the quantum proposition representing a partic-
> ular global property, as opposed to alternative possible global prop-
> erties, is already selected as true by the state. It is then simply a
> question of determining, by a suitable measurement, which of the
> alternative propositions is true, i.e., which of the alternative propo-
> sitions is represented by the state. This is quite different from the
> claim that the quantum state already contains the information cor-
> relating all values in the domain of the function with corresponding
> values in the range. There is, in principle, no way of extracting this
> information by any measurement, while the information about the
> global property encoded in a subspace can be extracted by a suitable
> measurement. It is this difference that makes possible the represen-
> tation of a global or disjunctive property of a function as a subspace
> containing the quantum state of a computer register in a quantum
> computation, while a classical computation represents such a prop-
> erty as a subset containing the classical state. The possibility of an
> exponential speedup arises because the classical state can end up in a
> particular subset only by ending up at a particular point in the sub-
> set, representing a particular pair of input-output values correlated
> by the function." ([90])

The crucial question in evaluating this characterization of the difference between quantum and classical computing, in relation to the question of parallelism, is precisely what is required of an *evaluation* of a function. Bub, in
effect, requires an evaluation to produce a *definite actual value* for a function
among a set of possible outcomes to be genuine. He considers computational
parallelism to occur *only when* more than such evaluations takes place in
parallel. For him, quantum parallelism would require the production of the
several different corresponding actual values during calculation rather than
only non-zero potentialities. This requirement is unnecessarily strict.

There are at least two problems with this requirement on the quantum
evaluation of functions, which become evident when one considers how quantum algorithms are designed. First, the difference between *possible* almost-
certain and *actual* outcomes is ignored; even the expected outcome will not
be actualized with absolute certainty, that is, its appearance is only a potentiality until measurement. Second, that algorithms are typically functionally
required to produce *intermediate* evaluations—that they usually involve sub-
routines or oracle evaluations—is overlooked. Although the distinction of the

first point may not matter under the quantum-logical interpretation of the theory, the second renders Bub's conception of evaluation improper for crucial algorithms. In order to be evaluations, intermediate steps would be required to include a *sequence of measurements* on key quantum registers during computing, which would render them inoperative; because the evolution of data requires pertinent superpositions to be sustained until the computation *ends* in a final read-out measurement, this would cause key algorithms to fail—for example, the Deutsch–Jozsa and Grover search algorithms.[13]

In the course of some algorithms, such as the Shor factoring algorithm, measurements of some subsystems (registers) are actually made in the course of its operation but these involve neither the production of function values nor collapsing superpositions in which various values are correlated in parallel.[14] It may therefore appear possible to finesse the above problem by denying that quantum algorithms do, in fact, involve intermediate *evaluations* despite the functionality imposed by their designers. Indeed, Bub believes this to be so.

> "Rather than 'computing all values of a function at once,' a quantum algorithm achieves an exponential speed-up over a classical algorithm precisely by avoiding the computation of any values of the function at all." ([90])

However, this is so only on a very particular and problematic view of the quantum state. Recall that there are two clear positions regarding the origin of the quantum speedup: (i) Bub's view that it comes from quantum logic (even though algorithms are designed to implement standard Boolean logic), and (ii) the standard view that it comes from quantum parallelism in the form of the computational-basis superpositions that are entangled. Having rejected (ii), Bub points to the nature of quantum-logical disjunction ∨, which can be viewed as involved in quantum computation on quantum-logical interpretations of the state vector. He argues that there need be no intermediate function evaluations because a proposition "can be true (or false) even if none of the disjuncts are either true or false." It is on such a view that Bub claims

> "The point of the procedure is precisely to avoid the evaluation of the function in the determination of the global property, in the sense of producing a value in the range of the function for a value in its domain, and it is this feature—impossible in the Boolean logic of classical computation—that leads to the speed-up relative to classical algorithms." ([90])

This claim stands or falls with the quantum-logical interpretation of the quantum state, which was seen in Section 2.1 to fail to provide adequate explanations: As Stachel has pointed out regarding this interpretation on Putnam's

---

[13] The Deutsch-Jozsa algorithm is described in Section 4.2.

[14] For a clear summary of the various stages of this algorithm, see [46].

comprehensive explication, which takes the quantum logical approach to its logical conclusion,

> "To summarize the nature of the quantum-logical explanation being offered: the logic is read off the Hilbert space. The physical laws have to be compatible with this logic. Measurement has to be compatible with these physical laws. Our knowledge has to be compatible with the measurements.
>
> If you fail to see in what sense a deeper explanation of quantum theory is achieved than follows from simply accepting the rules of quantum mechanics, I can join you in puzzlement." ([427], p. 305)

Bub claims, in effect, that quantum computer scientists design algorithms that *do not* evaluate functions during their operation, despite their design methodology but only provide correct answers in the end. It is hard to see why one should prefer Bub's view of the quantum state over the more intuitive one of the designers of the algorithms themselves, given its shortcomings vis-à-vis explanation. His explication of the operation of quantum algorithms is a forced one, given that the gates implemented during the algorithms are essentially Boolean in design and chosen precisely to evaluate in that way functions that are encoded in computational-basis states. In the course of its operation, a quantum algorithm typically 'consults,' that is, evaluates a quantum oracle function in the course of its operation, with a different effect on various components of the superposition of computational basis states conditioned on the evaluation. Bub simply denies that these 'consultations' are evaluations of functions at all. This approach, like the quantum logical interpretation on which it depends, fails to explain; in this case, it is the quantum speedup that is left unexplained. It merely remains internally consistent.

The standard view of quantum computation, according to which quantum interference of computational basis states is responsible for the speedup and evaluations of functions typically do occur coherently, is both natural *and* explanatory but does not require the ontological excess of the multi-verse view of Deutsch. Almost certain actual evaluations of functions, which accord with the intentions of algorithm designers, are understood to take place more efficiently than classically due to the interference between the computational basis states appearing in the intermediate steps, which *are* genuine evaluations carried out by parallel processing under a Boolean logical scheme. Given that interference is a direct consequence of the superposition principle, which is the most fundamental postulate of quantum theory and is shared by all interpretations in one way or another, there seems little reason to reject the standard explanation of quantum speedup.

Steane claims in his article "Quantum computation needs only one universe," that not only is the Many-worlds interpretation not required for the explanation of the efficacy of quantum computing, but that "quantum computation is... not well described by interpretations of quantum mechanics which invoke the concept of vast numbers of parallel universes." Although this is only so to the extent that the multi-verse theory itself is problematic, as shown in the previous chapter, he is correct in noting that the existence of quantum speedup is sometimes incorrectly "used as evidence that quantum physics is best understood in terms of vast numbers of parallel universes" [436]. In particular, Deutsch has claimed that the prime exemplar of quantum computing, the Shor algorithm, is *inexplicable* without the ontological commitment he sees as natural to the Collapse-Free interpretation.

> "With Shor's algorithm, the argument has been writ very large. To those who still cling to a single-universe world-view, I issue this challenge: *explain how Shor's algorithm works*. I do not merely mean predict that it will work... I mean provide an explanation. When [it] has factorized a number, using 10^{500} or so times the computational resources that can be seen to be present, where was the number factorize? There are only 10^{80} atoms in the entire visible universe, an utterly minuscule number compared to 10^{500}. So if the visible universe were the extent of physical reality, physical reality would not even remotely contain the resources required to factorize such a large number." ([130], p. 217)

Deutsch's challenge is readily met by the standard conception summarized further above. The threat of the challenge is incorrectly represented by Deutsch, because it is based on the conflation of computational resources with the number of actualized quantum states, a point considered further below. However, Steane himself makes a not dissimilar error.

Steane's position on the question of quantum speedup is that "Quantum computers cannot manipulate classical information more efficiently than classical ones..." as a consequence of Holevo's theorem. This would be true if quantum computation were simply a matter of *communicating* information. However, communication is not what constitutes computing; computing efficiency is not bounded by *communication* efficiency. In particular, quantum speedup is achieved by reducing the number of *computational steps*, such as the number of oracle function evaluations, required to produce a result rather than, for example, achieving greater mutual information between input and outputs as is the case in more efficient communication. On insufficient grounds, Steane views entanglement *alone* as responsible for the differences between quantum and classical computation, stating that "entanglement makes available types of computation processes which, while not exponentially larger than classical ones, are unavailable to classical systems. The essence of quantum computation is that it uses entanglement to generate and manipulate a physical representation of the correlations between logical entities, without the need to completely represent the logical entities themselves." ([436])

Similarly to Bub, Steane views quantum computations as merely involving "elementary processing operations which achieve some given degree of transformation of a body of information, such as evolving it from one state to an orthogonal state" [436]. Unlike Deutsch, Steane correctly notes that the number of qubits rather than the number of quantum symbols encoded in actualized states of two-level systems is pertinent to information processing. Nonetheless, he fails to recognize that accounting for computational resources involves more than simply the number of qubits involved. Rather, computation rates are primarily quantified by the number of necessary computational *steps* including oracle calls, a smaller number of which there clearly is under the mentioned quantum algorithms relative to classical alternatives. Computational speedup is the reduction of the number of required such steps. It is on *that* basis that speedup is seen to occur in quantum computing.

The number of required computational steps, such as oracle calls, for a successful computation under a given algorithm is found by counting the necessary transformations implemented by specific logic gates acting on the quantum registers. The usual explanation of quantum speedup is, in fact, simply that quantum mechanics allows a reduction of the number of computational steps, most importantly oracle calls, required for their operation. The number of oracle calls implemented in the course of the operation of an algorithm in no way depends on the number of worlds assumed to exist by an interpretation, contrary to the claim of Deutsch that the power of quantum computing is inexplicable without reference to many parallel universes. Thus, the standard explanation is in no way undermined by the arguments just considered.

4.6 Protocols and the Nature of Quantum Information

Quantum states enable agents to perform information processing tasks surpassing what they could achieve using only classical computing and communication systems. The superior capability of quantum communication involving *entangled* systems is most clearly evidenced by the viability of the quantum dense coding and quantum teleportation protocols.

Let us first consider the quantum dense-coding communication protocol. Holevo's theorem dictates that without previously shared entangled signs providing quantum correlated symbols, the transmission of a single two-level system from one agent to another can support at most the communication of *one* bit of information. Because communicating *two* units of information requires twice the communications resources needed to communicate a single unit, it would appear that a pair of two-level systems is required to communicate two bits. Indeed, were *only* individual two-level systems used for communication, this would be the case. However, previously shared bipartite quantum systems in Bell states are such that local operations on just one of the pertinent pair of two-level systems are sufficient for one of the four Bell states to be transformed to any other, allowing their correlations to be used as a communication resource. Quantum dense coding, illustrated in Figure 4.4, makes use of this property of Bell states of a compound system shared by two agents, typically referred to as Alice and Bob, in labs A and B, respectively, to communicate two bits of information by transferring only *one* two-level system [42].

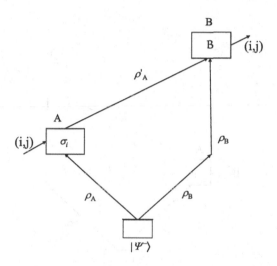

Fig. 4.4. Schematic of the communication of two bits using quantum dense coding. This protocol allows the communication of two bits of information transferring only a single two-level system from a pair in a previously shared Bell state.

In particular, shared entanglement enables Alice and Bob, in effect, to increase the capacity of a shared quantum channel via the quantum dense-coding protocol consisting of the following three steps.

(1) Alice and Bob share pair of two-level systems, one in each laboratory, in a previously agreed Bell basis state, say, the singlet state $|\Psi^-\rangle$.
(2) Alice performs either the identity ($\mathbb{I}$), basis-state flip (σ_1), phase flip (σ_3) or basis-state+phase flip transformation (σ_2) on her two-level

system, so that the full system is in a different one of the four Bell states, and sends it to Bob.

(3) Bob performs a Bell-state measurement on the pair of two-level systems then in his laboratory, obtaining two bits of information, corresponding to the identity of the new Bell state.

Step (1) can be realized using only a well-characterized source, which need not be in the possession of either agent. Typically, a photon subspace, such as the polarization subspace, is used as the system for carrying out the protocol. A two-level system capable of encoding only one qubit is communicated from Alice to Bob in step (2). The "Bell state measurement" of step (3) is a joint measurement, in this case localized within a single laboratory, that perfectly distinguishes any given Bell state from all other Bell states. Two bits are communicated from Alice to Bob.

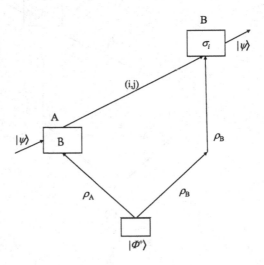

Fig. 4.5. Schematic of quantum state "teleportation," the transposition of an unknown quantum state $|\psi\rangle$ from one laboratory A to another B, using one e-bit of communication resources, represented by the encoding state $|\Phi^+\rangle$, and two classical bits, (i,j). A Bell-state measurement is indicated in laboratory A at left.

Like quantum dense coding, quantum state teleportation also cannot be achieved using only classical communication but can be accomplished using quantum states. Quantum (state) teleportation, illustrated in Figure 4.5, is the placing of a remotely located quantum system into the unknown quantum state identical to that of a locally possessed quantum system. This process is often described as 'the transmission of an 'unknown qubit state,' $|\psi\rangle$, from one agent to another. However, it is important to realize that this is *neither* the communication of a qubit *nor* the transfer of a two-level system from one of the mentioned laboratories to the other. Rather, an *unknown state* of a two-level system is transposed in the process, a state which, furthermore, has in no way been used by the 'sender' for encoding information to the 'receiver.' Indeed, there is no encoding or decoding, and hence no transmission of information in the course of carrying out the protocol, other than that required to carry out the protocol itself, although the state could, in principle, play a role in communication in some larger unmentioned context involving two *other* agents. The protocol can be used as part of a quantum computation.

As in the quantum dense coding protocol, the two agents involved first share a pair of two-level systems in a Bell state [34, 75, 76]. Let the unknown state in question be written $|\psi\rangle = a_0|0\rangle + a_1|1\rangle$, as in Equation 4.1, and, for specificity, the initially shared Bell state be $|\Phi^+\rangle = \frac{1}{\sqrt{2}}(|00\rangle + |11\rangle)$. Without loss of generality, let the first of the subsystems in the shared pair be in the possession of Alice and the second be in the possession of Bob. According to the protocol, Alice first measures, in the Bell basis, the state of the different pairing of subsystems, that constituted by the two systems in her laboratory, namely, that the state of which is to be transposed and the one two-level system of the pair which she and Bob jointly share that initially lies in her laboratory. Before this measurement, the three two-level systems involved are in the state

$$|\Psi\rangle = |\psi\rangle|\Phi^+\rangle = \frac{1}{\sqrt{2}}(a_0|0\rangle + a_1|1\rangle)(|00\rangle + |11\rangle) . \qquad (4.15)$$

This overall state can be conveniently rewritten as a sum of four terms wherein each term has the first two subsystems in a Bell state differing from the others:

$$|\Psi\rangle = \frac{1}{2}\big[(|00\rangle + |11\rangle) (a_0|0\rangle + a_1|1\rangle) + (|00\rangle - |11\rangle)(a_0|0\rangle - a_1|1\rangle)$$
$$+ (|10\rangle + |01\rangle) (a_1|0\rangle + a_0|1\rangle) + (|10\rangle - |01\rangle)(a_1|0\rangle - a_0|1\rangle)\big] .$$

Now, when Alice performs a Bell-state measurement on her subsystems, she obtains an outcome corresponding to the particular one of the four Bell states in which her two subsystems end up, as manifest in the second of the above forms of the state. The four *a priori* equiprobable possibilities of outcome amount to two bits of (classical) information. Alice then communicates the result of her joint measurement, that is, those two bits to Bob. Upon receiving the two bits, Bob finally performs one of four corresponding appropriate operations on the second subsystem of the composite system the two agents are jointly sharing, placing his two-level system in the desired, unknown state $|\psi\rangle = a_0|0\rangle + a_1|1\rangle$.

The character of protocols such as quantum dense coding and quantum teleportation, considered in the context of Bell inequality tests, have led Penrose to introduce a new term, "quanglement."

"There is a difficulty about [the term *quantum information*], namely the appearance of the word 'information.' In my view, the prefix 'quantum' does not do enough to soften the association with ordinary information, so I am proposing that we adopt a new term for it: QUANGLEMENT... The term suggests 'quantum mechanics' and it suggests 'entanglement.' " ([346], p. 603)

Indeed, Penrose reconsiders entanglement and identifies it as the essence of quantum information, emphasizing its role as a physical resource useful in communication. In particular, he specifies that there is "no way to send an ordinary signal by quanglement alone" ([346]), which is something true of entangled states but not of states of individual two-level systems.

What Penrose appears to doing in his recent explorations of quantum information is adopting the 'reverse-wave interpretation' of quantum subsystem behavior [276]. In particular, in regard to the specific protocols above, he suggests tracing trajectories in space-time corresponding to the propagation of entangled-system pairs conceived of as signal systems (*cf.* Figures 4.4 and 4.5). Moreover, he discusses the paradigmatic example of this conception of space-time propagation of quantum systems: a pair of two-level systems in a Bell state emitted from a non-linear crystal by spontaneous parametric down-conversion. In this example, he draws a single continuous spacetime trajectory from one emitted photon backward in time to the crystal, which is said to be reflected as from a mirror and then forward in time to the other photon. Leveraging this picture, Penrose argues that "... past-directed channels of quanglement can be used just as well as future-directed channels. If quanglement were transmittable information, then it would be possible to send messages into the past, which it clearly isn't." Finally, he adds that "quanglement links have the novel feature that they can zig-zag backwards and forwards in time, so as to achieve an effective 'spacelike propagation' " [346].

It is not entirely clear from the writings of Penrose that there is much new to his 'quanglement' beyond the above sort of heuristic crutch for understanding the behavior of information in the manner of a quasi-physical substance. However, in that case, as pointed out by Timpson in regard to signaling, that any reference to pseudo-substances must be understood as such.

" 'How does the information get from A to B?'... is a perfectly legitimate question if it is understood as a question about what the causal processes involved in the transmission of... information are, but note that it would be a mistake to take it as a question concerning how information, construed as a particular, or as some pseudo-substance, travels... Thus when considering an information transmission process, one that involves entanglement or otherwise, we should not feel it incumbent upon ourselves to provide a story about how some thing, denoted by 'the information', travels from A to B; nor, *a fortiori*, worry about whether this supposed thing took a spatio-temporally continuous path or not." ([451], p. 331)

Timpson's point applies equally well to Penrose's 'quanglement,' that is, entanglement. Moreover, Penrose remarks that

> "not being capable of carrying information, quanglement does not respect the normal restrictions of relativistic causality," ([346], p. 607)

another characteristic of entanglement. Fundamentally, what Penrose is doing is characterizing a sort of causal influence capable of acting backward in time as well as forward. This idea can already be found in the course of the ongoing attempt to ground the concept of entanglement; such influences have previously been carefully considered, for example, by Maudlin ([313], Ch. 5) in regard to entanglement and violations of Bell theorem. This idea is taken up again in Section 4.9, below.

Another element of Penrose's thinking, which seems also to have been shared at least at one time in the past by Richard Jozsa, is the view that a single two-level system state contains an *infinite* amount of information; this idea was explored, for example, in Jozsa's article, "Quantum information and its properties" (in [301]). In Penrose's view, quantum teleportation sends an infinite amount of information because it faithfully reproduces in Bob's laboratory a specific pure state of a two-level system previously present in Alice's laboratory. The basis for Penrose's view in this regard is simple. The communication of the description of a particular pure state of a two-level system would be that of a precise location in the Poincaré–Bloch sphere. The specification of this location with absolute precision would require an infinite amount of information. However, as we have already seen, what takes place in quantum teleportation is *not* the communication of such information at all; there is merely a complex and indirect transposition of what *could* serve as a quantum signal state—in that it *could* serve to encode a symbol in an appropriate, entirely absent context—in a quantum channel with a theoretical information capacity of one bit per use.

It is helpful here to consider also that the standard textbook of quantum computing and quantum information theory, written by Michael Nielsen and Isaac Chuang, comes perilously close to endorsing this view.

> "How much information is in a qubit? Paradoxically, there are an infinite number of points on the [Poincaré–Bloch] unit sphere, so that in principle one could store an entire text of Shakespeare in the infinite binary expansion of θ [the complement of the altitudinal angle in the sphere]. However, this conclusion turns out to be misleading, because of the behavior of a qubit when observed. " ([324], p. 15)

(Note: here the term *qubit*, as is standard in that literature, is used to refer to the two-level system rather than the information unit.) Fortunately, these authors then proceed to discuss the misleading nature of this idea.

"Recall that measurement of a qubit will given *only* 0 or 1. Furthermore, measurement *changes* the state of a qubit, collapsing it from its superposition of $|0\rangle$ and $|1\rangle$ to the specific state consistent with the measurement result... from a single measurement one obtains only a single bit of information about the state of the qubit, thus resolving the apparent paradox. It turns out that only if infinitely many identically prepared qubits were measured would one be able to determine α and β for a qubit in the [corresponding superposition state]. But an even more interesting question to ask might be: how much information is represented by a qubit *if we do not measure it*? This is a trick question, because how can one quantify information if it cannot be measured? Nevertheless, there is something conceptually important here, because when Nature evolves as a closed system of qubits, not performing 'measurements,' she apparently does keep track of all the continuous variables describing the state, like α and β. In a sense, in the state of a qubit, Nature conceals a great deal of 'hidden information.' " ([324], pp. 15-16)

Nielsen and Chuang thus do correctly note that measuring the state of one two-level system may yield *up to* one bit of information but does not *necessarily* do so. For example, the destination may be aware that the transmitter is capable of sending only the states $|\uparrow\rangle$ and $|\nearrow\rangle$, which are not perfectly discriminable, so that less than one bit can be communicated with each state, even when measurement is perfectly performed, as discussed previously in this chapter. The point is that one cannot assess the information contained in a signal state in isolation from its encoding, which may involve considerable context, as Weaver noted long ago.

The above "paradox" arises when the information *capacity* of a single-qubit channel is conflated with the state-specification of the two-level system that serves as its signal. This is related to the primary error committed in the article first formally introducing the qubit, in particular, in the paragraph considered in the introduction of this chapter. Nielsen and Chuang do correctly defuse the suggested paradox by noting that any functioning signal must be decoded upon arrival at its destination. However, the metaphorical language of the final comment of this passage regarding "hidden information" is evidence of the incoherency of the notion of *essentially inaccessible* information with which they continue to flirt. An analogous paradox can be presented in the classical case as well, for example, by considering a large number of other uncontrolled degrees of freedom of a classical system, such as those of the particles within a piece of paper imprinted with either a one or a zero, and how *those* might behave during classical signal transmission. Indeed, the "hidden information" being referred to in the above text would be the information that *could* be carried by hidden variables model of a two-level system *should* hidden variables be used for encoding by supernatural agents, which is an incoherent notion under naturalism.

The above portrayal of state teleporation by Penrose is misleading precisely because *no* information is communicated from Alice to Bob in the process of entanglement distribution, a fact which, rather surprisingly, he acknowledges. In the case of the quantum teleportation protocol that he considers as an example, no information beyond the two bits of classical information that are 'consumed' in the process of carrying out the protocol are communicated, so that $|\psi\rangle$ is left unchanged in the end. If "quanglement" is nothing over above entanglement, this term is merely a new contraction of the phrase "quantum entanglement." Despite his stated intention, Penrose insufficiently distinguishes quantum information from classical information, which can be understood as due to a reliance on a picture of both as pseudo-substances. For his part, Jozsa takes care to mention that such 'information'—*i.e.* the description of a precise preparation of $|\psi\rangle$—is fundamentally *inaccessible* but, nonetheless, also misses this point. If this 'information' is inaccessible, *there can be no communication of that information*. Indeed, there is only the net transposition of an *unknown* pure two-level-system state from one place to another. Moreover, classical information conforming with the requirements of causality is communicated in the course of carrying out the protocol.

Fuchs has criticized Penrose's understanding of the behavior of the quantum state in the quantum teleportation process in order to argue for the essential subjectivity of the quantum state function, as follows.

> "Roger Penrose argues in his book *The Emperors New Mind* that when a system 'has' a state $|\psi\rangle$ there ought to be some property in the system (in and of itself) that corresponds to its '$|\psi\rangle$-ness.' For how else could the system be prepared to reveal a **YES** in the case that Alice actually checks it? But there is a crucial oversight... If Alice fails to reveal her information to anyone else in the world, there is no one else who can predict the qubit's ultimate revelation with certainty. More importantly, there is nothing in quantum mechanics that gives the qubit the power to stand up and say **YES** all by itself: If Alice does not take the time to walk over to it and interact with it, there is no revelation. There is only the confidence in Alice's mind that, should she interact with it, she could predict the consequence of that interaction." ([181])

The observer's ability to make predictions is independent of the question of the *reality* of the quantum state. It is an epistemic question. Even in a classical situation, in the absence of information corresponding to the encoding, that is, "Alice's information," no information *can* be transmitted as a matter of communications theory. It is not a matter of the reality of the quantum state or of quantum mechanics more generally, but only one of the basic principles of communication theory.

The most fundamental points in relation to all of the above discussions are (i) that information can be communicated only if the receiver knows the encoding method involved, and (ii) that signal states must at least be accessible in principle to encode information. A third noteworthy point was made by Bohr, who pointed out in 1929 that, given a quantum system that has been prepared,

> "...a subsequent measurement to a certain degree deprives the information given by a previous measurement of its significance for predicting the future course of the phenomena...these facts...set a limit to the *extent* of the information obtainable..." ([57], p. 18; [354], p. 15)

4.7 Informational Interpretations of Quantum Mechanics

As seen above, due to the superposition principle, quantum information has features not possessed by its classical counterpart, most importantly those associated with entanglement; quantum signal state correlations arising from state superposition in the tensor-product Hilbert space can behave in ways that classical signal correlations cannot. It was evident early on, for example, to EPR and Schrödinger, even before the explorations of Bohm, Bell, and others motivated by their work, that special insight into the physical world can be achieved by considering special situations involving entangled systems. These situations have served as grounds for rejecting some interpretations of the theory, although they have not proven definitive in determining the *best* interpretation. A similar attitude now prevails toward situations considered by quantum information science, because symbol entanglement is an information processing resource with correlations between physical states serving to enhance signaling.

The consideration of questions of an information-theoretic nature provides a new perspective on the physical correlations that can be used to encode information. Indeed, valuable insights into the behavior of quantum systems have been made in recent years by exploiting the tools of information theory. The fact that quantum states enable communication and information processing tasks to be accomplished that cannot be carried out, at least as efficiently, using corresponding classical states lends further support to the view that information and the foundations of quantum mechanics are intimately related. As a result, some have argued that quantum information theory is not only useful but *the essential key* to understanding quantum mechanics. This claim is plausible in some versions and implausible in others. It appears quite plausible, in light of recent results discussed in the final section of this chapter, that in order for important information-theoretic concepts to sustain their fundamental character in the quantum context, the strength of correlations between signs must be physically constrained in specific ways that the consideration of information theoretic situations may better reveal. However, when the above claim is based on the idea that information is *physical* or *more fundamental* than the long-central physical correspondent of the philosophical concept of substance, namely, matter, it is highly suspect.

An influential version of the latter idea involves the assumption that an ontological reduction of physical objects to information obtains. This reductionist claim is associated with what has been called the *Information interpretation of quantum mechanics* [442]. "In the information interpretation of quantum mechanics, information is the most fundamental, basic entity." This slowly developing approach is one of increasing interest, and involves the following picture.

> "Every quantized system is associated with a definite discrete amount of information. This information content remains constant at all times and is permutated one-to-one throughout the system evolution. What is interpreted as measurement is a particular type of information transfer over a fictitious interface. The concept of a many-to-one state reduction is not a fundamental one but results from the practical impossibility to reconstruct the original state after the measurement." ([442])

The last two statements characterize state reduction 'for all practical purposes,' which is typically associated with single-universe variants of the Collapse-Free interpretation including state decoherence, and are also compatible with subjectivism. However, this Informational interpretation differs fundamentally from the Radical Bayesian interpretation, which suggests that the fundamental referent of quantum information theory is knowledge, in that the Informational interpretation considers quantum mechanics to describe *objectified* information, with a metaphysics more closely resembling logical atomism, and does not deny quantum mechanics the status of a physical theory.

One motivation for the idea of the conservation of information involved in this interpretation is that the no-cloning theorem discussed above and the following no-deleting theorem, both of which are based on the superposition principle, seem to recommended it. Given the superposition principle, these theorems follow immediately from the Schrödinger state evolution, which is unitary. The *quantum no-deleting theorem* is the statement that it is impossible to delete one copy of a pure qubit state [334]. Perfect reversible deletion of non-orthogonal quantum states was shown to be impossible by the following argument. Consider two copies of an unknown two-level system pure state $|\psi\rangle$. No linear transformation exists from $\mathcal{H}_A \otimes \mathcal{H}_B \otimes \mathcal{H}_C$ to itself such that $|\psi\rangle|\psi\rangle|C\rangle \mapsto |\psi\rangle|B\rangle|C'\rangle$, where $|\psi\rangle$ is the state to be deleted, $|B\rangle$ is a blank state, and $|C'\rangle$ is a state that is independent of $|\psi\rangle$; the only linear transformation capable of performing a transformation of the above form is one that violates this final requirement by *swapping* the unknown state $|\psi\rangle$ with the ancilla state $|C\rangle$; that is, that has $|C'\rangle = |\psi\rangle$. Conservation of information is also a natural corollary of the single-universe variants of Collapse-Free interpretations.

The primary idea of reducing physics to information was far earlier and more influentially advocated by Wheeler in his *it from bit* thesis:

"Every **it**, every particle, every field of force, even the spacetime continuum itself, derives its way of action and its very existence entirely,
even if in some contexts indirectly, from the detector-elicited answers
to yes or no questions, binary choices, **bits**. Otherwise stated, all
things physical, all **its**... must in the end submit to an information-
theoretic description." ([494])

This famous claim can itself be seen as originating in the view of Bohr that "no
elementary phenomenon is a phenomenon until it is a registered (observed)
phenomenon," that is, that "all things physical are information-theoretic in
origin and this is a *participatory universe*" [494]. This idea of Bohr was appreciated and forcefully driven to the extreme by Wheeler. Similarly to the
presumption of Zeno and his predecessors that all things have spatial extension, those claiming physics is reducible to information must assume all things
must have what one might call *informational magnitude*. Indeed, for Wheeler
quanta exist only because "what we call existence is an information-theoretic
entity" [494].

Wheeler set out a basic agenda, namely, to find "what, if anything, has to
be added to distinguishability and complementarity to obtain *all* of standard
quantum theory" [494]. The example offered by Wheeler of a physical entity
most genuinely so reducible is the black hole, which is parameterized by its
area.

"[The area] expressed in units of the basic Bekenstein–Hawking area
$4(\hbar G/c^3) \log_e 2$, is given by the **bit** count, N, of that black hole. Here
N represents the number of bits of information it would have taken
to distinguish the initial configuration of particles and fields that fell
in to make this particular black hole from the 2^N alternative quantum configurations that would have produced a black hole externally
identical to it." ([493] and [494], p. 755)

However, the black hole is perhaps the *least* exemplary of physical entities.
Indeed, despite their appearance at the center of most if not all spiral galaxies,
black holes are entities for which it is most often said "physics breaks down"
[209]. The theoretical motivation behind Wheeler's thesis is a set of results in
loop quantum gravity, which is, unfortunately, beyond the scope of this book.
His deeper claim is that general relativity is reducible to quantum gravity
as an approximation, and that space and time are "secondary ideas." On
that assumption, Wheeler's position vis-à-vis matter may be more justified.
Ultimately, for him, the "it from bit" thesis constituted the basis of a research
program rather than an interpretation of quantum mechanics *per se*. It has
been others who have taken up this picture in that fashion.

The idea that physics is reducible to information is problematic for at
least two reasons. One difficulty is that it is far from clear that all physical
things have anything intrinsic corresponding to informational magnitudes,
much less that they are "submitting to information-theoretical descriptions"

in all their aspects. That is, on this approach, physical objects are required to have information-theoretic characterizations that are also the most complete descriptions that can be given of them. A second, insurmountable difficulty is that any information-theoretic description of an object is, by definition, *entirely different* from the *existent* it describes. A physical entity is not a simulacrum and cannot be equated with its own description; that this issue could have been ignored is a symptom of the influence of postmodernism, one negative effect of which has been characterized as follows.

"Our world, Jean Baudrillard tells us, has been launched into hyper-space in a kind of postmodern apocalypse. The airless atmosphere has asphyxiated the referent, leaving us satellites in aimless orbit around an empty center. We breathe an ether of floating images that no longer bear a relation to any reality whatsoever. That, according to Baudrillard, is simulation: the substitution of signs of the real for the real." ([312])

In the current case, the orbit appears to be the "self-excited circuit" [492].

Česlav Bruckner and Anton Zeilinger were inspired to adopt an information-theoretical interpretation of quantum mechanics by the following comment of Feynman, not unrelated to the picture of the relationship between physics and information offered by Wheeler, although inverted in that Feynman's concern can only be one in the presence of a computation realized *in* space.

"It always bothers me that, according to the laws as we understand them today, it takes a computing machine an infinite number of logical operations to figure our what goes on in no matter how tiny a region of space, and no matter how tiny a region of time. How can all that be going on in that tiny space?" ([170])

For the question itself to matter, the size of the physical region in which the pertinent phenomena occur must be related to the information resources required to simulate them using a computational device. It is not clear there need be such a scaling.

In any event, Feynman's concern as understood by advocates of the Informational interpretation is somewhat different. Bruckner and Zeilinger have offered two intuitive postulates characterizing their position: (1) the amount of information carried by any system is finite, and (2) the amount of information carried is lesser the smaller the system in terms of the number of its parts, rather than its spatial extent *per se*. These authors claim that their postulates "solve Feynman's problem" and that one arrives "at a natural limit when a system only represents one bit of information" ([82], p. 57). The second postulate is at best unremarkable, provided one take "amount of information" to have the usual meaning of "the *maximum* average amount of information obtainable by witnessing an event" and one understands "carried by" as the standard *façon de parler* discussed at the outset of this chapter. In the absence of a specified encoding it is demonstrably false, as can be seen by reconsidering Weaver's example of transmitting a bit: One can transmit a bit using the printed text of the King James Version of the Bible, but one can also transmit *many more bits* using, for example, any given single chapter of that same printed text. That is, the actual information content of a given signal depends on the encoding convention in use. An additional assumption that would avoid this problem is needed, for example, that there is a natural encoding such as that provided by the quantum logic of Birkhoff and von Neumann which, however, is based on the Hilbert space formalism.

Zeilinger's aim is to resolve the conceptual difficulties in the foundations of quantum mechanics by demonstrating the measurement problem to be a *Scheinproblem*. Bruckner and Zeilinger have so far failed to provide a robust conceptual framework in which that problem is seen to evaporate, in particular given its dependency on the Hilbert-space formalism, although Zeilinger has made a serious attempt to do so. The world view offered by him is based on what he refers to as a "foundational principle" for quantum mechanics that accords with the above postulates. Zeilinger is also motivated by the fact that

"both the special and the general theory of relativity are based
on firm foundational principles, while [in] quantum mechanics... we
have a number of coexisting interpretations utilizing mutually contradictory concepts." ([515])

He argues that this fact is something that "contains an important message," namely that "a generally accepted foundational principle for quantum mechanics has not yet been identified," and adds that "the purpose of [his] paper is not to compare and analyze [existing] interpretations, but to go significantly beyond them" [515]. As should now be evident, this approach bears similarities, at least at the rhetorical level, to other new interpretations inspired by quantum information science.

The 'foundational principle' offered by Zeilinger is grounded in a sort of logical atomism akin to that introduced by Bertrand Russell [275] but, importantly, without Russell's assumption that everything that is physical is ultimately composed of *matter*. "An elementary system carries 1 bit of information," because "an elementary system represents the truth value of one proposition." Zeilinger remarks of his foundational principle that "this might also be interpreted as a definition of what is the most elementary system" and that it

"underline[s] that notions such as that a system 'represents' the truth
value of a proposition or that it 'carries' one bit of information only
implies a statement concerning what can be said about possible measurement results." ([515])

An obvious way of approaching this picture is to compare it with the ideas of the early Wittgenstein, in which "the world is everything that is the case" and "the totality of facts, not of things" ([503], Propositions I and I.I).

Presumably because, on the face of it, the foundational principle is a tautology, Zeilinger is driven to offer an analysis supporting it using the two postulates given above. To obtain an 'elementary system,' one decomposes "a system which may be represented by numerous propositions into constituent systems"; "the limit" of this decomposition "is reached when an individual system finally represents the truth value to one single proposition only." Thus, he can be seen as espousing logical atomism in a specific form in which all facts are propositions represented by quantum state projectors, à la Birkhoff and von Neumann. In his expositions of the approach, Zeilinger focuses on propositions associated with the spin components of a spin-$\frac{1}{2}$ particle.

> "The spin of the particle carries the answer to one question only, namely, the question, What is its spin along the z-axis?... Since this is the only information the spin carries, measurement along any other direction must necessarily contain an element of randomness. We remark that this kind of randomness must then be irreducible, that is, it cannot be reduced to hidden properties of the system, otherwise the system would carry more than a single bit of information." ([515])

The reasoning here is circular. Moreover, perfectly good hidden-variables models for the spin-$\frac{1}{2}$ system *do* exist, such as that offered by Bell long ago [23]. Only an analysis of measurement in a world described by such models can answer the question of whether the spin-$\frac{1}{2}$ system under a hidden-variables model could or could not be used to encode additional information. Zeilinger further claims that

> "We have thus found a reason for the irreducible randomness in quantum measurement. It is the simple fact that an elementary system cannot carry enough information to provide definite answers to all questions that could be asked [of it]." ([515])

However, the pertinent fact, captured by Holevo's theorem, follows from the mathematical formalism of quantum mechanics itself rather than from the foundational principle on offer. Were the principle truly foundational, one would expect standard postulates of quantum mechanics themselves to emerge non-problematically from it or be corrected by it. The measurement problem is not shown to be illusory under the interpretation, as claimed.

In Relativity, the foundational Principle of Equivalence motivates the mathematical formalism which, in turn, is used to provide physical explanations. By contrast, this interpretation only rationalizes a form of logical atomism through the purported foundational principle precisely by reference to the predictions of the mathematical formalism of quantum mechanics that are confirmed by experiment. The irreducible nature of quantum probability can equally well understood, for example, from the point of view of the Basic or Copenhagen interpretations. Zeilinger argues that this atomistic "viewpoint... lends support to Bohr's notion of complementarity," noting that

> "when the measurement direction is orthogonal to the eigenstate direction... for the new measurement situation the system does not contain any information whatsoever, and the result is completely random." ([515])

More particularly,

> "Measurements of a particle's spin along orthogonal directions are complementary, and the reason is, again, the fact that an elementary system carries only one bit of information." ([515])

It is true that, for example, Pauli had commented on the irreducible nature of randomness in quantum mechanics as follows, some 45 years earlier, although for him this was a natural result instead of the Copenhagen approach.

> "The non-deterministic character of the natural laws postulated by quantum mechanics rests precisely upon [the] possibilities of a free choice of experimental procedures complementary one with the other." ([339])

However, Bohr's notion of complementarity is more fundamental than this new principle in that it applies directly to physical systems rather than indirectly via information that they *could* be used to communicate under a special convention. Thus, the principle of complementarity is a more satisfactory foundational principle than the informational one Zeilinger has offered. The 'foundational principle' is therefore better seen as a *consequence* of the principle of complementarity than as a notion supporting or underlying it.

Finally, it is noteworthy in this regard that in 1937, Bohr commented on similarities of quantum mechanics to relativity, as follows.

> "Not withstanding all differences, a certain analogy between the postulate of relativity and the point of view of complementarity can be seen in this, that according to the former the laws which in consequence of the finite velocity of light appear in different forms depending on the choice of the frame of reference, are equivalent to one another, whereas, according to the latter the results obtained by different measuring arrangements apparently contradictory because of the finite size of the quantum of action, are logically compatible." ([59], p. 291)

Thus, the complementarity interpretation can plausibly be seen as contextualizing quantum measurement theory by providing a logical analogy to that of another physical theory. The Informational interpretation only provides informational descriptions of various physical situations but adds nothing to the theory.

A more common view of the relation of information to quantum mechanics is the position that there is a reduction of information theory to quantum physics, rather than the other way around. A particularly popular version of this is the following. "[T]he theory of information is not purely a mathematical concept, but... the properties of its basic units are dictated by the laws of physics" [353]. An intuitive support for this view is the fact, as Neil Gershenfeld concisely put it in his textbook *The physics of information technology*, that the physics of computation, which he (at certain points) identifies with information theory itself, provides "an explanation of how noise and energy limit the amount of information that can be represented in a physical system, which in turn provides insight into how to efficiently manipulate information in the system" ([186], p. 36).

A strongly reductionist version can be found to be advocated in the influential lecture notes of John Preskill, who has provided the following characterization of the role of quantum information science within physics.

"Why is a physicist teaching a course about information? In fact, the *physics of information and computation* has been a recognized discipline for at least several decades. This is natural. Information, after all, is something that is encoded in the state of a physical system; a computation is something that can be carried out on an actual physically realizable device. So the study of information and computation should be linked to the study of the underlying physical processes. Certainly, from an engineering perspective, mastery of principles of physics and materials science is needed to develop state of the art computing hardware." ([362], p. 7)

Like Gershenfeld's characterization, this *prima facie* is a clear and informative rendition of the relationship. However, Preskill's notion in the above statement is that information can be understood *only* as encoded in a physical system. This is more evident after noting a statement that later follows. "The moral we draw [from the major achievements in the physics of computation] is that 'information is physical' " ([362], p. 10).

The most influential version of this position is that of Landauer, earlier presented in his essay "The physical nature of information" [288, 291].

"Information is not a disembodied abstract entity; it is always tied to a physical representation. It is represented by engraving on a stone tablet, a spin, a charge, a hole in a punched card, a mark on paper, or some other equivalent. This ties the handling of information to all the possibilities and restrictions of our real physical world, its laws of physics and its storehouse of available parts. This view was implicit in Szilard's discussion of Maxwell's demon... The acceptance of the view, however, that information is a physical entity, has been slow... Indeed, our assertion that information is physical amounts to an assertion that mathematics and computer science are a part of physics... Mathematicians, in particular, have long assumed that mathematics was there first, and that physics needed that to describe the universe." ([291])

The error of this statement is the invalid inference from a being typically associated with a physical representation in the physical context, such as the technological one, to being *essentially* physical. In the case of information conceived of as an entity—something that itself may be problematic, *cf.* [450]—it is only the case that information *may be constrained by physical law* in some respects (indeed, there are "restrictions") *when* physical systems are used to communicate, which again is only typical. Consider information acquired by minds. Although human consciousness has been clearly shown to be 'tied' to the human body, especially to the brain, this does not in itself suffice to show that consciousness reduces to physics, however attractive that view might be to engineers, physicians, physicists, and others whose preferred manner of understanding of the world is rooted in physics.

Moreover, although physics does *constrain* information in the technological context—which is what enables the achievements of the physics of computer systems to which Preskill refers in his lectures, such as the possibility of reversible computation—the entirety of the behavior of information is not dictated by physics alone. This can be seen, for example, by noting that, although the physical characteristics of a particular source and other elements of a physical communication system constrain an agent's ability to transmit the information in question to a receiver, *precisely how much* information is *in fact* communicated by a signal is ultimately dictated by the choices of the sending and receiving agents using the source and system, rather than the merely the physics of its signals. Again, this was pointed out by Weaver as quoted in the introduction to this chapter: the sending agent using a communication system is free to choose whatever encoding he or she wishes for use with a given set of signals. Furthermore, the receiving agent must be informed as to this choice of encoding in order to receive *any information at all* from the sender. Moreover, an agent may be free to choose the sort of source used, for example, a classical source or a quantum source to encode the very same information in a given channel, so long as the channel has the required communication capacity; in that case, the two physical theories describing the signals, the classical and the quantum, are *different* theories.

One might still argue that classical mechanics has been successfully reduced to quantum mechanics. However, even with this granted, physicalists must successfully argue that the agents *choices* are not actually free but entirely dictated by quantum physics. This is very far from having been established. An important element of mechanics is that agents have freedom of choice in relation to experimental arrangements in measurement. For example, Bohr stated that

> "The freedom of experimentation, presupposed in classical physics, is of course retained and corresponds to the free choice of experimental arrangement for which the mathematical structure of the quantum mechanical formalism offers the appropriate latitude." ([57], p. 73)

Pauli argued more strongly.

"The non-deterministic character of the natural laws postulated by quantum mechanics rests precisely upon these possibilities of a free choice of experimental procedures complementary one with the other." ([339])

Similarly, von Neumann's account of measurement involves choices that are not assumed to be physically determined, although the case that such choices are determined *might* be made within the Collapse-Free approach, if the branching 'of worlds' could be put on a firm footing. Neither of the physicists of computation quoted above have cited any particular interpretation of quantum mechanics to bolster their claims. The thesis is presumably not assumed to rest on any particular set of quantum interpretational assumptions.

Rather than asking whether there is a reduction of one theory or ontology to the other, assuming that question is well defined in the first place, a more constructive approach to the relationship between information and mechanics might be to ask whether there is *supervenience* either of information magnitudes on physical magnitudes or vice-versa. By addressing this question, rather than that of reduction, one equally well engages the issues of interest while avoiding strong assumptions of the sort critically assessed above. Supervenience is a sort of dependency that can exist between sets of properties possessed by quite different sorts of entity and so could more plausibly hold in the situations considered above; the relation of supervenience is neither symmetric nor dependent on a strong reductive relationship. Supervenience can be defined as follows. A set of properties **A** *supervenes* on a set of properties **B** if and only if two entities a and b that share all properties in **B** necessarily also share all properties in **A**, in which case the properties in **B** are 'base properties' and the properties in **A** are the 'supervenient properties.' Because supervenience is weaker than reduction, it is *a priori* more plausible to assert that quantum information supervenes on physics or vice-versa.

If information-theoretic properties supervene on physical properties, then any two information-theoretic properties that are indistinct must necessarily also be physically indistinct.[15] Nonetheless, as argued above and further below, the relationship between the physical properties and the information-theoretical properties described in the physics literature, when properly explicated, also does not indicate supervenience one way or the other: different signal states may communicate the same information and different information may involve the same signal states. However, the recognition of a weaker relationship, namely, that in the context of computation and communication involving physical signs physics provides the pertinent probability distributions merely *constraining* signals, has fostered discoveries about the quantum world through the study of quantum information. Putting aside questions of reduction and supervenience, one can productively move on to consider in-

---

[15] *Cf.*, for example, [106], Chapter 2.

sights into the nature of the quantum world already achieved by the study of quantum information, for example, using thermodynamical methods. Importantly, the behavior of signals used to communicate qubits and those used to communicate bits are constrained by *Landauer's principle*, which is essentially a recontextualization of *Szilard's limit* [519]. This principle dictates that the erasure of information using a physical computing system is irreversible, with an associated minimum physical energy cost of $kT \ln 2$ per bit in an environment of temperature T, because there is an associated dissipation of a minimum energy of $kT \ln 2$ into the environment, kT being the mean thermal energy [289, 290].

Szilard's limit has been used to resolve the (classical) 'Maxwell demon' paradox [274]. The basis of the paradox is that the second law of thermodynamics can be stated as "the entropy of a closed system never decreases" but the following thought experiment is also conceivable. Consider a machine consisting of a box within which there is a barrier with an internal door and which contains a classical gas of molecules that is under the examination of a finite intelligent 'demon' capable of controlling the door at will. The demon would appear to be able to increase the mean kinetic energy of molecules on one side of the barrier relative to that on the other and, hence, the temperature difference between the two sides, by selectively opening and closing the door based on the speeds of the individual molecules approaching the door. This would result in a decrease of entropy in the box. An apparent paradox thus arises because the demon could perform useful work with no energy expense by cycling the machine while carrying out this process.

During the same era in which the mathematical foundations of quantum mechanics were being clarified, the physical situation involved in the paradox was described by Szilard in a manner simpler yet more rigorous than that of Maxwell's original description, as follows [443]. Szilard considered Maxwell's box to be a cylinder containing two equal chambers and to contain a one-molecule fluid and the chamber in which the molecule is contained to be both measured and (physically) recorded by the demon, and finally that the diaphragm+door to be replaced by a piston. The result of the observation of the molecular speed by the demon is used by him to connect the piston to a mechanical load on which a specifiable amount of work W can be done. This load is continually varied so as precisely to match the average force of the fluid on the piston in a way that renders the process a quasi-static and reversible thermodynamic one. In this situation, pressure of the fluid moves the piston to one end of the container and so brings the gas back to its initial volume, and an energy Q equal to the work W is given to the gas through heat transfer from a constant temperature bath [293]. The net result is that the gas has not only precisely its initial volume but also its initial temperature, whereas the entropy of the heat bath has been reduced. The paradox can then be resolved: There is a paradox in this situation only if the machine, minus the demon, is a *closed* physical system, which it is not, because the demon must be a physical system as well as an intelligent one in order to be capable of interacting with the door.

The apparent paradox is resolved by examining the physical behavior of the pertinent truly closed system, namely, that consisting of both the box *and* the demon, as follows. One first notes that the demon, which is a physical system by virtue of its ability to open and close the door, must perform physical measurements on the molecules and *physically* record them in order to obtain their velocities while carrying out the required molecule-sorting process. The pertinent entropy is that of the total system of machine and the demon's physical memory. One then also notes that the demon's memory must be *erased* for the entire procedure on the system, including the demon, to be genuinely cyclic. This erasure comes at the thermodynamical cost of the finite entropy increase, as discussed above. The apparent paradox is thereby resolved, because there is a corresponding *increase* of entropy in the process, removing the appearance of conflict with the second law of thermodynamics.

The increase in entropy in the total system will, in fact, always be at least as much as the decrease of entropy of the gas in the process. Szilard stated this as follows.

> "One may reasonably assume that a measurement procedure is fundamentally associated with a certain definite average entropy production, and that this restores concordance with the second law. The amount of entropy generated by the measurement may, of course, always be greater than this fundamental amount, but not smaller."
> ([443])

This fundamental amount he identified as $k \ln 2$, namely, the minimum associated with obtaining one bit of information by measurement, which is the smallest amount of information a measurement can provide the demon when it is considered part of the natural encoding procedure associated with this cylinder–piston–molecule apparatus. Because the initial volume occupied by the single molecule is changed by a factor of two in the process, the associated entropy change is $k \ln 2$, because $\Delta S_{\text{thermo}} = \Delta Q/T$, where S_{thermo} is the thermodynamical entropy and Q is heat energy. This analysis is widely recognized as seminal. Indeed, Jordan, for his part, viewed Szilard's treatment as "one of the greatest achievements of modern theoretical physics... the tendency in Szilard's views is to acknowledge also a microphysical applicability of thermodynamics" [264], which, notably, is the same move exploited by von Neumann in his theory of quantum measurement.

Szilard thereby identified the physical 'correspondent' of the bit of information, albeit in the limited sense in which such a correspondence is possible, as argued in the introduction to this chapter regarding the relation of the quantum two-level system and quantum bit and applicable *mutatis mutandis* to that between the classical two-state system and the classical bit, although Szilard did not specifically identify the role of *erasure* in resolving the Maxwell demon paradox. It is in this analysis that he first provided Szilard's limit; he identified the smallest free energy cost associated with a gain of information, the basic unit of which is the bit, by a computing machine in a context in which the associated (physical) memory system must ultimately be erased. Although the latter point was one not obviously fully appreciated by Szilard, it was later fully clarified by Charles Bennett in 1982 along the above lines [30]. Computers have, since Szilard's analysis, often been understood from the thermodynamical point of view, as mechanisms that consume free energy in producing mathematical work and produce waste heat in the process [289].[16]

Although Szilard's analysis takes place entirely within classical physics, it clearly describes the sort of situation now considered to be significant in the domain of quantum mechanics. Żurek argued in the mid-1980s that it is only in semi-classical or quantum treatments that such an analysis is uncontrovertibly consistent [519]. One objection to the application of equilibrium thermodynamics in the classical scenario is that it restricts the validity of results to situations where the thermodynamic limit applies; the thermodynamic limit is that of very large particle number and volume while the particle *density* remains constant, whereas Szilard's idealization is employed in precisely the opposite limit. Thus, Jauch and Baron argued that in a classical analysis an inconsistency arises from the simultaneous employment of dynamical and thermodynamical idealizations that are incompatible [259]. Żurek argued that this objection is overcome by considering a large collection, such as a quantum ensemble, of such Maxwell box experiments.

Żurek also noted that a *more difficult* problem exists, namely, that a classical gas particle is localized, and so will not fill the full box volume after the insertion of Szilard's piston, as a quantum description would require. Indeed, because the piston acts only as a finite barrier, the wave-function of the particle is still present on both sides of it, the particle location being indeterminate until the molecule is measured, in a quantum analysis [519]. More recently, Seth Lloyd published an article involving the resolution of the paradox associated with a fully quantum Maxwell demon scenario that makes use of entanglement [299].

Summarizing what has been established thus far in this chapter, both reductionist positions, that physics is reducible to information and that information is reducible to physics, are implausible. Nonetheless, because physics constrains information processing, the two bear a close relationship that can be helpful in the investigation of both. Accordingly, quantum information science, in which there has been not only tremendous growth but also many surprising discoveries, promises to offer new insight into the foundations of quantum theory, as shown in the remainder of the chapter.

---

[16] This clearly regards only to computers performing computations in a manner that is not fully reversible.

4.8 Entanglement 'Thermodynamics'

Perhaps unsurprisingly in light of the above results, analogies between entanglement theory and thermodynamics have been usefully explored in quantum information science. Seminally, Popescu and Rohrlich showed that any process using collective local operations and classical communication that preserves the degree of state entanglement must be a reversible one [358]. This was established by an argument analogous to the well known thermodynamic characterization of the ideal efficiency of an engine in the Carnot cycle.

> "When Einstein searched for a universal formal principle from which to derive a new mechanics (namely, special relativity) he took for inspiration a general principle of thermodynamics: The laws of nature are such that it is impossible to construct a *perpetuum mobile*. This general principle (the second law) enabled Carnot to show that all reversible heat engines operating between given temperatures T_1 and T_2 are equally efficient... we can draw an analogy with entanglement, as follows: The laws of physics are such that it is impossible to create (or increase) entanglement between remote quantum systems by local operations." ([359])

The central condition on entanglement measures, given by Equation 1.51 and sometimes called *the fundamental postulate of entanglement*, was seen to be analogous to the second law of thermodynamics [110]: The "fundamental postulate" is that no net increase of entanglement between systems in distinct laboratories can occur *solely* as a result of local operations and classical communication. Together with the condition of partial additivity condition on entanglement measures, which applies to collections of states, the postulate provides a 'unit of entanglement,' the *e-bit*: "*ebit* denotes a single entangled bit" ([359], p. 43). By considering these two conditions and exploiting techniques of thermodynamics, Popescu and Rohrlich demonstrated that the von Neumann entropy is the *unique* measure of ordinary bipartite entanglement in a specific sense by approaching the problem of quantifying entanglement of k pure states through that of finding defining a measure of entanglement for n singlet states, $|\Psi^-\rangle$ [358]. This inspired later, more speculative attempts to use thermodynamical analogies, leading to what has come to be called "entanglement thermodynamics."

The argument of Popescu and Rohrlich runs as follows. The allowed local transformations of quantum states of a bipartite quantum system, with one subsystem in laboratory A and one in laboratory B, are reversible only in the limit where the number of copies of a state becomes arbitrarily large. Furthermore, there is no way to define total entanglement for an infinite number of state copies, because it would then clearly take an infinite value, something physically precluded except when entanglement is defined intensively.

"Here too, thermodynamics provides the formal principle: the thermodynamic limit requires us to define intensive quantities. Likewise, the measure of entanglement must be intensive." ([359], p. 43)

This requires that the measure of entanglement for n singlets must be proportional to n. In that case, the entanglement of a collection of k systems in an arbitrary pure state $|\Psi\rangle_{AB}$ approaches that of n systems each in the singlet state $|\Psi^-\rangle$, that is,

$$E(|\Psi\rangle_{AB}) = \lim_{n,k\to\infty} \left(\frac{n}{k}\right) E(|\Psi^-\rangle_{AB}) , \qquad (4.16)$$

which is identified as the "entropy of entanglement" of the state $|\Psi\rangle_{AB}$ [33]. Any such measure of pure-state entanglement is, therefore, determined up to a constant factor, namely, the amount of entanglement associated with the singlet state, which is then taken to be that of the fundamental unit, the e-bit. This also constrains entanglement manipulation analogously to that of heat in thermodynamics.

Again, following on the heals of this result, there have been attempts to construct a full blown "entanglement thermodynamics." More conservative explorations of relationships between thermodynamics and quantum information analogy have also taken place, for example [5]. Under the broader analogy, entanglement plays the role analogous to heat in traditional thermodynamics, the distillation of pure entangled states plays the role of extracting work from heat. The bound entanglement is given by Equation 1.56, namely $B(\rho) \equiv E_f(\rho) - D(\rho)$. This expression is seen to be formally similar to the Gibbs–Helmholtz equation of thermodynamics, namely, $TS = U - A$, where U is the internal energy and A the free energy, with TS serving as the "bound energy."

Recall the three basic laws of thermodynamics, which can be given the following simple forms. (1) Heat is a form of energy. (2) It is impossible for any cyclic process to occur the sole effect of which is the extraction of heat from a reservoir and the performance of an equivalent amount of work. (3) The entropy of a system approaches a constant value as the temperature approaches zero. These laws allow the reversible transformation of work into heat and vice versa. One can, therefore, formulate the assumptions of thermodynamics as follows, "There is a form of energy (heat) that cannot be used to do work, that nonetheless can be used to store work though work can be stored in heat only if there is some heat to begin with, in which case work can be stored reversibly" [242].

Now, consider two agents in laboratories A and B initially sharing a collection of pairs of subsystems described by an n-fold product of the bipartite system state ρ, $\rho^{\otimes n} = \rho \otimes \cdots \otimes \rho$, with n a large number, who collectively and locally operate on the members of each shared pair, communicate using a classical information channel if they desire, and can arrange their local subsystems into subensembles ρ_i represented with probabilities p_i. The "fundamental postulate" of entanglement theory, the putatitive 'second law of entanglement thermodynamics,' dictates that the entanglement remaining at the end of such a CLOCC transformation, on average, cannot exceed the initial shared entanglement. The combination of the entangled-state distribution process and the entanglement distillation process, which accumulates pure entangled states from mixed states, is viewed as functioning analogously to the process of *cycling an engine* that obtains work from heat; bound entangled states, being those entangled states from which *no* pure entanglement can be distilled (*cf.* Section 1.10), are similarly viewed as analogous to thermodynamic systems from which no *work* can be drawn and are seen as containing "fully disordered entanglement."

The following "laws of entanglement thermodynamics" are then suggested by analogy to the above traditional thermodynamic laws [242]. (1) The entanglement of formation is conserved. (2) The disorder of entanglement can only increase. (3) One cannot distill singlet states with perfect fidelity. The "law" (1) corresponds to condition (ii) of Section 1.10. There is an analogy to reversible work extraction, although in general one needs more entanglement (in e-bits) to create a state than can be drawn from it. In traditional thermodynamics, the second law dictates that any thermodynamical system has more energy than can be extracted from it, except when one of the reservoirs is at zero temperature; the same holds in this "thermodynamics of entanglement" where for a general mixed state ρ, $D(\rho) < E_f(\rho)$. However, attempts to continue further with this treatment of entanglement in order properly to complete the analogy run into difficulties [110, 251]. In particular, it requires the completion of the correspondence between fundamental quantities in the two theories. For example, if a thermodynamic system gains a quantity ΔQ of heat energy, there will be an entropy increase of $\Delta S = \Delta Q/T$. Given that the role of entropy is played by $S(\rho)$ in quantum entanglement theory, it is by no means clear what quantity is to play the role of temperature, T. A well-defined 'entanglement temperature,' $\bar{T}(\rho)$, for mixed states (when $S(\rho) > 0$) of the form $\bar{T}(\rho) = B(\rho)/S(\rho)$, is required if the 'entanglement entropy' is to be taken to be $S(\rho)$, as is suggested by the fact that this results in the equality of the entanglement of formation and entanglement of distillation for pure states (*cf.* [110]). For example, the third law of thermodynamics is expressed in terms of the behavior of entropy with respect to temperature. However, the temperature analogue is absent from the above statement of 'laws of entanglement thermodynamics.' The lack of a well-defined such quantity brings this approach strongly into question.

In addition to the difficulty of completing "entanglement thermodynamics," the argument for the uniqueness of the quantum entanglement measure based on a *mutatis mutandis* argument has be seen to induce an unwarranted dependence on the choice of unit—the introduction of the Bell singlet state as providing an "e-bit" of entanglement—manifest in the *ratio problem*: ratios of entanglement measures, such as the entanglement of formation or distillable entanglement of two different states, depend in general on the particular state chosen as the basic unit of entanglement when the degree of entanglement is referenced to it [323].

The definition of an entanglement unit is particularly problematic in the multi-party context; in multiple-component systems their exist *different sorts of entanglement* that are not quantifiable in terms of the e-bit. The argument for the uniqueness in the case of two pure states of a pair of two-level systems will not work in the multi-party context because reversible interconversions cannot be performed between all pairs of such states. Moreover, it has been shown that no unique measure of entanglement exists in the case of *mixed* states [321]. By contrast, the thermodynamic entropy does have a unique measure without such a ratio problem [189].

4.9 Information and Entanglement

The quantitative study of entanglement has usefully exploited suggestive relationships between, on the one hand, the behavior information in the possession of agents with the ability to perform local actions on quantum systems and with the ability to classically communicate and, on the other, the behavior of heat. Despite the fact that such thermodynamical concepts have proven useful in the quantum information context as analogues to concepts involving entanglement, there are clear limitations to the thermodynamic analogy, as just described. Nonetheless, the investigation of the relationship between information and entanglement by making more direct use of results of the mathematics of correlation and entropy has been productive. Indeed, as pointed out by Leah Henderson, the "second law of entanglement thermodynamics" itself follows directly from basic properties of relative entropy ([225], *cf.* [295]). The remainder of this chapter focuses on this more direct approach to the study of entanglement and on the relationship between entanglement and communication complexity.

The significance of entanglement for quantum communication has recently been characterized in the following way by Wootters. "[It is] remarkable that, even though entanglement by itself does not constitute a communication channel, the presence of entanglement allows modes of communication that are not possible without it" ([506], p. 229). Indeed, *prima facie*, this fact appears surprising. However, careful study of the character of and relationship between differing degrees of correlation does much to remove the apparent mystery; even *classical* shared randomness serves as a communication resource. As shown below, recent investigations have gone some distance toward revealing the essential character of quantum behavior. If there turns out to be something to entanglement beyond (i) that the uniquely strong correlations associated with it occur only in quantum systems and (ii) that pure entangled states violate local causality, then that something may well turn out to have nothing in particular to do with communication theory. In any event, the investigation of the standard theory of quantum mechanics through the theory of communication is one of the most incisive currently being pursued.

Although the claim that entanglement allows for genuinely new modes of communication may be an overstatement, entanglement does allow a number of tasks to be accomplished *more efficiently* than is possible when only the transmission of classical bits is allowed and quantum entanglement between physical systems is absent, as already seen in this chapter. Without prior-shared quantum entanglement or prior-shared classical randomness, agents in distant laboratories must communicate to establish distant physical correlations or to evaluate distributed functions; previously shared quantum entanglement reduces the amount of communication required to carry out such tasks compared with situations involving generic shared random strings. It is, therefore, not surprising that demonstrations of the failure of local causality given pure entangled states can be related to communication complexity.

The communication complexity of a function is the minimum number of classical bits that must be broadcast for every party involved to come to know the value of the function. A two-party quantum communication protocol exists that operates more efficiently than any local classical protocol for determining a binary function of two input bits x_0 and y_0 separately received by agents in distinct labs, who thereafter are not allowed to communicate. A binary function can be determined when the two agents output two bits a and b, such that $a \oplus b = x_0 \wedge y_0$ with as high a probability as possible, where "$\oplus$" indicates addition mod 2 ("XOR") and "$\wedge$" indicates Boolean conjunction ("AND"). Let the two agents, A and B, begin with the binary strings $\mathbf{a} = a_1 a_2$ and $\mathbf{b} = b_1 b_2$, respectively. Consider also the set of four conditions consisting of the three conditions $a_i \oplus b_j = 0$, for the cases when i and j are not both unity, together with the condition that $a_1 \oplus b_1 = 1$. With some thought, one realizes that these four conditions cannot *all* be simultaneously deterministically satisfied; they can, at best, be satisfied with a probability of up to $\frac{3}{4}$ when the strings are chosen randomly. Let the a_i and b_j be randomly distributed independently of inputs. There is a quantum protocol in which one agent sends one two-level system of an entangled pair to the other agent, the outcome of a measurement on which can be used often correctly to evaluate $f(\mathbf{a}, \mathbf{b}) = a_1 \oplus b_1 \oplus (a_2 \oplus b_2)$, even though the resulting evaluation will sometimes be incorrect. The receiver can compute $f(\mathbf{a}, \mathbf{b})$ by obtaining appropriate data regarding the string possessed by the sender, namely, a_1 or the binary sum of its two bits. This can be done quantum mechanically with probability $p(1-p) = \cos^2(\frac{\pi}{8}) > \frac{3}{4}$ if the two agents make joint measurements on the pair of two-level systems in a Bell state and perform appropriate local rotations on the subsystems in both laboratories [91].

An exact procedure for this more efficient protocol is the following. Let the agents, Alice and Bob, each possess one of two subsystems initially in the shared Bell spin-singlet state $|\Phi^-\rangle_{AB}$. Let a two-bit string $x_0 y_0$ be taken as input to be used in the determination of the function $f(x, y) = x_1 \oplus y_1 \oplus (x_0 \wedge y_0)$. In the case that $x_0 = 0$, Alice rotates its subsystem by $-\frac{\pi}{16}$ in the plane involved in the joint measurements shown in Figure 1.2 (of Section 1.8); otherwise, a rotation of $\frac{3\pi}{16}$ is applied. Alice then performs a spin measurement, obtaining outcome bit a. Bob proceeds similarly, obtaining the outcome bit b. The result of these two steps is a superposition of two Bell states, $|\Phi^-\rangle_{AB}$ and $|\Psi^+\rangle_{AB}$; the probability amplitude for the former is $\cos(\theta^A + \theta^B)$ and for the latter $\sin(\theta^A + \theta^B)$, where the θ^X ($X = A, B$) are the angles of those rotations locally performed conditionally on the specific inputs. The probability of $a \oplus b = 0$, which is that of the two subsystems both ending up in their original states is, therefore, $p(a \oplus b) = \cos^2(\theta^A + \theta^B)$, that is, the square of the probability amplitude in the case that the outcomes are anti-correlated (so that the XOR of the two is unity). Notice then that one has $\theta^A + \theta^B = \frac{\pi}{8}$ in every case. Finally, Alice classically communicates $a \oplus x_1$ to Bob, who communicates $b \oplus y_1$ to A. After that communication round, both parties can individually determine $f(a, b)$, because they can then determine the bit $(a \oplus x_1) \oplus (b \oplus y_1) = x_1 \oplus y_1 \oplus (a \oplus b)$, which will be $f(x, y)$ with probability $\cos^2(\frac{\pi}{8}) = 0.854$ [77], which exceeds the greatest classically achievable value of $\frac{3}{4}$. More generally, one can consider n parties in possession of partial input data for some n-variable function. It has been shown that there is a class of quantum communication complexity protocols capable of increasing the efficiency of solution of such problems beyond what is classically possible if and only if a Bell-type inequality for three-level systems is violated [83].

Returning specifically to entangled pairs of two-level systems, recall the CHSH inequality $|S| \leq 2$, where S is defined in terms of expectation values of joint measurements, and that the maximum value of $|S|$ beyond 2 has been taken as a measure of how 'quantum mechanical' a bipartite system is; once $|S|$ exceeds 2, the behavior of a system is no longer classical (Bell local) in nature but can be described quantum mechanically.[17] Recall also that the maximal violation of the inequality by an entangled state is $|S| = 2\sqrt{2}$, the Tsirel'son bound. Although Bell himself did not argue that causal violation of this inequality distinguishes quantum mechanics from other conceivable theories besides classical mechanics, the question of whether it does is an important one. Indeed, Shimony engaged the question by asking whether "*non-locality* plus *no signaling* plus *something else simple and fundamental*" suffices to uniquely single out quantum mechanics from the set of conceivable mechanical theories providing correlated local-measurement outcomes [407].

Shimony's question led Popescu and Rohrlich to consider, as a starting point of investigation, whether quantum mechanics might be uniquely distinguished by the conjunction of non-locality and causality alone. In other words, "Is quantum mechanics the *only* causal theory—*i.e.* theory under which signaling is constrained from above by the speed of light—that violates the Bell

---

[17] Note, however, this criterion is questionable in the case of *mixed* states, in that, for example, a Werner state, which is obtained by mixing a fully mixed state and a maximally entangled state of a pair of two-level systems, can be entangled without violating the CHSH inequality [489], as discussed in Chapter 1

inequality?" [357]. They introduced a schema they called the *non-local box*, now referred to as the *PR box*, to model the generic class of 'non-local' theory. They answered the question in the negative, by showing that a theory providing the values of this schema provides correlations that *exceed* the Tsirel'son bound. This opened the way to the consideration of the relationship between mechanical theories and information in greater generality than had previously been done. Indeed, the communication complexity results obtained if these *non-physical* boxes were available has become of considerable interest.[18]

As the above example indicates, entanglement has been shown to serve as an information-theoretic resource that surpasses the local classical resource of prior-shared random bit strings. Further insight into it has been gained by investigating the quantitative relationships between it, considered via the correlations exhibited by signs in shared Bell singlet states, that is, *e-bits* and other information-theoretic resources; useful comparisons between shared random classical data, shared secret data, prior-shared entanglement, and non-local boxes have been drawn by studying whether and how the information-theoretic resources can simulate one another. Such relationships have operational realizations in quantum protocols. The question of which probabilistic 'machines' or 'boxes,' or combinations thereof, are capable of capturing the behavior of others, that is, are able to simulate them is, therefore, pivotal. The term *box* is shorthand for *black box* in the traditional sense of an abstract rule that may formalize a mechanism. Such boxes formalize communication resources, being so in the sense that they must be resupplied after use for communication. Finding the number of uses of various sorts of box that are needed to provide the joint probabilities arising from the use of prior-shared Bell states by means *not* quantum mechanical in nature serves to connect quantum mechanics and information theory [454], enabling the systematic classification of physical theory in relation to communication.

It is helpful here also to recall the implications of the violation of each of the conditions, parameter independence (PI) and outcome independence (OI), underlying Bell's locality condition ([419], p. 118). Violating PI, something *not* occurring in quantum mechanics, allows one bit to be perfectly and superluminally communicated using an ensemble of composite systems all in the same complete state and making the same choice of measurements for each member of the ensemble. Violating OI, as quantum mechanics *does do*, implies influences into the past—although in the case of quantum mechanics this does not allow for superluminal *communication*, because the outcomes of measurements cannot be controlled by the agents making the measurements. From the perspective of these conditions, Bell's theorem, as in its CHSH form, implies that (Bell-)local theories cannot simulate the behavior of outcomes of measurements on a singlet without *some* communication.

The CHSH inequality can be reformulated in a particularly convenient form by considering $p(m_x^A \oplus m_y^B \equiv x \wedge y)$, the probability for the (mod 2) sum

---

[18] See, for example, [20, 21, 104, 454, 461].

of measurement outcomes m_x^A, $m_y^B \in \mathbb{Z}_2$ to equal the product of "measurement setting" parameters x and y in $\mathbb{Z}_2 \equiv \{0,1\}$: Any local hidden-variables theory describing a joint system shared by two agents A and B obeys the inequality in the form

$$s = \sum_{x,y \in \mathbb{Z}_2} p(m_x^A \oplus m_y^B \equiv x \wedge y) \leq 3 . \tag{4.17}$$

The corresponding Tsirel'son-type bound is then that the maximum quantum mechanical value for the left-hand-side of this inequality is $s = 3.41 = 2 + \sqrt{2}$. The sum provided by quantum probabilities exceeds the constraints provided by the condition of Bell locality by this specific amount. The non-local boxes are found to exceed it by an even greater amount, reaching the value 4.

Correlation polytopes are natural structures for illustrating such relationships. The geometrical representation of joint probabilities within these polytopes is a technique with a long history in the study of entanglement [352, 455]. In particular, the probabilities arising in (Bell-)local theories as constrained by Bell-type inequalities provide specific boundaries between quantum and other correlations. For an arbitrary number n of independent classical events occurring with marginal probabilities $(p_1, p_2, \ldots, p_n)$ and joint probabilities $(p_{12}, \ldots)$, these probabilities can be taken together to form a vector, $p = (p_1, p_2, \ldots, p_n, p_{12}, \ldots)$ in Euclidean space. Because the p_i, $i = 1, \ldots, n$ are assumed independent, each of them can reach both 0 and 1. The combined values of $p_1, p_2, \ldots, p_n$ of these extreme cases and the associated joint probabilities $p_{ij} = p_i p_j$ can be interpreted as truth values; they correspond to a two-valued (or dispersionless) measure.

Every convex polytope has an equivalent description as the convex hull of extreme points or as an intersection of a finite number of half-spaces each given by a linear inequality, which is useful for representing constraints. The relationship between prior-shared non-local boxes and better known information-theoretic resources can be illustrated by reference to this structure. Consider the joint probabilities $p(x, y | A, B)$ of obtaining the pair of outputs x and y conditionally on both inputs A and B, which by definition satisfy the conditions of positivity $p(x, y | A, B) \geq 0$ for all a, b, X, and Y and summing to one, $\sum_{x,y} p(x, y | A, B) = 1$ for all A and B. Two classes of boxes can be distinguished. Signaling boxes are those allowing signaling through the agents' choices of input, acting like two-way classical channels; non-signaling boxes are those for which such communication between agents via the choice of A and B is not allowed. In the latter case, the marginal probability $p(x|A)$ is independent of B and $p(y|B)$ is independent of A, that is, $\sum_y p(x, y | A, B) = \sum_y p(x, y | A, B') = p(x|A)$ for all x, A, B, B' and similarly $\sum_x p(x, y | A, B) = \sum_x p(x, y | A', B) = p(y|B)$ for all y, B, A, A'; their joint probabilities form a convex polytope, $\mathcal{P}$.

Of the non-signaling boxes, one can consider boxes which satisfy locality and those which don't, where a box is a local if and only if its probabilities can be simulated by independent agents initially sharing only a vector of random variables $\mathbf{v}$, that is, random classical data. Thus, for local boxes, $p(x, y | A, B) =$

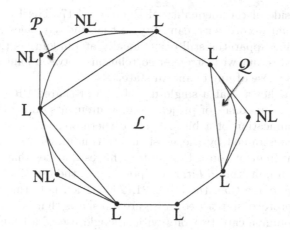

Fig. 4.6. Joint probability loci [20]. The polytope $\mathcal{P}$ of joint probabilities constrained by the no-signaling condition, corresponding to the "non-signaling boxes." Bell-type inequalities define the dashed lines and bound the local polytope $\mathcal{L}$; the vertices labeled L correspond to local correlations and those labeled NL to non-local correlations. The quantum mechanical probabilities lie only in the region $\mathcal{Q}$, the boundary of which includes smooth curves.

$\sum_{\mathbf{v}} p_{\mathbf{v}} p_{\mathbf{v}}(x|A) p_{\mathbf{v}}(y|B)$, where $p_{\mathbf{v}}$ is the probability of occurrence of the vector of values $\mathbf{v}$. The set of joint probabilities given by these boxes is again a convex polytope, the local polytope $\mathcal{L}$, the vertices of which are deterministic, that is, the marginal conditional probabilities take *only* the values 0 and 1; the boundary of this polytope corresponds to the cases wherein either the positivity conditions or the Bell-type inequalities are satisfied. The former are also facets of $\mathcal{P}$, whereas the latter are facets of $\mathcal{L}$ interior to $\mathcal{P}$ [20]. By contrast, the set of correlations obtainable by measurements on bipartite quantum states $\mathcal{Q}$, for a fixed number of measurement settings and outcomes, is convex [287, 311, 455, 456] but does *not* form a polytope since the number of extrema is infinite, providing smooth interior boundaries as well as facets shared with $\mathcal{P}$. The sets just described are related as $\mathcal{L} \subset \mathcal{Q} \subset \mathcal{P}$, as shown in Fig. 4.6.

Popescu and Rohrlich have found that non-local causal correlations are exceptionally powerful. Spatially separated agents sharing only random classical data that are not allowed to communicate are unable to simulate the non-signaling *non-local* boxes, $\bar{\mathcal{L}}$. The PR box, defined by its causal provision of maximal correlations, gives the following conditional probabilities. If the two-bit string xy is a member of $\{00, 01, 10\}$ then $p(m_x^A = 0, m_y^B = 0) = \frac{1}{2}$ and $p(m_x^A = 1, m_y^B = 1) = \frac{1}{2}$, and when xy takes the value 11 then $p(m_x^A = 0, m_y^B = 1) = \frac{1}{2}$ and $p(m_x^A = 1, m_y^B = 0) = \frac{1}{2}$; all other pertinent probabilities are zero. For this schema, the value 4 can be reached by the left-hand side of the inequality of Equation 4.17. The PR box provides probabilities that accord with causality in that the outcomes on both sides of the measuring apparatus still occur locally at random, as they do in the case of Bell states, but with stronger correlations between joint measurement outcomes than those of *any* quantum state.

It has been shown that a single use of a prior-shared PR box is capable of simulating all outcomes of projective measurements on a Bell state *without* the communication of a bit [104]. Furthermore, there are sets of joint probabilities constrained by the no-signaling conditions that cannot be obtained by measurements on a Bell state. This establishes that the non-local box information unit, the *nl-bit*, corresponding to the use of a single PR box is a stronger resource than the e-bit [21, 260]. Then note that the quantum dense coding protocol illustrated in Figure 4.5 shows that an e-bit allows one to perfectly communicate two bits with a single use of a channel that perfectly communicates one qubit, where "perfectly communicates" is shorthand for "communicates the encoding quantum state with perfect fidelity." Moreover, it has been demonstrated that one bit of (superluminal) communication and prior-shared classical randomness together are sufficient to produce *all* Bell-inequality-violating measurement correlations which are associated with 1 e-bit [104]. Finally, of all the information-theoretic resources so far conceived, one bit of super-luminal communication is the strongest resource because, unlike a PR non-local box, it is not unconstrained by causality. Thus, in terms of resource strength, an ordering relation has been established: 1 e-bit $\prec$ 1 nl-bit $\prec$ 1 bit (superluminal), in order of increasing strength.

The PR box and superluminal-signaling box provide statistics that would be shocking were they to be physically realized. Providing PR non-local boxes to two spacelike separated agents would allow them to perform *all* distributed computations with perfect accuracy given a *trivial* amount of communication, namely, 1 bit; the 1 bit is necessary for 'preserving causality' [460]. This result reduces all possible distributed functions in the standard inner-product form: The correlations provided by non-local boxes allow the inner-product functions to be solved with just 1 bit of communication, with a maximum of an exponential amount of prior-shared PR boxes. Such a collapse of the classes of communication problems is remarkable, because the classification of computational complexity problem is fundamental in *computer science*. "[T]heir absence... goes squarely against the world view and experience of probably all researchers in the field of complexity theory" [461]. By contrast, some communication complexity problems using e-bits require *unlimited* prior-shared entanglement and a number of communicated bits [109].

In this way, there naturally arises the question of whether what distinguishes quantum mechanics is the physical constraint corresponding to the non-triviality of *communication complexity theory* when quantum resources are allowed, which is an information-theoretic constraint, rather than the physical constraint of *causality*. Although the above result suggests that non-local PR boxes themselves describe correlations that can't be realized on the grounds of the theory of communication rather than only physical experience, researchers in theoretical computer science have had little time to contemplate even previously shared *quantum* correlations, much less those correlations constrained by causality alone. It may be that changes in the foundation of their subject based on these newly contemplated cases will reduce the surprise evoked by the associated results.

The original question of what physical condition, together the constraint of local causality, can provide non-local correlations of the strength predicted by quantum mechanics but *only* that strength has yet to be precisely answered. Thus far, a bound has been demonstrated assuming non-trivial computational complexity, but one that only constrains correlations to a strength *beyond* the specific strength of quantum correlations.

"In any world in which it is possible, without communication, to implement an approximation to the [non-local PR box] that works correctly with probability greater than $\frac{3+\sqrt{6}}{6} \approx .91$, every Boolean function has trivial probabilistic communication complexity." ([77])

Recall that a Bell state approximates a PR non-local box with probability 0.854. This leaves a gap of about 5 percent. Should this difference be closed, the physical condition the violation of which would correspond to the trivialization of communication complexity hierarchy might then be identifiable.

4.10 The Great Arc

A series of steps involving quantum probability that form what can be called the *Great Arc* in the study of the foundations of quantum mechanics can be discerned. Each of these steps has influenced the interpretation of the theory, the fundamental principle of which is that of state superposition. Each of these has contributed to a deepening of our understanding of quantum mechanics but has yet entirely to clarify its foundations. Related to this is that the quantum measurement problem has yet to be adequately resolved. Interpretations of the theory advocates of which have claimed explain away the measurement problem only do so at the price of introducing equally difficult problems. The most recent step along this arc has been the explicit consideration of the information-theoretic implications of quantum mechanics.

One of the first steps in the probing of the foundations of quantum theory was the raising of the question of how the probabilities appearing in quantum mechanics are to be understood, which was addressed by Born and Pauli in their interpretation of quantum amplitudes via the Born rule. Another was further reflection on the adequacy of the wave-function description, as addressed by Einstein in his rough division of interpretations into those presenting the quantum formalism as complete and those presenting it as incomplete and by von Neumann in his imperfect but influential argument against hidden variables. These initial steps regard the question of whether, as Einstein put it,

God plays dice with the world, that is, whether the probabilities of quantum theory are irreducible. In the fuller interpretations that emerged, there are inelegant features: either there is a stochastic "collapse of the wavepacket" only during measurement, with or without the mind playing a functional role in the process, or there is an essential dependence on classical instruments or subjectivity in understanding physical processes during measurements.

A third step forward was a more pointed probing of the completeness of the theory and its conflict with local causality, as engaged by Einstein, Podolsky, and Rosen. EPR also brought the question of the relationship between quantum mechanics and realism closer to the foreground. A fourth was Gleason's theorem demonstrating that the formalism of quantum mechanics *is* complete in a particular sense. A fifth was the exploration of the extent to which local hidden variables models can be well defined and whether they can be empirically tested, addressed by Bell, CHSH, and others. The results associated with these steps, the difficulties associated with a naive approach to quantum propositions, and the experimental demonstrations of Aspect and later workers showing local causality to fail, albeit with certain loopholes left open, led some to reject realism on less than sufficient grounds.

A sixth major step was the discovery that the Bell locality condition is logically decomposable into simpler conditions, outcome independence and parameter independence, as shown by Jarrett and further considered by Shimony, allowing one to determine that outcome independence is the condition of the two that is violated in the quantum world. A seventh has been the attempt to find simple conditions uniquely distinguishing quantum mechanics from other theories on the basis of the sorts of joint correlations predicted under these circumstances, pursued by Popescu and Rohrlich and then others, that has benefited from the consideration of elements of communication complexity theory. These last steps have pointed out that, like realism, causality can be retained despite the counterintuitive non-local character of quantum physics.

Given that quantum theory involves probability, state preparation and state measurement, which are essential elements of signaling, and that communication is based on the establishment of correlations using these, it is clear that information theory will remain of considerable relevance to the investigation of the foundations of quantum physics. As seen in the previous section, the perspective provided by the recent focus on information has contributed to what is the most detailed picture yet of the broader implications of quantum theory. Although there are good reasons to reject recently proffered information-centered conceptions of physics, it appears that the requirement that the mechanics of quantum signals allow for the realization of communication tasks consistently with basic information-theoretic principles, assuming that physics is causal, may aid in the construction of a more satisfying picture of the quantum world.

The results considered in this chapter illustrate the way in which recent foundational investigations have captured quantum physics in a new light. In particular, they show that the quantitative consideration of information-theoretic questions in relation to the physical constraints on quantum signals has illuminated the most fundamental aspects of the theory. Like the preceding results just summarized, they suggest that the foundations of quantum mechanics are best investigated through a constructive and balanced combination of conceptual and mathematical analyses.

A

Appendix

A.1 Mathematical Elements

The central mathematical structure of quantum mechanics is Hilbert space, which is a specific sort of vector space. The Dirac notation commonly used to describe Hilbert-space calculations in the physics literature is introduced below, after this structure is described in standard mathematical notation.

A *vector space* V over a field F is a set of vectors together with operations of scalar-multiplication and addition satisfying the following. For all elements $a, b \in F$ and $u, v, w \in V$ (with the zero vector denoted $\mathbf{0}$), scalar multiples and vector sums are elements of V such that:

(i) There is a unique scalar *zero element* $0 \in F$ such that $0u = \mathbf{0}$ for all u;
(ii) $\mathbf{0} + u = u$;
(iii) $a(u + v) = au + av$ and $(a + b)u = au + bu$; and
(iv) $u + v = v + u$, and $u + (v + w) = (u + v) + w$.

A (scalar) *inner product* (u, v) is assumed for all pairs $u, v \in V$ such that:

(i) $(u, v) = (v, u)^*$, where * indicates complex conjugation;
(ii) $(u, u) \geq 0$, with $(u, u) = 0$ if and only if $u = \mathbf{0}$;
(iii) $(u, v + w) = (u, v) + (u, w)$; and
(iv) $(u, av) = a(u, v)$.

The *norm* of a vector, $||u|| \equiv \sqrt{(u, u)}$, satisfies:

(i) $||u|| \geq 0$ for all $u \in V$, with $||u|| = 0$ if and only if $u = \mathbf{0}$;
(ii) $||u + v|| \leq ||u|| + ||v||$, for all $u, v \in V$; and
(iii) $||au|| = |a| \, ||u||$, for all $a \in F$ and all $u \in V$.

The field F used in quantum mechanics is usually taken to be that of the complex numbers, $\mathbb{C}$. The vectors for which $||u|| = 1$ are the *unit vectors*. Vectors u and v are *orthogonal* ($u \perp v$) if and only if $(u, v) = 0$.

A function of two vector arguments that is more general than the inner product, which serves similar purposes but is applicable in broader contexts, is the *Hermitian form*, $h(u, v)$, which satisfies:

(i) $h(u, v) = h(v, u)^*$;

(ii) $h(u, av) = ah(u, v)$, for all $a \in F$, $u, v \in V$; and

(iii) $h(u + v, w) = h(u, w) + h(v, w)$, for all $u, v, w \in V$.

A *positive Hermitian form* is an Hermitian form that also satisfies the condition $h(u, u) \geq 0$ for every $u \in V$. An Hermitian form is *positive-definite* if $h(u, u) = 0$ implies $u = \mathbf{0}$; a *positive-definite* Hermitian form is an inner product. A *basis* for a vector space is a set of mutually orthogonal vectors such that every $u \in V$ can be written as a linear combination of its elements; a basis is *orthonormal* if it is composed entirely of unit vectors. A set S of vectors in V is a *subspace* of V if S is a vector space in the same sense as V itself is. The one-dimensional subspaces of V are called *rays*.

A *Hilbert space*, $\mathcal{H}$, is a complete vector space with an inner product for which $(u, u) \geq 0$ for all $u \in \mathcal{H}$. A map $A : \mathcal{H} \rightarrow \mathcal{H}$, $u \mapsto Au$ is a *linear operator* if, for all $u, v \in \mathcal{H}$ and scalars $a \in F$: (i) $A(u + v) = Au + Av$; and (ii) $A(au) = a(Au)$. If, instead of (ii), one has $A(au) = a^*(Au)$, then A is an *anti-linear operator*.[1] A vector space with an inner product, such as a Hilbert space, can be attributed a norm $||\cdot|| = (v, v)^{1/2}$, which provides a distance via $d(v, w) = ||v - w||$. Such a vector space is *separable* if there exists a countable subset in the space that is everywhere dense, that is, for every vector there is an element of the space within a distance ϵ of it for every positive real ϵ; the space is *complete* if every Cauchy sequence—namely, every sequence such that for every $\epsilon > 0$ there is a number $N(\epsilon)$ such that $||v_m - v_n|| < \epsilon$ if $m, n > N(\epsilon)$— has a limit in the space. (For finite-dimensional spaces, one usually considers the norm topology, although weak topologies may be required to define needed limits and to give proper definitions of continuity.) A *subspace* of a Hilbert space $\mathcal{H}$ is a closed linear manifold, that is, a linear manifold containing its limit points; a linear manifold in $\mathcal{H}$ is a collection of vectors such that the scalar multiples and sums of all its vectors are in it.

A *bounded* linear operator is a linear transformation L between normed vector spaces V and W such that the ratio of the norm of $L(v)$ to the norm of v is bounded by the *same* number, for all non-zero vectors $v \in V$. The set of bounded linear operators on a Hilbert space $\mathcal{H}$ is designated $B(\mathcal{H})$. The sum of two operators A and B, $A + B$, is another operator defined by $(A + B)v = Av + Bv$, for all $v \in \mathcal{H}$; multiplication of an operator by a scalar a is defined by $(aA)v = a(Av)$, for all $v \in \mathcal{H}$; multiplication of two operators is defined by $(AB)v = A(Bv)$, for all $v \in \mathcal{H}$. The *zero operator*, $\mathbb{O}$, and the *unit operator*, $\mathbb{I}$, are defined by $\mathbb{O}v = \mathbf{0}$ and $\mathbb{I}v = v$, respectively. An operator B is the *inverse* of another operator A whenever $AB = BA = \mathbb{I}$; it can be written $B = A^{-1}$. An ordering relation $A \geq B$ for self-adjoint bounded linear operators is defined by $A - B \geq \mathbb{O}$.

A non-zero vector $v \in \mathcal{H}$ is an *eigenvector* of the linear operator A if $Av = \lambda v$, for any scalar λ, which is said to be the *eigenvalue* of A corresponding to

---

[1] Together, the linear and anti-linear operators are fundamental to quantum mechanics.

v; one can then write $(A - \lambda\mathbb{I})v = \mathbf{0}$. By considering the linear operator A in its matrix representation, the solutions to this *eigenvalue problem* can be found by solving the *characteristic* equation $\det(A - \lambda\mathbb{I}) = 0$, the left-hand side of which, in cases where A has a *finite set* (*spectrum*) of eigenvalues, is an nth-degree polynomial in λ.

The *adjoint*, $A^\dagger$ of the operator A is defined by the property that $(A^\dagger v, w) = (v, Aw)$ for every $v, w \in \mathcal{H}$. Any operator for which $A^\dagger = A$ is said to be (Hermitian) *self-adjoint* and has the two following properties. (i) All eigenvalues are real. (ii) Any two eigenvectors v_1 and v_2 with corresponding eigenvalues λ_1 and λ_2, respectively, are orthogonal to each other when λ_1 and λ_2 are non-identical. A linear operator O is *unitary* if $OO^\dagger = O^\dagger O = \mathbb{I}$, in which case $O^\dagger = O^{-1}$. Unitary operators are usually designated by the symbol U and have the following properties.

(i) The rows of U form an orthonormal basis.
(ii) The columns of U form an orthonormal basis.
(iii) U preserves inner products, that is, $(v, w) = (Uv, Uw)$ for all $v, w \in \mathcal{H}$.
(iv) U preserves norms and angles.
(v) The eigenvalues of U are of the form $e^{i\theta}$.

The matrix representing any unitary transformation U on a Hilbert space of countable dimension d can be diagonalized as above, to take the form

$$\begin{pmatrix} e^{i\theta_1} & 0 & \cdots & 0 \\ 0 & \ddots & \ddots & 0 \\ \vdots & \ddots & \ddots & \vdots \\ 0 & \cdots & 0 & e^{i\theta_d} \end{pmatrix}.$$

A bounded linear operator $\mathcal{O} \in B(\mathcal{H})$ is *positive*, $\mathcal{O} \geq \mathbb{O}$, if $\langle\psi|\mathcal{O}|\psi\rangle \geq 0$ for all $|\psi\rangle \in \mathcal{H}$. The set of positive operators is convex. Positive operators are Hermitian, always admit a positive square root, and, if invertible, have a unique decomposition into *polar form*, by which is meant that such an operator $\mathcal{O}$ can be written $\mathcal{O} = |\mathcal{O}|\mathcal{U}$ with $\mathcal{U}$ a unitary operator and $|\mathcal{O}| = \sqrt{\mathcal{O}\mathcal{O}^\dagger}$, analogous to the polar form of a complex number.

An Hermitian operator A is a *projection operator* (projector) if and only if $A^2 = A$, in which case it is usually denoted $P(S)$, where S is a subspace of $\mathcal{H}$, often a ray. A bounded linear operator is a *trace-class* operator if its *trace*, $\text{tr}A \equiv \sum_k (Av_k, v_k)$, is absolutely convergent for any orthonormal basis $\{v_k\}$ of $\mathcal{H}$. The trace is a linear functional over the space of trace-class operators, that is, $\text{tr}(aA+bB) = a\,\text{tr}A+b\,\text{tr}B$, for A, B trace-class, and is *cyclic*, *i.e.* $\text{tr}(AB) = \text{tr}(BA)$. The bilinear map $\langle A, B\rangle \equiv \text{tr}(A^\dagger B)$ is an inner product on the trace-class operators, and provides the *Hilbert–Schmidt norm*. By contrast, the *spectral norm* of an operator described by an $n \times n$ complex matrix A is $\max\{|\lambda| \,|\lambda \in \text{Spec}(A)\}$, where $\text{Spec}(A)$ is the eigenvalue spectrum of A; it is the square root of the *spectral radius* of A, which is the largest eigenvalue of $A^\dagger A$.

A.2 The Standard Postulates

A set of postulates for standard quantum mechanics is given here, which similar to those given by von Neumann.[2] In the standard approach to quantum mechanics, the pure states of systems are elements of complex Hilbert spaces.

The first postulate is often referred to as the *superposition principle*.

Postulate I:
Each physical system is represented by a Hilbert space and described by physical quantities and a state represented by linear operators in that space.

The Hilbert space is usually taken to be complex projective Hilbert space. The treatment of the motion of quantum particles in space requires the use of an infinite-dimensional separable Hilbert spaces.

The second postulate is known as the Born rule.

Postulate II:
Each physical quantity of a quantum system is represented by a positive Hermitian operator O, the expectation value of which is given by $\text{tr}(\rho O)$, where ρ is the bounded positive Hermitian trace-class operator representing the state of the system.

The third postulate is commonly referred to as the Dirac–von Neumann projection postulate.

Postulate III:
When a physical quantity of a system initially prepared in a state represented by the statistical operator ρ is measured, the state of the system immediately after this measurement is represented by the statistical operator $\rho' = P_k \rho P_k / \text{tr}(\rho P_k)$, where P_k is the projection operator onto the subspace corresponding to measurement outcome k, with a probability given by the expectation value of P_k for ρ.

This postulate and alternative versions of the projection postulate for different situations are discussed in greater detail Chapter 2.

The fourth postulate provides a prescription for representing systems.

Postulate IV:
Each physical system composed of two or more subsystems is represented by the Hilbert space that is the tensor product of the Hilbert spaces representing its subsystems; the operators representing its physical quantities act in this product space.

The tensor product structure is described in Appendix A.

---

[2] A related formulation is given by Arno Bohm, whose textbook takes great to incorporate all situations in non-relativistic quantum mechanics through the use of variants of some postulates [50].

The fifth postulate is most commonly referred to as the *Schrödinger evolution*.

Postulate V:

The temporal evolution of the state of each closed physical system, that is, each physical system not interacting with anything outside of itself, takes place according to $\rho(t) = U(t)\rho(0)U^{\dagger}(t)$ (in the "Schrödinger picture"), where t is the time parameter and $U = e^{-itH/\hbar}$ is a *unitary operator*, H being the generator of time-translations.

This postulate provides the natural temporal evolution of closed systems, which provides a linear transformation of state-vectors.

A.3 The Dirac Notation

The Hilbert-space structures described in the previous section can all be written in *Dirac notation*, which we now introduce. The state of a physical system described by a state-vector v is written as a *ket*, $|v\rangle$, and corresponds to a pure statistical operator; the corresponding Hermitian adjoint is given by a *bra*, $\langle v|$. The inner product (v, w) of two such vectors is written as the *braket* $\langle v|w\rangle$, and is a complex scalar. Operators acting from the left on a ket yield a ket and acting from the right on a bra yield a bra. A *ketbra*, $|v\rangle\langle w|$, is an operator that, when acting on a ket $|u\rangle$, yields

$$|v\rangle\langle w|(|u\rangle) = \langle w|u\rangle|v\rangle = (w, u)|v\rangle \ . \tag{A.1}$$

Every statistical operator ρ can be written as a linear combination of (pure) *projector ketbras*, $P(|u_i\rangle) \equiv |u_i\rangle\langle u_i|$, having weights p_i. The inner product of vectors, $\langle v|w\rangle$ taking $|v\rangle = \sum_i \alpha_i|i\rangle$ and $|w\rangle = \sum_i \beta_i|i\rangle$, is therefore

$$\langle v|w\rangle \doteq (\alpha_1^* \ \alpha_2^* \cdots) \begin{pmatrix} \beta_1 \\ \beta_2 \\ \vdots \end{pmatrix} . \tag{A.2}$$

The row vector $(\alpha_1^*\alpha_2^* \cdots)$ represents $\langle v|$ and the column vector $(\beta_1\beta_2 \cdots)^{\mathrm{T}}$ represents $|w\rangle$. The general ketbra $|v\rangle\langle w|$ can be written as the outer product

$$|v\rangle\langle w| \doteq \begin{pmatrix} \alpha_1 \\ \alpha_2 \\ \vdots \end{pmatrix} (\beta_1^* \ \beta_2^* \cdots). \tag{A.3}$$

Recalling that the projector $P(|v\rangle) = |v\rangle\langle v|$, one thus has, for any $|w\rangle$, $P(|v\rangle)|w\rangle = (\langle v|w\rangle)|v\rangle$; see Figure 2.1. Note that $P^2 = |v\rangle\langle v|v\rangle\langle v| = P$ because $|v\rangle$ has norm 1, that is, projectors are idempotent.

The matrix for an operator O and basis states $|i\rangle$ and $|j\rangle$ has elements $\langle j|O|i\rangle \in \mathbb{C}$. The representation of an operator by a collection of matrix elements is relative to the choice of eigenbasis. Hermitian operators (observables) correspond to physical magnitudes and have matrices with real diagonal elements O_{ii} and complex off-diagonal elements such that $O_{ij} = O_{ji}^*$. The matrix representation of a statistical operator ρ, such as is necessary to describe mixed quantum states, is known as a *density matrix* and is designated by the same symbol. When the state is pure, the density matrix is of rank one. Recalling the spectral representation, one can define a function of an operator O by

$$f(O) = \sum_n f(o_n)P(|o_n\rangle) , \qquad (A.4)$$

under appropriate conditions.

The *tensor product* of two vectors $|v\rangle$ and $|w\rangle$ is written $|v\rangle \otimes |w\rangle$. The tensor product space, written $V \otimes W$, is the linear space formed by such products of vectors: given bases $|v_1\rangle, \ldots, |v_k\rangle$ and $|w_1\rangle, \ldots, |w_l\rangle$ for two vector spaces V and W, respectively, a corresponding basis for $V \otimes W$ is given by

$$\{|v_i\rangle \otimes |w_j\rangle : 1 \leq i \leq k, 1 \leq j \leq l\},$$

and $\dim(V \otimes W) = kl$. Any vector $|\Psi\rangle \in V \otimes W$ can be written in the form

$$|\Psi\rangle = \sum_{ij} \alpha_{ij}|v_i\rangle|w_j\rangle , \qquad (A.5)$$

where the α_{ij} are corresponding scalar components. Every linear operator O in such a tensor product space $V \otimes W$, where

$$O(v_i \otimes w_j) \equiv (O_1 v_i) \otimes (O_2 w_j) , \qquad (A.6)$$

is a linear combination of direct products of linear operators, namely,

$$O = \sum_i O_1^{(i)} \otimes O_2^{(i)} . \qquad (A.7)$$

A.4 The Classification of Entangled States

Group theory has provided a basis for contemporary conceptions of entanglement that have been developed in quantum information science which have an operational basis.[3] This is largely due to the fact that a specific group of transformations can be associated with each of the classes of operations that can be considered. In particular, the general approach to the classification of entangled states is to examine the inherent transformational properties of

[3] Note that the term 'operational' in this context is more restricted than the standard philosophical use of the term.

states and to identify equivalence classes of states under these transformations, which themselves correspond to well known group structures, in particular based on the accessibility of states from each other by local operations as
mentioned above. A set G, together with a product map $G \times G \to G$, forms
a *group* if it satisfies the following conditions.

(1) Multiplication is *associative*: $a(bc) = (ab)c$ for all $a, b, c \in G$.
(2) An *identity element* $e \in G$ exists, for which $eg = ge = g$ for all $g \in G$.
(3) An *inverse* $g^{-1} \in G$ exists for every $g \in G$, such that $g^{-1}g = gg^{-1} = e$.

A map θ between two groups G and H is a *group homomorphism* if $\theta(g_1 g_2) = \theta(g_1)\theta(g_2)$ for all $g_1, g_2 \in G$. An *action* of a group G on another set S is
given by a map $G \times S \to S$ such that $g_2(g_1 s) = (g_1 g_2)s$ and $es = s$ for any
$s \in S$, that is, a homomorphism from the group into the group of one-to-one
transformations of S. A *unitary representation* of a group on a vector space
assigns unitary operators U on the space such that $U(gh) = U(g)U(h)$ for
all $g, h \in G$; a mere *projective representation* is a representation for which
$U(gh) = \omega(g, h)U(g)U(h)$, where $\omega(g, h)$ is a phase. A representation is *irreducible* if there is no vector subspace that is mapped to itself by every element
of the representation. Representations are *equivalent* if there is an isomor-
phism M between them such that $MU(g) = U(g)M$. The *orbit* $\mathcal{O} \equiv G \cdot S$
of an element m of a set S under the action of a group G is the subset of S
given by $\{gm | g \in G\}$, g ranging over all elements of G. Orbits stratify the set
of quantum states of a system.

The orbit of a statistical operator ρ under the group $U(n)$ of unitary operators of dimension n is determined by its spectrum. An orbit $\mathcal{O}$ can therefore
be specified by a representative diagonal matrix, the eigenvalues of which are
ordered from greatest to smallest. Lower bounds on the number of parameters needed to describe equivalence classes have been provided that show the
insufficiency of the total set of state descriptions of local systems for specifying the state description of the *compound* system they comprise. The extra
parameters are known as 'hidden non-localities.' One can find equivalence
classes under local unitary transformations (LUTs) of the statistical operator
and equivalently under (local) rotations of the Stokes tensor, the tensor of expectation values of the tensor product of Pauli matrices, because compound
states are equivalent in their global behavior if they can be transformed into
each other by such operations. Let S be the vector subspace kept fixed by a
subgroup (the *stabilizer*, $\mathcal{S}$) of elements of G_n, the n-subsystem Pauli group
formed (essentially) by the tensor product of Pauli matrices. This is equivalent to invariance under the choice of local Hilbert space basis. Generally, each
group $\mathcal{G}$ of transformations acts transitively on an orbit $\mathcal{O} = \mathcal{G}/\mathcal{S}$, where $\mathcal{S}$ is
the stabilizer subgroup of the orbit. Because states of N qubits are equivalent
in entanglement when they lie on the same orbit under LUTs of the statistical operator, each such orbit corresponds to a single entanglement class with
characteristic invariant quantities.

The orbits have specific dimensionalities, $\dim\mathcal{O}$, given by the dimension $\dim\mathcal{S}$ of the stabilizer subgroups of states on the orbit and the dimension $\dim\mathcal{G}$ of the group in question: $\dim\mathcal{O} = \dim\mathcal{G} - \dim\mathcal{S}$. For LUTs, $\mathcal{G}$, being local, has elements of the form $U_1 \otimes U_2 \otimes \cdots \otimes U_N$ so that each unitary transformation U_i acts on a Hilbert space corresponding to a component of the total system in the possession of a single party in its local laboratory. The dimension of the orbit is just the number of real parameters required to specific the location of a state in the orbit. The Hilbert space of pure states of N parties, each in possession of a single qubit is $\mathcal{H}^{(N)} = \mathbb{C}^2 \otimes \mathbb{C}^2 \otimes \cdots \otimes \mathbb{C}^2$. Any pure state of the compound system is therefore described by $2(2^N - 1)$ real parameters, because there are 2^N *complex* parameters and so 2^{N+1} real parameters describing any state on this space, of which normalization reduces the number of real parameters by one, as does the freedom of global phase. The number of parameters describing a state thus grows *exponentially* with the number of components, N. This fact has particularly significant ramifications for quantum computation, discussed in Chapter 4.

More broadly, one can consider equivalence classes of multipartite states under accessibility via SLOCC. In particular, on can find equivalence classes of multipartite states via the criterion of mutual accessibility via invertible local operations: a *local invertible operator* (ILO) is an operator that can be written in tensor product form where each factor has a well-defined inverse and acts in a single-party Hilbert subspace. Recall that SLOCC transformations are local quantum operations together with classical communication that transform states with some finite *probability* of success, rather than with certainty. The number of state parameters that can be altered by a multiparty ILO grows linearly in the number of parties, being $6N$; a single-qubit ILO described by a four-complex-component matrix is required to have a non-zero determinant scalable to unity because multiplication by a scalar does not affect accessibility, and depends only on six real parameters [148].

Two pure states are of the same class in this sense if the parties involved have a chance of successfully converting one state into another under SLOCC, that is, if $|\Psi'\rangle = M_1 \otimes M_2 \otimes \cdots \otimes M_N|\Psi\rangle$, where $M_i \in SL(d,\mathbb{C})$ is an ILO acting on the d_i-dimensional Hilbert space of subsystem i [38]. It is difficult to find canonical states on the orbits of these multipartite states because the set of equivalence classes of multi-qubit states under SLOCC, in the space of orbits $\mathcal{H}^{(N)}/(SL(2,\mathbb{C}) \times SL(2,\mathbb{C}) \times \cdots \times SL(2,\mathbb{C}))$, depends on at least $[2(2^N - 1) - 6N]$ parameters [148]. For $N = 2,3$ there is a finite number of equivalence classes, but there may be an *infinite* number for $N > 3$. The situation when one party possesses *more than one qubit* is worse, even in the case of three parties. In the case of two parties, there is a maximally entangled state from which all states may be reached with certainty; in the case of three parties, there is generally no such state [297].

One way of proceeding in the study of entanglement despite these difficulties is to consider symmetrical states, which can be generated via special operations. In particular, the LOCC protocols known as *twirling operations*, have been applied to bipartite states of systems described by finite-dimensional Hilbert spaces, to make limited further progress in this line of investigation *cf.* [271]. The twirling operations, written in the form of group averages, namely,

$$\mathcal{T}_1 = \int_{U(d)} dU \otimes U^* \rho \, (U \otimes U^*)^\dagger \, , \mathcal{T}_2 = \int_{U(d)} dU \otimes U \rho \, (U \otimes U)^\dagger \, ,$$

have the following effects on states ρ.

$$\mathcal{T}_1(\rho) \to \rho_{\text{iso}}(F) = \tfrac{1-F}{d^2-1}\mathbb{I}_{d^2} + \tfrac{Fd^2-1}{d^2-1}P(|\Psi_d^+\rangle) \, , \mathcal{T}_1(\rho) \to \rho_{\text{W}}(\epsilon) = \tfrac{1}{d(d+\epsilon)}V \, ,$$

where $F = \text{tr}\big(\rho P(|\Psi_d^+\rangle)\big)$, $|\Psi_d^+\rangle = \frac{1}{\sqrt{d}}\sum_{i=}^{d-1}|i\rangle|i\rangle$ is a maximally entangled bipartite state in the product of two d-dimensional Hilbert spaces, V is the swap operator which takes $|\phi\rangle \otimes |\psi\rangle \to |\psi\rangle \otimes |\phi\rangle$, dU is the normalized left-invariant Haar measure on U, and ρ_{W} is a Werner state. One of these operations, $\mathcal{T}_1$, maps the quantum state ρ into an isotropic state, which is separable if and only if $F \leq \frac{1}{d}$, which characterizes the PPT property; the Werner states are separable iff they have the PPT property. If a state $\rho \in \mathcal{H}_d \otimes \mathcal{H}_d$ has a PPT, then it is non-distillable, that is, a bound state [237].

A.5 Elements of Traditional and Quantum Logic

A *Boolean algebra* B_n is an algebraic structure given by the collection of 2^n subsets of the set $I = \{1, 2, \ldots, n\}$ and three operations under which it is closed: the two binary operations of union ($\vee$) and intersection ($\wedge$), and a unary operation, complementation ($\neg$). In addition to there being complements (and hence the null set $\emptyset$ being an element), one assumes
   (1) Commutativity: $S \vee T = T \vee S$ and $S \wedge T = T \wedge S$;
   (2) Associativity: $S \vee (T \vee U) = (S \vee T) \vee U$ and $S \wedge (T \wedge U) = (S \wedge T) \wedge U$;
   (3) Distributivity: $S \wedge (T \vee U) = (S \wedge T) \vee (S \wedge U)$ and $S \vee (T \wedge U) = (S \vee T) \wedge (S \vee U)$; and
   (4) $\neg \emptyset = I$, $\neg I = \emptyset$, $S \wedge \neg S = \emptyset$, $S \vee \neg S = I$, $\neg(\neg S) = S$,
for all its elements S, T, U. The algebra B_1 is the *propositional calculus* arising from the set $I = \{1\}$, which is also used in digital circuit theory, where $\emptyset$ corresponds to FALSE, I to TRUE, $\vee$ to OR, $\wedge$ to AND, and $\neg$ to NOT. The 'basic' logic operation XOR corresponds to $(S \vee T) \wedge (\neg S \vee \neg T)$. A σ-*algebra* is a nonempty collection S of subsets of a set X such that

   (1) The empty set $\emptyset$ is in S;
   (2) If A is in S, then the complement of A in X is in S; and
   (3) If A_n is a sequence of elements of S, then the union of the elements of the sequence is also in S.

A *poset* (partially ordered set) P is a set S together with a binary (partial ordering) relation, $\leq$, that is

(1) Reflexive: $a \leq a$;

(2) Antisymmetric: $a \leq b$ and $b \leq a$ implies that $a = b$, for all $a, b \in S$; and

(3) Transitive: $a \leq b$ and $b \leq c$ implies that $a \leq c$, for all $a, b, c \in S$.

The *least upper bound* (lub) of two elements, a and b, under $\leq$ is written $a \vee b$; the *greatest lower bound* (glb) is written $a \wedge b$. An *orthomodular poset* is a poset, with a unary operation $^\perp$, fulfilling the following four conditions.

(1) $0 \leq a \leq 1$ for all $a \in P$, 0 being the zero element and 1 the unit;

(2) For all $a, b \in P$, $(a^\perp)^\perp = a$, $a \leq b \Rightarrow b^\perp \leq a^\perp$, $a \vee a^\perp = 1$;

(3) If $a \leq b^\perp$ then $a \vee b \in P$; and

(4) If $a \leq b$, then there is an element $c \in P$ such that $c \leq a^\perp$ and $b = a \vee c$.

Condition (2) ensures that the operation $^\perp : P \rightarrow P$, corresponding to set-theoretic complementation, is an orthocomplementation; (4) is the orthomodular law. Two elements a and b of an orthomodular poset are said to be *orthogonal* if $a \leq b^\perp$.

A *lattice* is a poset for which there exists both a lub and a glb for every pair of elements. A lattice contains both a zero element, 0, and an identity element, 1, if $0 \leq a$ and $a \leq 1$ for every one of its elements a. A lattice is a *complemented lattice* if there exists a complement, $a^\perp$, for every one of its elements, a— that is, if for every a there exists an element $a^\perp$, such that $a \vee a^\perp = 1$ and $a \wedge a^\perp = 0$. A lattice is a *distributive lattice* if for all triplets of elements a, b, c, $a \wedge (b \vee c) = (a \wedge b) \vee (a \wedge c)$ and $a \vee (b \wedge c) = (a \vee b) \wedge (a \vee c)$. An *orthomodular lattice* is an orthomodular poset that is a lattice. A *Boolean lattice* (or *Boolean algebra*) is a lattice that is both complemented and distributive. Every element of a Boolean lattice has a unique complement that is an orthocomplement. An *orthomodular lattice* is an orthomodular poset that is a lattice. Elements a and b of an orthomodular poset are *orthogonal* if $a \leq b$. Given two orthomodular posets P_1 and P_2, P_1 is *orthorepresentable* in P_2 if there exists a mapping, called the *orthoembedding*, $h: P_1 \rightarrow P_2$, such that, for every $a, b \in P_1$:

(1) $h(0) = 0$,

(2) $h(a^\perp) = h(a)^\perp$,

(3) $a \leq b$ if and only if $h(a) \leq h(b)$, and

(4) $h(a \vee b) = h(a) \vee h(b)$ whenever $a \perp b$.

An orthomodular poset P_1 is *representable* in another orthomodular poset P_2 if there exists a mapping $h : P_1 \rightarrow P_2$ such that h is an *orthoembedding* for which $h(a \vee b) = h(a) \vee h(b)$ for every $a, b \in P_1$. The set $h(P_1)$ is then an *orthorepresentation* of P_1 in P_2. A *Boolean subalgebra* of an orthomodular poset is a suborthoposet that is then a Boolean algebra.

A.6 C*-Algebras

A *Banach space* is a normed space in which every Cauchy sequence converges. The observables of a quantum system form a real Banach (*i.e.* complete normed linear) space $\mathcal{A}$ such that $A \in \mathcal{A}$. The powers of $A, A^n \in \mathcal{A}$ for n= 0, 1, ..., are well-defined and such that the usual rules for operating with polynomials in a single variable hold. A *Banach *-algebra* is an algebra of operators that form a Banach algebra with respect to the operator norm and has defined on it an *involution* $A \to A^*$, satisfying i) $(A + B)^* = A^* + B^*$, ii) $(cA)^* = c^* A^*, \forall c \in \mathcal{C}$, iii) $(AB)^* = A^* B^*$, and iv) $(A)^{**} = A$, for all $A, B \in \mathcal{A}$. In the context of quantum mechanics, one identifies * with †, (Hermitian) adjoint. A *-subalgebra is sometimes called an *operator *-algebra*. The set of linear operators on Hilbert space fulfills these conditions. A Banach algebra with involution, $\mathcal{A}$ can be given a representation on a Hilbert space H, by a linear map $\pi : \mathcal{A} \to B(H)$, into the bounded linear operators on II such that i) $\pi(AB) = \pi(A)\pi(B)$, and ii) $\pi(A^*) = \pi(A)^*$, for all $A \in \mathcal{A}$. The Hilbert space is then called the *representation space*. Every norm-closed *-subalgebra of the algebra of all bounded operators on a Hilbert space, with the norm induced by the inner product is a Banach *-algebra; it also fulfills the relations $||A^*|| = ||A||$ and $||A^*A|| = ||A^*|| \, ||A||$, so that $||A^*A|| = ||A||$. A *C*-algebra* is Banach algebra for which $||A^*A|| = ||A||^2$, for all its elements. Quantum mechanics can thus be viewed as a C*-system where the C*-algebra is that of the set of all *bounded* self-adjoint operators acting on finite-dimensional Hilbert space. A *W*-algebra* $\mathcal{A}$ is a C*-algebra, the dual space of a Banach space $\mathcal{A}_*$, so that $(\mathcal{A}_*)^* = \mathcal{A}$.

The *C*-algebraic approach to quantum mechanics* is based on the abstract algebra $\mathcal{A}$, the self-adjoint elements (the *observables*) of which form a Jordan algebra. Quantum states are normalized positive linear function ρ on the C*-algebra $\mathcal{A}$. Every state ρ assigns to each $A \in \mathcal{A}$ a complex number $\rho(A)$ in such a way that we have a ρ which is i) $\rho(1) = 1$, normalized, ii) $\rho(A^*A) \geq 0$, positive, and iii) $\rho(aA + bB) = a\rho(B) + b\rho(B)$, for every $A, B \in \mathcal{A}$, for all $a, b \in \mathcal{C}$. The algebra of observables can be taken to be the algebra $\mathcal{B}(H)$ of all bounded observables acting on a complex Hilbert space H. Every normalized vector $\Psi \in H$ defines a state $\rho_\Psi(A) \equiv \langle \Psi | A\Psi \rangle$, for all $A \in \mathcal{A}$. Such a state is known as a *vector state*. Generally, every density operator $\rho_D \in \mathcal{B}(H)$. defines a state $\rho_D(A) \equiv tr(DA)$, for all $A \in \mathcal{A}$. Such states are *normal states*. A normal state is pure if and only if D is a projection, *i.e.* $D^2 = D$; it is therefore pure if and only if it is a vector state.

Note that most C*-algebras have an uncountable number of unitarily inequivalent representations. It was hoped that all quantum-mechanical theories could be formulated independent of representation, but this has not been the case thus far. Another difficulty for C*-algebraic quantum mechanics is that it presupposes a statistical interpretation, and in any non-classical system (*i.e.* when the algebra of observables does not commute) a mixed statistical state cannot determine the ensemble since the set of states.

References

1. Abramson, N., *Information theory and coding* (McGraw-Hill; New York, 1963).
2. Aerts, D., "The stuff the world is made of: Physics and reality," in D. Aerts *et al.* (Eds.), *Einstein meets Magritte: An interdisciplinary reflection* (Kluwer Academic; Dordrecht, 1999).
3. Alexandrov, A. D., "On the meaning of the wave function," Doklady Akademii Nauk SSSR **85**, 292 (1952).
4. Allahverdyan, A. E., A. Khrennikov, and Th. M. Nieuwenhuizen, "Brownian entanglement," Phys. Rev. A **72**, 032102 (2005).
5. Anders, J., J. Hajdušek, D. Markham, and V. Vedral, "'How much of one-way computation is *just* thermodynamics?" Found. Phys. **38**, 506 (2008).
6. Ann, K., and G. S. Jaeger, "Disentanglement and decoherence in two-spin and three-spin systems under dephasing," Phys. Rev. B **75**, 115307 (2007).
7. Ann, K., and G. S. Jaeger, "Generic tripartite Bell non-locality sudden death under local phase noise," Physics Letters A (to appear; http://dx.doi.org/10.1016/j.physleta.2008.10.003).
8. Anscombe, F. R., "Quiet contributor: The civic career and times of John W. Tukey," (special Tukey issue) Stat. Sci. **18** (3), 287 (2003).
9. Arndt, M., O. Nairz, J. Vos-Andrae, C. Keller, G. van der Zouw, and A. Zeilinger, "Wave-particle duality of C_{60} molecules," Nature **401**, 680 (1999).
10. Arrighi, P., and C. Patricot, "Conal representation of quantum states and non-trace-preserving quantum operations," Phys. Rev. A **68**, 042310 (2003).
11. Aspect, A., J. Dalibard, and G. Roger, "Experimental realization of Einstein–Podolsky–Rosen–Bohm gedankenexperiment: A new violation of Bell's inequalities," Phys. Rev. Lett. **49**, 91 (1982).
12. Aspect, A., P. Grangier, and G. Roger, "Experimental test of realistic theories via Bell's inequality," Phys. Rev. Lett. **47**, 460 (1981).
13. Ayer, A. J. (Ed.), *Logical positivism* (The Free Press; New York, 1959).
14. Bacciagaluppi, G. and M. Hemmo, "Making sense of approximate decoherence," in *PSA: Proceedings of the Biennial Meeting of the Philosophy of Science Association, Vol. 1994, Volume One: Contributed Papers* (1994), p. 345.
15. Balaguer, M., "Platonism in metaphysics," The Stanford Encyclopedia of Philosophy (Spring 2004 Edition), Edward N. Zalta (Ed.), URL = http://plato.stanford.edu/archives/win2003/entries/platonism/.

16. Ballentine, L., "The statistical interpretation of quantum mechanics," Rev. Mod. Phys. **42**, 385 (1970).
17. Barnum, H., E. Knill, and N. Linden, "On quantum fidelities and channel capacities," IEEE Trans. Inform. Theory **46**, 1317 (2000).
18. Barnum, H., and N. Linden, "Monotones and invariants for multi-particle quantum states," J. Phys. A **34**, 6787 (2001).
19. Barrett, J. A., *The quantum mechanics of minds and worlds* (Oxford University Press; Oxford, 1999).
20. Barrett, J., N. Linden, S. Massar, S. Pironio, S. Popescu, and D. Roberts, "Nonlocal correlations as an information-theoretic resource," Phys. Rev. A **71**, 022101 (2005).
21. Barrett, J., and S. Pironio, "Popescu–Rohrlich correlations as a unit of non-locality," Phys. Rev. Lett. **95**, 140401 (2005).
22. Bell, J. S., "On the Einstein–Podolsky–Rosen paradox," Physics **1**, 195 (1964).
23. Bell, J. S., "On the problem of hidden variables in quantum mechanics," Rev. Mod. Phys. **38**, 447 (1966).
24. Bell, J. S., *Speakable and unspeakable in quantum mechanics* (Cambridge University Press; Cambridge, 1987).
25. Bell, J. S., "Against 'measurement'," in A. J. Miller (Ed.), *Sixty-two years of uncertainty* (Plenum; New York, 1990), pp. 17-32.
26. Beller, M., "The rhetoric of antirealism and the Copenhagen spirit," Phil. Sci. **63**, 183 (1996).
27. Beller, M., *Quantum dialogues* (University of Chicago Press; Chicago, 1999).
28. Benioff, P., "The computer as a physical system: A microscopic quantum mechanical Hamiltonian model of computers as represented by Turing machines," J. Stat. Phys. **22**, 563 (1980).
29. Benioff, P., "Models of quantum Turing machines," Fortschr. Phys. **46**, 423 (1998).
30. Bennett, C. H., "The thermodynamics of computation: A review," Int. J. Theor. Phys. **21**, 905 (1982).
31. Bennett, C. H., "Time/space trade-offs for reversible computation," SIAM J. Comput. **18**, 766 (1989).
32. Bennett, C. H., "Quantum cryptography using any two nonorthogonal states," Phys. Rev. Lett. **68**, 3121 (1992).
33. Bennett, C. H., H. Bernstein, S. Popescu, and B. Schumacher, "Concentrating partial entanglement by local operations," Phys. Rev. A **53**, 2046 (1996).
34. Bennett, C. H., G. Brassard, C. Crépeau, R. Jozsa, A. Peres, and W. K. Wootters, "Teleporting an unknown quantum state via dual classical and Einstein–Podolsky–Rosen channels," Phys. Rev. Lett. **70**, 1895 (1993).
35. Bennett, C. H., G. Brassard, S. Popescu, B. Schumacher, J. Smolin, and W. Wootters, "Purification of noisy entanglement and faithful teleportation via noisy channels," Phys. Rev. Lett. **76**, 722 (1996).
36. Bennett, C. H., D. P. DiVincenzo, and J. A. Smolin, "Capacities of quantum erasure channels," Phys. Rev. Lett. **78**, 3217 (1997).
37. Bennett, C. H., D. P. DiVincenzo, J. A. Smolin, and W. K. Wootters, "Mixed-state entanglement and quantum error correction," Phys. Rev. A **54**, 3824 (1996).
38. Bennett, C. H., S. Popescu, D. Rohrlich, J. A. Smolin, and A. V. Thapliyal, "Exact and asymptotic measures of multipartite pure state entanglement," Phys. Rev. A **63**, 012307 (2001).

39. Bennett, C. H., and P. W. Shor, "Quantum information theory," IEEE Trans. Inform. Theory **44**, 2724 (1998).
40. Bennett, C. H., P. W. Shor, J. A. Smolin and A. V. Thapliyal, "Entanglement-assisted classical capacity of noisy quantum channels," Phys. Rev. Lett. **83**, 3081 (1999).
41. Bennett, C. H., P. W. Shor, J. A. Smolin, and A. V. Thapliyal, "Entanglement-assisted capacity of a quantum channel and reverse Shannon theorem," IEEE Trans. Inform. Theory **48**, 2637 (2002).
42. Bennett, C. H., and S. J. Wiesner, "Communication via one- and two-particle operations on Einstein–Podolsky–Rosen States," Phys. Rev. Lett. **69**, 2881 (1992).
43. Bergou, J., M. Hillery, and Y. Sun, "From unambiguous quantum state discrimination to quantum state filtering," Fortschr. Phys. **51**, 74 (2003).
44. Bernstein, H. J., D. M. Greenberger, M. A. Horne, and A. Zeilinger, "Bell's theorem without inequalities for two spinless particles," Phys. Rev. A **47**, 78 (1993).
45. Bernstein, J., *Quantum profiles* (Princeton University Press; Princeton, 1991).
46. Beth, T., and M. Rötteler, in G. Alber, T. Beth, M. Horodecki, P. Horodecki, R. Horodecki, M. Rötteler, H. Weinfurter, R. Werner, and A. Zeilinger (Eds.), *Quantum information*, STMP **173**, VII–X (Springer-Verlag; Berlin, 2001), p. 96.
47. Bethe, H., "Quantum theory," Rev. Mod. Phys. **71**, S1 (1999).
48. Birkhoff, G., and J. von Neumann, "The logic of quantum mechanics," Ann. Math. **37**, 823 (1936).
49. Bitbol, M., *Schrödinger's philosophy of quantum mechanics* (Kluwer; Dordrecht, 1996).
50. Bohm, A., *Quantum mechanics: Foundations and applications* (Springer-Verlag; Berlin, 1979).
51. Bohm, D., *Quantum theory* (Prentice-Hall, Inc.; Englewood Cliffs NJ, 1951).
52. Bohm, D., "A suggested interpretation of quantum theory in terms of 'hidden' variables I.," Phys. Rev. **85**, 166 (1952); II. Phys. Rev. **85**, 180 (1952).
53. Bohm, D., *Causality and chance in modern physics* (Routledge and Kegan Paul; London, 1957).
54. Bohm, D., and Y. Aharonov, "Discussion of experimental proof for the paradox of Einstein, Rosen and Podolsky," Phys. Rev. **108**, 1070 (1957).
55. Bohr, N., "The quantum postulate and the recent development of atomic theory," in *Atti del congresso internationale dei fisici, Como, 11-20 Septembre 1927* (Zanachelli; Bologna, 1928), p. 565.
56. Bohr, N., "Das Quantenpostulat die neuere Entwicklung der Atomistik," Naturwissenschaften **16**, 245 (1928); "The quantum postulate and the recent development of atomic theory," Nature **121**, 580 (1928).
57. Bohr, N., *Atomic theory and the description of nature* (Cambridge University Press; Cambridge, 1934).
58. Bohr, N., "Can the quantum mechanical description of reality be considered complete?" Phys. Rev. **48**, 696 (1935).
59. Bohr, N., "Causality and complementarity," Phil. Sci. **4**, 289 (1937).
60. Bohr, N., "The causality problem in modern physics," in *New theories in physics* (International Institute of Intellectual Cooperation; Paris, 1939), p. 11.

61. Bohr, N., "On the notions of causality and complementarity," Dialectica **2**, 312 (1948).
62. Bohr, N., "Discussions with Einstein on epistemological problems in atomic physics," in P. A. Schilpp (Ed.), *Albert Einstein: Philosopher-scientist. The library of living philosophers, Volume 7, Part I* (Open Court; Evanston, IL, 1949), p. 201.
63. Bohr, N., *Essays 1932-1957 on atomic physics and human knowledge* (Wiley; New York, 1958).
64. Bohr, N., *Essays 1958-1962 on atomic physics and human knowledge* (Wiley; New York, 1963).
65. Bokulich, A., and G. Jaeger (Eds.), *Philosophy of quantum information and entanglement* (Cambridge University Press; Cambridge, 2009).
66. Borel, É., *Le Hasard* (F. Alcan; Paris, 1914).
67. Born, M., "Zur Quantenmechanik der Stoßvorgänge," Z. Phys. **37**, 863 (1926).
68. Born, M., *Natural philosophy of causality and chance* (Clarendon Press; London, 1949).
69. Born, M., "Ist die klassische Mechanik tatsächlich deterministisch?" Physikalische Blätter **2**, 49 (1955).
70. Born, M., "Physics and metaphysics," in *Physics in my generation* (Pergamon Press; London, 1956), p. 93.
71. Born, M., "Quantum mechanics of collision processes," in G. Ludwig (1968), *Wave mechanics* (Pergamon Press; Oxford, 1968), p. 207.
72. Born, M., and I. Born (translator), *The Born–Einstein letters* (Walker and Co.; London, 1971).
73. Born, M., W. Heisenberg, and P. Jordan, "Zur Quantenmechanik II," Zeit. Phys. **35**, 557 (1926).
74. Born, M., and P. Jordan, "Zur Quantenmechanik," Zeit. Phys. **34**, 858 (1926).
75. Boschi, D., S. Branca, F. De Martini, L. Hardy, and S. Popescu, "Experimental realization of teleporting an unknown pure quantum state via dual classical and Einstein–Podolsky–Rosen channels," Phys. Rev. Lett. **80**, 1121 (1998).
76. Bouwmeester, D., J. W. Pan, K. Mattle, M. Eibl, H. Weinfurter, and A. Zeilinger, "Experimental quantum teleportation," Nature **390**, 575 (1997).
77. Brassard, G., H. Buhrmann, N. Linden, A. A. Méthot, A. Tapp, and F. Unger, "A limit on nonlocality in any world in which communication complexity is not trivial," http://xxx.lanl.gov quant-ph/0508042 (4 Aug., 2005).
78. Braunstein, S. L., "Entanglement in quantum information processing," in Akulin, V. M., A. Sarfati, G. Kurizki, and S. Pellegrin (Eds.), *Decoherence, entanglement and information protection in complex quantum systems* (Springer; Dordrecht, 2005).
79. Braunstein, S. L., and Pati, A. K. (Eds.), *Quantum information with continuous variables* (Springer-Verlag; Berlin, 2003).
80. Braunstein, S. L., and P. van Loock, "Quantum information with continuous variables," Rev. Mod. Phys. **77**, 513 (2005).
81. Brenner, M., and H. A. Stone, "Modern classical physics through the work of G. I. Taylor," Physics Today, June 2000.
82. Bruckner, Č., and A. Zeilinger, "Quantum physics as a science of information," in A. Elitzur *et al.* (Eds.), *Quo vadis quantum mechanics?* (Springer; Heidelberg, 2005), p. 47.

83. Bruckner, Č., M. Żukowski, and A. Zeilinger, "Quantum communication complexity protocol with two entangled qubits," Phys. Rev. Lett. **89**, 197901 (2002).

84. Bub, J., "Comment on the Daneri–Loinger–Prosperi quantum theory of measurement," in T. Bastin (Ed.), *Quantum theory and beyond* (Cambridge University Press; Cambridge, 1971), p. 65.

85. Bub, J., *The interpretation of quantum mechanics* (D. Reidel Publishing Co.; Dordrecht, 1974).

86. Bub, J., "Von Neumann's projection postulate as a possible conditionalization rule in quantum mechanics," J. Phil. Logic **6**, 381 (1977).

87. Bub, J., *Interpreting the quantum world* (Cambridge University Press; Cambridge, 1997).

88. Bub, J., "Why the quantum?" Stud. Hist. Phil. Mod. Phys. **35**, 241 (2004).

89. Bub, J., "Quantum mechanics is about quantum information," Found. Phys. **35**, 541 (2005).

90. Bub, Jeffrey, "Quantum computation: Where does the speedup come from?" in [65] (2009).

91. Buhrman, H., R. Cleve, and W. van Dam, "Quantum entanglement and communication complexity," SIAM J. Comput. **30**, 1829 (2001).

92. Buhrman, H., W. van Dam, P. Høyer, and A. Tapp, "Multiparty quantum communication complexity," Phys. Rev. A **60**, 2737 (1999).

93. Bunge, M. (Ed.), *Quantum theory and reality* (Springer; New York, 1967).

94. Busch, P., "Quantum states and generalized observables: A simple proof of Gleason's theorem," Phys. Rev. Lett **91**, 120403 (2003).

95. Busch, P., M. Grabowski, and P. J. Lahti, *Operational quantum physics* (Springer-Verlag; Berlin, 1995).

96. Busch, P., T. Heinonen, and P. J. Lahti, "Heisenberg's uncertainty principle: Three faces, two rôles," http://arxiv.org/abs/quant-ph/0609185 (2006).

97. Busch, P., P. J. Lahti, and P. Mittelstaedt, *The quantum theory of measurement, Second, Revised edition* (Springer-Verlag; Berlin, 1996).

98. Busch, P. and C. Shilladay, "Complementarity and uncertainty in Mach-Zehnder interferometry and beyond," Phys. Rep. **435**, 1 (2006).

99. Callebaut, W., *Taking the naturalistic turn, or how real philosophy of science is done* (University of Chicago Press; Chicago, 1993).

100. Cassinelli, G., E. De Vito, and A. Levrero, "On the decompositions of a quantum state," J. Math. Anal. App. **210**, 472 (1997).

101. Caves, C., and C. A. Fuchs, "Quantum information: How much information is in a state vector?" in A. Mann and M. Revzen, *Sixty years of EPR*, Ann. Phys. Soc., Israel, 1996; also http://xxx.lanl.gov quant-ph/9601025 (1996).

102. Caves, C., C. A. Fuchs, and R. Schack, "Quantum probabilities as Bayesian probabilities," Phys. Rev. A **65**, 022305 (2002).

103. Caves, C., C. A. Fuchs, and R. Schack, "Subjective probability and quantum certainty," Stud. Hist. Phil. Mod. Phys. B **38**, 255 (2007).

104. Cerf, N. J., N. Gisin, S. Massar, and S. Popescu, "Simulating maximal quantum entanglement without communication," Phys. Rev. Lett. **94**, 220403 (2005).

105. Cerf, N. J., N. Gisin, and S. Popescu, "Simulating maximal quantum entanglement without communication," Phys. Rev. Lett. **94**, 220403 (2005).

106. Chalmers, D., *The conscious mind* (Oxford University Press; New York, 1996).

107. Clauser, J. F., M. Horne, A. Shimony, and R. A. Holt, "Proposed experiments to test local hidden-variable theories," Phys. Rev. Lett. **23**, 880 (1973).
108. Cleve, R., and H. Buhrman, "Substituting quantum entanglement for communication," Phys. Rev. A **56**, 1201 (1997).
109. Cleve, R., W. van Dam, M. Nielsen, and A. Tapp, "Quantum entanglement and the communication complexity of the inner product function," *Lecture notes in computer science* **1509**, 61 (1999).
110. Clifton, R., "The subtleties of entanglement and its role in quantum information theory," Phil. Sci. **69**, S150 (2002).
111. Clifton, R., J. Bub, and H. Halvorson, "Characterizing quantum theory in terms of information-theoretic constraints," Found. Phys. **33**, 1561 (2003).
112. Clifton, R., and H. Halvorson, "Bipartite-mixed-states of infinite-dimensional systems are generically nonseparable," Phys. Rev. A **61**, 02108 (2000).
113. Coffman, V., J. Kundu, and W. K. Wootters, "Distributed entanglement," Phys. Rev. A **61**, 052306 (2000).
114. Collins, D., N. Gisin, N. Linden, S. Massar, and S. Popescu, "Bell inequalities for arbitrarily high-dimensional systems," Phys. Rev. Lett. **88**, 040404 (2002).
115. Cooper, L. N., and D. van Vechten, "On the interpretation of measurement within quantum theory," Am. J. Phys. **37**, 1212 (1969).
116. Cushing, J. T., A. Fine, and S. Goldstein (Eds.), *Bohmian mechanics and quantum theory: An appraisal* (Kluwer; Dordrecht, 1995).
117. Dalla Chiara, M., R. Giuntini, and R. Greechie, *Reasoning in quantum theory: Sharp and unsharp quantum logics* (Springer-Verlag; Berlin, 2004).
118. Davies, E. B., *Quantum theory of open systems* (Academic Press; London, 1976).
119. Davies, P. C. W., and J. R. Brown (Eds.), *The ghost in the atom* (Cambridge University Press; Cambridge, 1986).
120. Davis, M., *The universal computer* (W. W. Norton and Co.; New York, 2000).
121. de Broglie, L., "La mécanique ondulatoire et la structure atomique de la matière et du rayonnement," J. Physique et du Radium **8**, 225 (1927).
122. de Broglie, L., "La nouvelle mécanique des quanta," in H. Lorentz (Ed.), *Rapports et discussions du cinquième conseil de physique Solvay* (Gauthier-Villars; Paris, 1928), p. 105.
123. de Laplace, P.-S., *Essai philosophique sur les probabilitiés* (Mme Ve Courcier; Paris, 1814); English translation by E. W. Truscott and F. L. Emory, *A philosophical essay on probabilities* (Dover; New York, 1951).
124. Deletete, R., and R. Guy, "Einstein and EPR," Phil. Sci. **58**, 377 (1991).
125. d'Espagnat, B. *Conceptual foundations of quantum mechanics* (Benjamin; Menlo Park, CA, 1971).
126. d'Espagnat, B. *On physics and philosophy* (Princeton University Press; Princeton, 2006).
127. Deutsch, D., "Quantum mechanics as a universal physical theory," Int. J. Theor. Phys. **24**, 1 (1985).
128. Deutsch, D., "Three connections between Everett's interpretation and experiment," in R. Penrose and C. J. Isham (Eds.), *Quantum concepts in space and time* (Clarendon Press; Oxford, 1986).
129. Deutsch, D., "Quantum computational networks," Proc. Roy. Soc. London A **425**, 73 (1989).
130. Deutsch, D., *The fabric of reality* (Penguin; London, 1997).

131. Deutsch, D., and P. Hayden, "Information flow in entangled quantum systems," http://xxx.lanl.gov quant-ph/9906007 (1999).
132. Deutsch, D., and R. Jozsa, "Rapid solution of problems by quantum computation," Proc. Roy. Soc. London A **439**, 553 (1992).
133. Devitt, M., *Realism and truth* (Princeton University Press; Princeton, 1984).
134. DeWitt, B. S., "Quantum mechanics and reality," Physics Today **23**, 30 (Sept. 1970).
135. DeWitt, B. S., and N. Graham, *The many-worlds interpretation of quantum mechanics* (Princeton University Press; Princeton, 1973).
136. Dickson, W. M., "An empirical reply to empiricism: Protective measurement opens the door for quantum realism," Phil. Sci. **62**, 122 (1995).
137. Dickson, W. M., *Quantum chance and non-locality* (Cambridge University Press; Cambridge, 1998).
138. Dieks, D., "Communication by EPR devices," Phys. Lett. A **92**, 271 (1982).
139. Dirac, P. A. M., *The principles of quantum mechanics* (Clarendon Press; Oxford, 1930).
140. Dirac, P. A. M., "The Lagrangian in quantum mechanics," Physik. Zeits. Sowjetunion **3**, 64 (1933); republished as an appendix in [168].
141. Dirac, P. A. M., "On the analogy between classical and quantum mechanics," Rev. Mod. Phys. **17**, 195 (1945).
142. Dirac, P. A. M., *The principles of quantum mechanics*, Fourth edition (Clarendon Press; Oxford, 1958).
143. Doherty, A. C., P. A. Parillo, and F. M. Spedalieri, "Distinguishing separable and entangled states," Phys. Rev. Lett. **88**, 187904 (2002).
144. Duan, L.-M., G. Giedke, J. I. Cirac, and P. Zoller, "Inseparability criterion for continuous variable systems," Phys. Rev. Lett. **84**, 2722 (2000).
145. Dummett, M., *Truth and other enigmas* (Harvard University Press; Cambridge MA, 1978).
146. Dummett, M., "Realism," Synthese **52**, 55 (1982).
147. Dür, W., and J. I. Cirac, "Classification of multiqubit states: Separability and distillability properties," Phys. Rev. A. **61**, 042314 (2000).
148. Dür, W., G. Vidal, and J. I. Cirac, "Three qubits can be entangled in two inequivalent ways," Phys. Rev. A. **62**, 062314 (2000).
149. Duwell, A., "How to teach an old dog new tricks: Quantum information, quantum computation, and the philosophy of physics," Ph.D. Thesis, University of Pittsburgh, 2004.
150. Einstein, A., "Quanten-Mechanik und Wirklichkeit," Dialectica **2**, 320 (1948).
151. Einstein, A., "Autobiographical notes," in P. A. Schilpp (Ed.), *Albert Einstein: Philosopher-scientist. The library of living philosophers, Volume 7, Part I* (Open Court; Evanston, IL, 1949), p. 1.
152. Einstein, A., "Remarks to the essays appearing in this collective volume," in P. A. Schilpp (Ed.), *Albert Einstein: Philosopher-scientist. The library of living philosophers, Volume 7, Part II* (Open Court; Evanston, IL, 1949), p. 663.
153. Einstein, A., B. Podolsky, and N. Rosen, "Can quantum-mechanical description of physical reality be considered complete?", Phys. Rev. **47**, 777 (1935).
154. Eisert, J. "Entanglement in quantum theory," Ph.D. Thesis, University of Potsdam, Potsdam, Germany, 2001.
155. Eisert, J., and H. J. Briegel, "Schmidt measure as a tool for quantifying entanglement," Phys. Rev. A **64**, 022306 (2001).

156. Eisert, J., C. Simon, and M. B. Plenio, "On the quantification of entanglement in infinite-dimensional quantum systems," J. Phys. A **39** 3911 (2002).

157. Ekert, A. K., "Quantum interferometers as quantum computers," Physica Scripta **T76**, 218 (1998).

158. Elby, A., "Why 'modal' interpretations of quantum mechanics don't solve the measurement problem," Found. Phys. Lett. **6**, 737 (1993).

159. Elby, A., and J. Bub, "Triorthogonal uniqueness theorem and its relevance to the interpretation of quantum mechanics," Phys. Rev. A **49**, 4213 (1994).

160. Englert, B.-G., and J. A. Bergou, "Quantitative quantum erasure," *Opt. Comm.* **179**, 337 (2000).

161. Englert, B.-G., J. Schwinger, and M. Scully, "Is spin coherence like Humpty-Dumpty? I. Simplified treatment," Found. Phys. **18**, 1045 (1988).

162. Enzer, D. G., P. G. Hadley, R. J. Hughes, C. G. Peterson, and P. G. Kwiat, "Entangled-photon six-state quantum cryptography," New J. Physics **4**, 45.1 (2002).

163. Everett, H. III, "On the foundations of quantum mechanics," Ph. D. thesis, Princeton University (1957).

164. Everett, H. III, " 'Relative state' formulation of quantum mechanics," Rev. Mod. Phys. **29**, 454 (1957).

165. Everett, H. III, "The theory of the universal wave function," in B. S. DeWitt and N. Graham (Eds.), *The many-worlds interpretation of quantum mechanics* (Princeton University Press; Princeton, 1973), p. 3.

166. Février, P., "Les relations d'incertitude de Heisenberg et la logique," Académie des Sciences (Paris) Comptes Rendus **204**, 481 (1937).

167. Feyerabend, P. K., "Problems of microphysics," in R. Colodny (Ed.), *Frontiers of science and philosophy, v. 1* (George Allen and Unwyn Ltd.; London, 1962), pp. 189-283.

168. Feynman, R. P., "The principle of least action in quantum mechanics," Ph.D. Thesis, Princeton University, 1942; reprinted in L. M. Brown (Ed.), *Feynman's thesis* (World Scientific; Singapore, 2005).

169. Feynman, R. P., "Space-time approach to non-relativistic quantum mechanics," Rev. Mod. Phys. **20**, 367 (1948).

170. Feynman, R. P., *The character of physical law* (MIT Press; Cambridge MA 1965).

171. Feynman, R. P., *The Feynman lectures on physics* [A. J. G. Hey and R. W. Allen (Eds.)] (Addison-Wesley; Reading, MA, 1965).

172. Feynman, R. P., *Feynman lectures on computation* (Addison-Wesley; Reading, MA, 1996).

173. Feynman, R. P., "Negative probability," in B. J. Hiley and F. D. Peat (Eds.), *Quantum implications* (Routledge and Kegan Paul; London, 1987), p. 235.

174. Fine, A., *The shaky game* (University of Chicago Press; Chicago, 1986).

175. Fleming, G., and H. Bennett, "Hyperplane dependence in relativistic quantum mechanics," Found. Phys. **19**, 231 (1989).

176. Folse, H. J., *The philosophy of Niels Bohr* (North-Holland; Amsterdam, 1985).

177. Forrest, P., *Quantum metaphysics* (Basil Blackwell Ltd.; Oxford, 1988).

178. Friedman, M. (Ed.), *Reconsidering logical positivism* (Cambridge University Press; Cambridge, 1999).

179. Fuchs, C., "Information gain vs. state disturbance in quantum theory," Fortschr. Phys. **46**, 535 (1998).

180. Fuchs, C., "Quantum foundations in the light of quantum information," http://xxx.lanl.gov quant-ph/0106166 (2001).
181. Fuchs, C., "Quantum mechanics as quantum information (and only a little more)," http://xxx.lanl.gov quant-ph/0205039 (2002).
182. Fuchs, C., and A. Peres, "Quantum theory needs no interpretation," Physics Today, **53** (3), 70 (2000).
183. Fuchs, C., and A. Peres, "Quantum theory—interpretation, formulation, inspiration: Fuchs and Peres reply," Physics Today **53** (9), 14 (2000).
184. Gell-Mann, M., and Hartle, J. B.. "Quantum mechanics in the light of quantum cosmology," in W. H. Żurek (Ed.), *Complexity, entropy, and the physics of information*," Proceedings of the Santa Fe Institute Studies in the Sciences of Complexity 8 (Addison-Wesley; Redwood City, CA, 1990), p. 425. Also [185].
185. Gell-Mann, M., and Hartle, J. B., "Quantum mechanics in the light of quantum cosmology," in S. Kobayashi, H. Ezawa, Y. Murayama, and S. Nomura (Eds.), Proceedings of the 3rd international symposium on the foundations of quantum mechanics in the light of new technology (Physical Society of Japan; Tokyo, 1990), p. 321.
186. Gershenfeld, N., *The physics of information technology* (Cambridge University Press; Cambridge, 2000).
187. Gibbins, P., *Particles and paradoxes* (Cambridge University Press; Cambridge, 1987).
188. Giedke, G., B. Kraus, M. Lewenstein, and J. I. Cirac, "Entanglement criterion for all bipartite quantum states," Phys. Rev. Lett. **87**, 167904 (2001).
189. Giles, R., *Mathematical foundations of thermodynamics. International series of monographs on pure and applied mathematics, Volume 53* (Pergammon; Oxford, 1964).
190. Gillespie, D. T., "Untenability of simple statistical interpretations of quantum measurement probabilities," Am. J. Phys. **54**, 4887 (1996).
191. Gisin, N., "Stochastic quantum dynamics and relativity," Helvetica Physica Acta **62**, 363 (1989).
192. Gisin, N., "Bell's inequality holds for all non-product states," Phys. Lett. A **154**, 201 (1991).
193. Gleason, A. M., "Measures on the closed subspaces of a Hilbert space," J. Math. Mech. **6**, 885 (1957).
194. Greenberger, D. M., M. A. Horne, A. Shimony, and A. Zeilinger, "Bell's theorem without inequalities," Am. J. Phys. **58**, 1131 (1990).
195. Greenberger, D. M., M. A. Horne, and A. Zeilinger, "Going beyond Bell's theorem," in M. Kafatos (Ed.), *Bell's theorem, quantum theory and conceptions of the universe* (Kluwer Academic; Dordrecht, 1989), p. 79.
196. Greenberger, D. M., M. A. Horne, and A. Zeilinger, "Multiparticle interferometry and the superposition principle," Physics Today, August 1993, p. 22.
197. Griffiths, R., "Consistent histories and the interpretation of quantum mechanics," J. Stat. Phys. **36**, 219 (1984).
198. Groenwald, H. J., "Quantal observation in statistical interpretation," in T. Bastin (Ed.), *Quantum theory and beyond* (Cambridge University Press; Cambridge, 1971), p. 43.
199. Groisman, B., S. Popescu, and A. Winter, "On the quantum, classical and total amount of correlations in a quantum state," Phys. Rev. A **72**, 032317 (2005).

200. Gudder, S., "On hidden-variables theories," J. Math. Phys. **11**, 431 (1970).
201. Hacking, I. (Ed.), *Scientific revolutions* (Oxford University Press; New York, 1981).
202. Hall, M. J. W., "Universal geometric approach to uncertainty, entropy, and information," Phys. Rev. A **59**, 2602 (1999).
203. Hájek, A., "Interpretations of probability," in E. Zalka (Ed.), The Stanford encyclopedia of philosophy, http://plato.stanford.edu .
204. Halvorson, H., "Generalization of the Hughston-Jozsa-Wootters theorem to hyperfinite von Neumann algebras," Los Alamos Archive preprint quant-ph/031001 (2003).
205. Hanbury–Brown, R., and R. Q. Twiss, "Correlation between photons in two coherent beams of light," Nature **177**, 27 (1956).
206. Hardy, L., "Quantum mechanics, local realistic theories, and Lorentz-invariant realistic theories," Phys. Rev. Lett. **68**, 2981 (1992).
207. Hartle, J. B., "Quantum mechanics of individual systems," Am. J. Phys. **36**, 704 (1968).
208. Hartley, R. V., "Transmission of information," Bell System Tech. J. **7**, 535 (1928).
209. Hawking, S. W., "Breakdown of predictability in gravitational collapse," Phys. Rev. D **14**, 112 (1976).
210. Hayden, P. M., M. Horodečki, and B. M. Terhal, "The asymptotic entanglement cost of preparing a quantum state," J. Phys. A **34**, 6891 (2001).
211. Healey, R., *The philosophy of quantum mechanics: An interactive interpretation* (Cambridge University Press; Cambridge, 1989).
212. Hegerfeldt, G. C., "Remark on causality and particle localization," Phys. Rev. D **10**, 3320 (1974).
213. Hegerfeldt, G. C., and S. N. M. Ruijsenjaars, "Remarks on causality, localization, and spreading of wave packets," Phys. Rev. D **22**, 377 (1980).
214. Hegerfeldt, G. C., "Causality problems for Fermi's two-atom system," Phys. Rev. Lett. **72**, 596 (1994).
215. Heisenberg, W., "Über quantentheoretische Umdeutung kinematischer und mechanischer Beziehungen," Z. Physik **33**, 879 (1925).
216. Heisenberg, W., "Über den anschaulichen Inhalt der quantentheoretischen Kinematik und Mechanik," Z. Physik **43**, 172 (1927).
217. Heisenberg, W. *Physical principles of the quantum theory* (University of Chicago Press; Chicago, 1930).
218. Heisenberg, W., "The representation of nature in contemporary physics," Daedalus **87**, 95 (1958).
219. Heisenberg, W., *Physics and philosophy* (Harper and Row; New York, 1958).
220. Heisenberg, W., *The physicist's conception of nature* (Harcourt, Brace & Co.; New York, 1958).
221. Heisenberg, W., "Quantum theory and its interpretation," in S. Rozental (Ed.), *Niels Bohr: His life and Work* (North-Holland Publishing Company; Amsterdam, 1967), p. 94.
222. Heisenberg, W., *Physics and beyond* (Allen and Unwin; London, 1971).
223. Heisenberg, W., "Remarks on the origin of the relations of uncertainty," in W. C. Price and S. S. Chissick (Eds.), *The uncertainty principle and foundations of quantum mechanics* (J. Wiley and Sons; London, 1977), p. 6.
224. Heisenberg, W., *Encounters with Einstein* (Princeton University Press; Princeton, 1983).

225. Henderson, L., "Measuring quantum entanglement," in T. Placek and J. Butterfield (Eds.), *Non-locality and modality* (Kluwer Academic; Dordrecht, 2002), p. 137.

226. Hendry, J., *The creation of quantum mechanics and the Bohr-Pauli dialogue* (D. Riedel; Dordrecht, 1984).

227. Herzog, T. J., P. G. Kwiat, H. Weinfurter, and A. Zeilinger, "Complementarity and the quantum eraser," Phys. Rev. Lett. **75**, 3034 (1995).

228. Heywood, P., and M. L. G. Redhead, "Nonlocality and the Kochen–Specker paradox," Found. Phys. **13**, 481 (1983).

229. Hill, S., and W. Wootters, "Entanglement of a pair of quantum bits," Phys. Rev. Lett. **78**, 5022 (1997).

230. Holevo, A. S., "Somes estimates for the information content transmitted by a quantum communication channel," Probl. Pered. Inform. **9**, 3 (1973) [Probl. Inf. Transm. (USSR) **9**, 177 (1973)].

231. Holevo, A. S., "Statistical decisions in quantum theory," J. Multivar. Anal. **3**, 337 (1973).

232. Home, D., and M. A. B. Whitaker, "Ensemble interpretation of quantum mechanics: A modern perspective," Phys. Rep. **210**, 223 (1992).

233. Honner, J., *The description of nature* (Oxford University Press; Oxford, 1987).

234. Hooker, C., *A realistic theory of science* (State University of New York Press; Albany, 1987).

235. Horne, M., A. Shimony, and A. Zeilinger, "Two-particle interferometry," Nature **347**, 429 (1990).

236. Horodecki, P., "Separability criterion and inseparable mixed states with positive partial transposition," Phys. Lett. A **232**, 333 (1997).

237. Horodecki, P., "Bound entanglement," in Bruss, D. and Leuchs, G. (Eds.), *Lectures on quantum information* (Wiley-VCH; Weinheim, 2007), p. 209.

238. Horodecki, M., P. Horodecki, and R. Horodecki, "Separability of mixed states: necessary and sufficient conditions," Phys. Lett. A **223**, 1 (1996).

239. Horodecki, M., P. Horodecki, and R. Horodecki, "Inseparable two spin-$\frac{1}{2}$ density matrices can be distilled to a singlet form," Phys. Rev. Lett. **78**, 574 (1997).

240. Horodecki, P., R. Horodecki and M. Horodecki, "Entanglement and thermodynamical analogies," http://xxx.lanl.gov quant-ph/9805072 (1998).

241. Horodecki, M., P. Horodecki, and J. Oppenheim, "Reversible transformations from pure to mixed states, and the unique measure of information," http://xxx.lanl.gov quant-ph/0212019 (2002).

242. Horodecki, M., J. Oppenheim and R. Horodecki, "Are the laws of entanglement thermodynamical?", http://xxx.lanl.gov quant-ph/0207177 (2002).

243. Howard, D., "Who invented the 'Copenhagen interpretation'? A study in mythology," Phil. Sci. **71**, 669 (2004).

244. Hughes, R. I. G., *The structure and interpretation of quantum mechanics* (Harvard University Press; Cambridge, MA, 1989).

245. Hughston, L. P., "Entropy, uncertainty, and nonlinearity," in R. Penrose and C. J. Isham (Eds.), *Quantum concepts in space and time* (Clarendon Press; Oxford, 1986).

246. Hughston, L. P., R. Jozsa, and W. K. Wootters, "A complete classification of quantum ensembles having a given density matrix," Phys. Lett. A **183**, 14 (1993).

247. Isham, C., *Lectures on quantum theory: Mathematical and structural foundations* (Imperial College Press; London, 1995).

248. Jackiw, R., and A. Shimony, "The depth and breadth of John Bell's physics," Phys. Perspect. **4**, 78 (2002).

249. Jaeger, G., "New quantum mechanical results in interferometry," Ph.D. Thesis, Boston University, 1995 (UMI Dissertation Services; Ann Arbor, 1994).

250. Jaeger, G., "Bohmian mechanics and quantum theory" Studies in History and Philosophy of Modern Physics **31**, 105 (2000).

251. Jaeger, G., *Quantum information: An overview* (Springer; New York, 2007).

252. Jaeger, G., M. A. Horne, and A. Shimony, "Complementarity of one-particle and two-particle interference," Phys. Rev. A **48**, 1023 (1995).

253. Jaeger, G., and A. Shimony, "Optimal distinction between two non-orthogonal quantum states," Phys. Lett. A **197**, 83 (1995).

254. Jaeger, G., A. Shimony, and L. Vaidman, "Two interferometric complementarities," Phys. Rev. A **51**, 54 (1995).

255. Jaeger, G., M. Teodorescu-Frumosu, A. V. Sergienko, B. E. A. Saleh, and M. C. Teich, "Multiphoton Stokes-parameter invariant for entangled states," Phys. Rev. A **67**, 032307 (2003).

256. Jammer, M., *The philosophy of quantum mechanics* (John Wiley and Sons; New York, 1974).

257. Jarrett, J., "Bell's inequality: A guide to the implications," in J. Cushing and E. McMullin (Eds.), *Philosophical consequences of quantum theory* (Univ. Notre Dame Press; Notre Dame, IN, 1989).

258. Jauch, J. M., *Foundations of quantum mechanics* (Addison-Wesley; Reading, MA, 1968).

259. Jauch, J. M., and J. G. Baron, "Entropy, information, and Szilard's paradox," Helv. Phys. Acta **45**, 220 (1972).

260. Jones, N., and L. Masanes, "Interconversion of nonlocal correlations," Phys. Rev. A **72**, 052312 (2005).

261. Joos, E., "Quantum theory and the appearance of a classical world," in Ann. N. Y. Acad. Sci. **480**, 242 (1986).

262. Joos, E., "Decoherence through interaction with the environment," in D. Giulini *et al.* (Eds.), *Decoherence and the appearance of a classical world in quantum theory* (Springer-Verlag; Berlin, 1996), p. 35.

263. Jordan, P., *Physics of the 20th Century* (Philosophical Library; New York, 1944).

264. Jordan, P., "On the process of measurement in quantum mechanics," Phil. Sci. **16**, 269 (1949).

265. Jozsa, R., "Fidelity for mixed quantum states," J. Mod. Opt. **41**, 2315 (1994).

266. Jozsa, R., and B. Schumacher, "A new proof of the quantum noiseless coding theorem," J. Mod. Opt. **41**, 2343 (1994).

267. Jozsa, R., and N. Linden, "On the role of entanglement in quantum computational speed-up," http://xxx.lanl.gov quant-ph/0201143 (2002).

268. Kemeny, J., "Fair bets and inductive probabilities," J. Symbolic Logic **20**, 263 (1955).

269. Kent, A., "Against many-worlds interpretations," Int. J. Mod. Phys. A **5**, 1745 (1990).

270. Kent, A., "Locality and reality revisited," in T. Placek and J. Butterfield (Eds.), *Non-locality and modality* (Kluwer Academic; 2002), p. 163.

271. Keyl, M., "Fundamentals of quantum information theory," Phys. Rep. **369**, 431 (2002).
272. Kirkpatrick, K. A., "The Schrödinger-HJW theorem," Found. Phys. Lett. **19**, 95 (2006).
273. Kirkpatrick, K. A., "Translation of G. Lüders' *Über die Zustandsänderung durch den Meßprozeß*," Ann. Phys. (Leipzig) **15**, 322 (2006).
274. Klein, M. J., "Maxwell, his demon, and the second law of thermodynamics," Am. Sci. **58**, 84 (1970).
275. Klement, K., "Russell's logical atomism," The Stanford encyclopedia of philosophy (Fall 2005 Edition), Edward N. Zalta (Ed.), URL = http://plato.stanford.edu/entries/logical-atomism/ .
276. Klyshko, D. N., "Quantum optics: quantum, classical and metaphysical aspects," Physics-Uspekhi **37**, 1097 (1994).
277. Kochen, S., and E. Specker, "The problem of hidden variables in quantum mechanics," J. Math. Mech. **17**, 59 (1967).
278. Körner, S. (Ed.), *Observation and interpretation; A symposium of philosophers and physicists* (Constable and Co. Ltd.; London, 1957).
279. Kraus, K., *et al.*, *States, effects, and operations. Springer lecture notes in physics. SLNP 190* (Springer-Verlag; Berlin, 1983).
280. Kremer, I., "Quantum communication," M.Sc. thesis, Computer Science Department, Hebrew University, 1995.
281. Krips, H., *The metaphysics of quantum theory* (Oxford University Press; Oxford, 1987).
282. Kronz, F., "Quantum theory: von Neumann vs. Dirac," The Stanford encyclopedia of philosophy (Fall 2008 edition), Edward N. Zalta (Ed.), URL=http://plato.stanford.edu/fall2008/entries/qt-nvd .
283. Kullback, S., and R. A. Leibler, "On information and sufficiency," Ann. Math. Stat. **22**, 79 (1951).
284. Kuhn, Thomas S., *The structure of scientific revolutions* (University of Chicago Press; Chicago, 1970).
285. Lahti, P., P. Busch, and P. Mittelstaedt, "Some important classes of quantum measurements and their information gain," J. Math. Phys. **32**, 2770 (1991).
286. Landau, L., "Das Dämpfungsproblem in der Wellenmechanik," Z. Phys. **45**, 430 (1927).
287. Landau, L. J., "Empirical two-point correlation functions," Found. Phys. **18**, 449 (1988).
288. Landauer, R., "Information is physical," Physics Today, May 1991, p. 23 (1991).
289. Landauer, R., "Irreversibility and heat generation in the computing process," IBM J. Res. Develop. **5**, 183 (1961).
290. Landauer, R., in E. D. Haidemenakis (Ed.), *Proceedings of the conference on fluctuation phenomena in classical and quantum systems*, Chiana, Crete, Greece, August 1969 (Gordon and Breach, Science Publishers Inc.; New York, 1970).
291. Landauer, R., "The physical nature of information," Phys. Lett. A **217**, 188 (1996).
292. Landé, A., *From dualism to unity in quantum physics* (Cambridge University Press; Cambridge, 1960).
293. Leff, H. S., and A. F. Rex (Eds.), *Maxwell's demon: Entropy, information, and computing* (Princeton University Press; Princeton, 1990), p. 15.

294. Lewenstein, M., and A. Sanpera, "Separability and entanglement of composite quantum systems," Phys. Rev. Lett. **80**, 2261 (1998).

295. Linblad, G. "Completely positive maps and entropy inequalities," Commun. Math. Phys. **40**, 147 (1975).

296. Linden, N., S. Massar, and S. Popescu, "Purifying noisy entanglement requires collective measurements," Phys. Rev. Lett. **81**, 3279 (1998).

297. Linden, N., and S. Popescu, "On multiparticle entanglement," Fortschr. Phys. **46**, 567 (1998).

298. Linden, N., S. Popescu, and W. K. Wootters, "The power of reduced quantum states," http://xxx.lanl.gov quant-ph/0207109 (2002).

299. Lloyd, S., "Quantum-mechanical Maxwell's demon," Phys. Rev. A **56**, 3374 (1997).

300. Lo, H.-K., and S. Popescu, "The classical communication cost of entanglement manipulation: Is entanglement an inter-convertible resource?" Phys. Rev. Lett. **83**, 1459 (1999).

301. Lo, H.-K., S. Popescu, and T. Spiller (Eds.), *Introduction to quantum computation and information* (World Scientific; Singapore, 1998).

302. London, F., and E. Bauer, "The theory of observation in quantum mechanics," in J. A. Wheeler and W. H. Żurek (Eds.), *Quantum theory and measurement* (Princeton University Press; Princeton, 1983), p. 217.

303. Lorentz, H. (Ed.), *Rapports et discussions du cinquième conseil de physique Solvay* (Gauthier-Villars; Paris, 1928).

304. Ludwig, G., *Grundlagen der Quantenmechanik* (Springer; Berlin, 1954).

305. Lüders, G., "Über die Zustandsänderung durch den Messprozess," Annalen der Physik **8**, 322 (1951).

306. MacKay, D. J. C., *Information theory, inference, and learning algorithms* (Cambridge University Press; Cambridge, 2003).

307. Mackey, G. W., *Mathematical foundations of quantum mechanics* (W. A. Benjamin; New York, 1963).

308. Margenau, H., "Quantum-mechanical description," Phys. Rev. **49**, 240 (1936).

309. Margenau, H., "Reality in quantum mechanics," Phil. Sci. **16**, 287 (1949).

310. Margenau, H., "Philosophical problems concerning the meaning of measurement in quantum physics," Phil. Sci. **25**, 23 (1958).

311. Masanes, L., "Tight Bell inequality for d-outcome measurements correlations," Quant. Inf. Comp. **3**, 345 (2003).

312. Masumi, B., "Realer than real: The simulacrum according to Deleuze and Guattari," Copyright **1**, 90 (1987).

313. Maudlin, T., *Quantum non-locality and relativity*, Second Edition (Blackwell Publishing; Malden MA, 2002).

314. Maudlin, T., "Space-time in the quantum world," in J. T. Cushing (Ed.), *Bohmian mechanics and quantum theory: An appraisal* (Kluwer Academic Publishers; Dordrecht, 1996), p. 285.

315. Mermin, N. D., "The great quantum muddle." Phil. Sci. **50**, 651 (1983).

316. Mermin, N. D., "Simple unified form for the major no-hidden-variables theorems," Phys. Rev. Lett. **65**, 3373 (1990).

317. Mermin, N. D., "What do these corrlations know about reality," Found. Phys. **29**, 571 (1999).

318. Mielnik, B., "Generalized quantum mechanics," Commun. Math. Phys. **37**, 221 (1974).

319. Mittelstaedt, P., *The interpretation of quantum mechanics and the measurement process* (Cambridge University Press; Cambridge, 1998).
320. Mittelstaedt, P., *Sprache und Realität in der modernen Physik* (B. I. Wissenschaftsverlag; Mannheim, 1986).
321. Morikoshi, F., M. Santos, and V. Vedral, "Accessibility of physical states and non-uniqueness of entanglement measure," J. Phys. A: Math. Gen. **37**, 5887 (2004).
322. Murdoch, D., *Niels Bohr's philosophy of physics* (Cambridge University Press; Cambridge, 1989).
323. Nielsen, M. A., "On the units of bipartite entanglement: Is sixteen ounces of entanglement always equal to one pound?" J. Phys. A: Math. Gen. **34**, 6987 (2001).
324. Nielsen, M. A., and I. L. Chuang, *Quantum computing and quantum information* (Cambridge University Press; Cambridge, 2000).
325. Nielsen, M. A., and J. Kempe, "Separable states are more disordered globally than locally," Phys. Rev. Lett. **86**, 5184 (2000).
326. Norsen, T., "Einstein's boxes," Am. J. Phys. **73**, 164 (2005).
327. Norsen, T., "Against 'Realism'," Found. Phys. **37**, 311 (2007).
328. Omnès, R., "Consistent interpretations of quantum mechanics," Rev. Mod. Phys. **64**, 339 (1992).
329. Omnès, R., *The interpretation of quantum mechanics* (Princeton University Press; Princeton, 1994).
330. Omnès, R., "General theory of the decoherence effect in quantum mechanics," Phys. Rev. A **56**, 3383 (1997).
331. Packel, E. W., and J. F. Traub, "Information-based complexity," Nature **328**, 29 (1987).
332. Pais, A., *Subtle is the lord* (Oxford University Press; New York, 1982).
333. Pan, J. W., D. Bouwmeester, H. Weinfurter, and A. Zeilinger, "Experimental entanglement swapping: Entangling photons that never interacted," Phys. Rev. Lett. **80**, 3891 (1998).
334. Pati, A. K., and S. L. Braunstein, "Impossibility of deleting an unknown quantum state," Nature **404**, 164 (2000).
335. Pauli, W., Letter to Eddington, Sept. 1923 in A. Hermann, K. V. Meyenn, and V. F. Weisskopf (Eds.), *Wissenschaftlicher Briefwechsel mit Bohr, Einstein, Heisenberg u.a.*, Band I: 1919–1929 (Springer; Berlin, 1979).
336. Pauli, W., "Quantentheorie," in H. Geiger and K. Scheel (Eds.), *Handbuch der Physik*, First Edition **23**, 1 (Springer-Verlag; Berlin, 1926).
337. Pauli, W., "Über Gasentartung und Paramagnetismus," Z. Phys. **41**, 81 (1927).
338. Pauli, W., "Die allgemeinen Prinzipien der Wellenmechanik," in Geiger, H. and K. Scheel (Eds.), *Handbuch der Physik*, Second Edition, **24**, 83 (Springer-Verlag; Berlin, 1933).
339. Pauli, W., "Wahrscheinlichkeit und Physik," Dialectica **8**, 112 (1954).
340. Pavičić, M., Bibliography on quantum logics and related structures, Int. J. Theor. Phys. **31**, 373 (1992).
341. Pearle, P., "Tales and tails and stuff and nonsense," in R. S. Cohen *et al.* (Eds.), *Experimental metaphysics* (Kluwer Academic Publishers; Dordrecht, 1997), p. 143.
342. Peierls, R., *Surprises in theoretical physics* (Princeton University Press; Princeton, 1979).

343. Peierls, R., "In defence of measurement," Physics World **4**, 19 (1991).
344. Penrose, R., "Gravity and state vector reduction," in R. Penrose and C. Isham (Eds.), *Quantum concepts in space and time* (Clarendon Press; Oxford, 1986).
345. Penrose, R., "Quantum physics and conscious thought," in B. J. Hiley and F. D. Peat (Eds.), *Quantum implications: Essays in honour of David Bohm* (Routledge & Kegan Paul Ltd.; London, 1987).
346. Penrose, R., *The road to reality* (BCA; Chatham, Kent, UK, 2004).
347. Peres, A., "Higher-order Schmidt decompositions," http://xxx.lanl.gov quant-ph/9504006 (1995).
348. Peres, A., "Separability criterion for density matrices," Phys. Rev. Lett. **77**, 1413 (1996).
349. Peres, A., "Clasical interventions in quantum systems," Phys. Rev. A **61**, 022116 (2000); *ibid.*, 022117 (2000).
350. Peres, A., and W. K. Wootters, "Optimal detection of quantum information," Phys. Rev. Lett. **66**, 1119 (1991).
351. Pitowsky, I., "Betting on the outcomes of measurements: A Bayesian theory of quantum probability," Stud. Hist. Phil. Mod. Phys. **34**, 395 (2003).
352. Pitowsky, I., *Quantum probability–quantum logic* (Springer-Verlag; Berlin, 1989).
353. Plenio, M. B., and V. Vitelli, "The physics of forgetting: Landauer's erasure principle and information theory," Cont. Phys. **42**, 25 (2001).
354. Plotnitsky, A., *Reading Bohr: Physics and philosophy* (Springer; Dordrecht, 2006).
355. Poincaré, H., *Théorie mathématique de la lumière, II* (Gauthier-Villars; Paris, 1892).
356. Popescu, S., and D. Rohrlich, "Generic quantum nonlocality," Phys. Lett. A **166**, 293 (1992).
357. Popescu, S., and D. Rohrlich, "Action and passion at a distance," in Cohen *et al.* (Eds.), *Potentiality, entanglement and passion-at-a-distance* (Kluwer; Dordrecht, 1997); also http://xxx.lanl.gov quant-ph/9605004 (3 May, 1996).
358. Popescu, S., and D. Rohrlich, "Thermodynamics and the measure of entanglement," Phys. Rev. A **56**, R3319 (1997).
359. Popescu, S., and D. Rohrlich, "The joy of entanglement," in [301], p. 29.
360. Popper, K. R., "The propensity interpretation of probability," Brit. J. Phil. Sci. **10**, 25 (1959).
361. Popper, K. R., *Quantum theory and the schism in physics* (Rowan and Littlefield; Totowa, NJ, 1982).
362. Preskill, J., *Physics 229: Advanced mathematical methods of physics— Quantum computation and information*, California Institute of Technology, 1998. http://www.theory.caltech.edu/people/preskill/ph229/ .
363. Price, H., "Probability in the Everett world: Comments on Wallace and Greaves," http://philsci-archive.pitt.edu/archive/00002654 .
364. Primas, H., *Chemistry, quantum mechanics and reductionism* (Springer; Berlin, 1983).
365. Przibram, K. (Ed.),*Letters on wave mechanics* (Philosophical Libary; New York, 1967).
366. Putnam, H., "Is logic empirical?" Boston Stud. Phil. Sci. **5**, 216 (1969).
367. Putnam, H., "Quantum mechanics and the observer," Erkenntnis **16**, 193 (1981).

368. Putnam, H., *Realism and reason* (Cambridge University Press; Cambridge, 1983).

369. Quine, W. V. O., "Epistemology naturalized," in *Ontological relativity or other essays* (Random House; New York, 1969), p. 69.

370. Raussendorf, R., and H. J. Briegel, "A one-way quantum computer," Phys. Rev. Lett. **86**, 5188 (2001).

371. Redhead, M. L. G., *Incompleteness, nonlocality and realism* (Oxford University Press; Oxford, 1987).

372. Redhead, M. L. G., *From physics to metaphysics* (Cambridge University Press; Cambridge, 1995).

373. Reichenbach, H., *Philosophic foundations of quantum mechanics* (University of California; Berkeley, 1944).

374. Richter, Th., "Interference and non-classical spatial intensity correlations," Quantum Opt. **3**, 115 (1991).

375. Robertson, H. P., "The uncertainty principle," Phys. Rev. **34**, 163 (1929).

376. Rosenfeld, L., "Strife about complementarity," Science Prog. **163**, 393 (1953).

377. Rosenfeld, L., "Misunderstandings about the foundations of quantum theory," in [278].

378. Rosenfeld, L., "Physics and metaphysics," Nature **181**, 658 (1958).

379. Rosenfeld, L., "Foundations of quantum theory and complementarity," Nature **190**, 384 (1961).

380. Rothstein, J., "Information, logic, and physics," Phil. Sci. **23**, 31 (1956); paper original presented before the American Physical Society in 1951.

381. Rothstein, J., "Information, measurement, and quantum mechanics," Science **25**, 510 (1951).

382. Rothstein, J., "Information and thermodynamics," Phys. Rev. **85**, 135 (1957).

383. Rothstein, J., "Information and organization as the language of the operational viewpoint," Phil. Sci. **29**, 406 (1962).

384. Ruelle, D., *Chance and chaos* (Princeton University Press; Princeton, 1991).

385. Rungta, P., and C. M. Caves, "Concurrence-based entanglement measures for isotropic states," Phys. Rev. A **67**, 012307 (2003).

386. Ryle, G., *The concept of mind* (The University of Chicago Press; Chicago, 1949).

387. Saunders, S., "Time, quantum mechanics, and decoherence," Synthese **102**, 235 (1995).

388. Saunders, S., "Time, quantum mechanics, and probability," Synthese **114**, 405 (1998).

389. Scheibe, E. (J. B. Sykes, translator), *The logical analysis of quantum mechanics* (Pergammon Press; Oxford, 1973).

390. Schlosshauer, M., "Decoherence, the measurement problem, and interpretations of quantum mechanics," Rev. Mod. Phys. **76**, 1267 (2005).

391. Schlosshauer, M., *Decoherence and the quantum-to-classical transition* (Springer; New York, 2007).

392. Schmidt, E. "Zur Theorie der linearen und nichtlinearen Integralgleichungen," Math. Annalen **63**, 433 (1906).

393. Schrödinger, E. "Quantisierung als Eigenwertproblem," Annalen der Physik **81**, 109 (1926).

394. Schrödinger, E., "Die gegenwaertige Situation in der Quantenmechanik," Die Naturwissenschaften **23**, 807 (1935).

395. Schrödinger, E., "Discussion of probability relations between separated systems," Proc. Cambridge Philos. Soc. **32**, 446 (1935).
396. Schumacher, B. W., "Quantum coding," Phys. Rev. A **51**, 2738 (1995).
397. Schumacher, B. W., "Sending entanglement through noisy quantum channels," Phys. Rev. A **54**, 2614 (1996).
398. Schumacher, B. W., and M. D. Westmoreland, "Quantum mutual information and the one-time pad," http://xxx.lanl.gov quant-ph/0604207 (2006).
399. Schumacher, B. W., M. D. Westmoreland, and W. K. Wootters, "Limitation on the amount of accessible information in a quantum channel," Phys. Rev. Lett. **76**, 3452 (1996).
400. Scully, M.O., and K. Drühl, "Quantum eraser - A proposed photon correlation experiment concerning observation and delayed choice in quantum mechanics." Phys. Rev. A **25**, 2208 (1982).
401. Scully, M. O., B.-G. Englert, and J. Schwinger, "Spin coherence and Humpty-Dumpty. III. The effects of observation," Phys. Rev. A **40**, 1775 (1989).
402. Scully, M. O., B.-G. Englert, and H. Walther, "Quantum optical tests of complementarity," Nature **351**, 111 (1991).
403. Scully, M., and M. S. Zubairy, *Quantum Optics* (Cambridge University Press, Cambridge, 1997), Section 20.1.
404. Seager, W., "A note on 'quantum erasure'," Phil. Sci. **63**, 81 (1996).
405. Shannon, C. E., "A mathematical theory of communication," Bell System Technical Journal **27**, 379 (1948); *ibid.*, 623 (1948).
406. Shannon, C. E., and W. Weaver, *The mathematical theory of communication* (University of Illinois; Urbana, IL, 1949).
407. Shimony, A., "Controllable and uncontrollable non-locality," in S. Kamefuchi *et al.* (Eds.), *Foundations of quantum mechanics in light of the new technology* (Physical Society of Japan; Tokyo, 1983), p. 225.
408. Shimony, A., "Contextual hidden variables theories and Bell's inequalities," Brit. J. Philos. Sci. **35**, 25 (1984).
409. Shimony, A., "Reply to Bell," *Dialectica* **39**, 107 (1985).
410. Shimony, A., "Conceptual foundations of quantum mechanics," in Paul Davies (Ed.), *The new physics* (Cambridge University Press; Cambridge, 1989), Ch. 13.
411. Shimony, A., *Search for a naturalistic world view, Volume I* (Cambridge University Press; Cambridge, 1993).
412. Shimony, A., *Search for a naturalistic world view, Volume II* (Cambridge University Press; Cambridge, 1993).
413. Shimony, A., in K. V. Laurikainen and C. Montonen (Eds.), *Symposia on the foundations of modern physics 1992* (World Scientific; Singapore, 1993).
414. Shimony, A., "Degree of entanglement," Ann. N.Y. Acad. Sci. **755**, 675 (1995).
415. Shimony, A., "The logic of EPR," Annales de la Fondation Louis de Broglie **26**, 399 (2001).
416. Shimony, A. "Some intellectual obligations of epistemological naturalism," in D. Malament (Ed.), *Reading natural philosophy* (Open Court; Peru, IL, 2002).
417. Shimony, A., "Bell's theorem," *The Stanford encyclopedia of philosophy* (Summer 2005 edition), E. N. Zalka (Ed.).
http://plato.stanford.edu/entries/bell-theorem .
418. Shimony, A., "Comment on Norsen's defense of Einstein's 'box argument'," Am. J. Phys. **73**, 177 (2005).

419. Shimony, A., "Aspects of nonlocality in quantum mechanics," in J. Evans and A. S. Thorndike (Eds.), *Quantum mechanics at the crossroads* (Springer; Heidelberg, 2007), p. 107.

420. Shimony, A., M. A. Horne, and J. F. Clauser, "An exchange on local beables," Dialectica **39**, 97; *ibid.*, 107 (1985).

421. Shor, P. W., J. A. Smolin, and B. M. Terhal, "Nonadditivity of bipartite distillable entanglement follows from conjecture on bound entangled Werner states," Phys. Rev. Lett. **86**, 2681 (2001).

422. Silverman, M. P., *Quantum superposition* (Springer-Verlag; Heidelberg, 2008).

423. Simon, D. R., "On the power of quantum computation," in S. Goldwasser (Ed.), *Proceedings of 35th annual symposium on the foundations of computer science* (IEEE Society Press; Los Alamitos CA, 1994), p. 116.

424. Sinha, S., and R. Sorkin, "A sum-over-histories account of an EPR(B) experiment," Found. Phys. Lett. **4**, 303 (1991).

425. Slater, J. C., *Solid state and molecular theory: A scientific biography* (Wiley; New York, 1975).

426. Srinivas, M. D.. "Collapse postulate for observables with continuous spectra," Comm. Math. Phys. **71**, 131 (1980).

427. Stachel, J., "Do quanta need a new logic?" in R. G. Colodny (Ed.), *From quarks to quasars* (Univ. Pittsburgh Press; Pittsburgh, 1986), p. 229.

428. Stachel, J., "Einstein and the quantum: Fifty years of struggle," in R. G. Colodny (Ed.), *From quarks to quasars* (Univ. Pittsburgh Press; Pittsburgh, 1986), p. 349.

429. Stachel, J., "Feynman paths and quantum entanglement," in R. Cohen *et al.* (Eds.), *Potentiality, entanglement and passion-at-a-distance* (Kluwer; Dordrecht, 1997), p. 245.

430. Stachel, J., "Structure, individuality, and quantum gravity," in D. Rickles *et al.* (Eds.), *The structural foundations of quantum gravity* (Clarendon Press; Oxford, 2006), Section 3.5.

431. Stachel, J., "Choice of variables and initial value problems in classical general relativity: Prolegomena to any future quantum gravity," in D. Oriti (Ed.), *Approaches to quantum gravity* (Cambridge University Press; Cambridge, to appear).

432. Stairs, A., "Quantum logic and the Lüders rule," Phil. Sci. **49**, 422 (1982).

433. Stairs, A., "Quantum logic, realism, and value-definiteness," Phil. Sci. **50**, 578 (1983).

434. Stapp, H. P., "The Copenhagen interpretation," Am. J. Phys. **40**, 1098 (1972).

435. Steane, A., "Quantum computing," Rep. Prog. Phys. **61**, 117 (1998).

436. Steane, A., "Quantum computing needs only one universe," Stud. Hist. Phi. Mod. Phys. **34**, 469 (2003).

437. Strauss, M., "Zur Begründung der statistischen Transformationstheorie der Quantenphysik," Berliner Berichte 1936, 382 (1936). [English translation, "The Logic of complementarity and the foundation of quantum theory," in M. Strauss, *Modern physics and its philosophy* (Reidel; Dordrecht 1972), p. 186.]

438. Suppe, F. (Ed.), *The structure of scientific theories* (University of Illinois Press; Urbana, 1977).

439. Suppes, P., "The probabilistic argument for a non-classical logic in quantum mechanics," Phil. Sci. **33**, 14 (1966).

440. Svozil, K., *Quantum logic* (Springer-Verlag; Berlin, 1998).
441. Svozil, K., "Quantum logic. A brief outline," http://xxx.lanl.gov quant-ph/9902042 (1999).
442. Svozil, K., "The information interpretation of quantum mechanics," Los Alamos Archive preprint quant-ph/0006033 (2000).
443. Szilard, L., "Über die Entropieverminderung in einem thermodynamischen System bei Eingriffen intelligenter Wesen," Z. Physik **53**, 840 (1929).
444. Taylor, G. I., "Interference fringes with feeble light," Proc. Camb. Philos. Soc. **15**, 114 (1909).
445. Teller, P., *An interpretive introduction to quantum field theory* (Princeton University Press; Princeton, 1995).
446. Thapliyal, A. V., "Multipartite pure-state entanglement," Phys. Rev. A **59**, 3336 (1999).
447. Thapliyal, A. V., reported in [38].
448. Thew, R. T., K. Nemoto, A. G. White, W. J. White, and W. J. Munro, "Qudit quantum-state tomography," Phys. Rev. A **66**, 012303 (2002).
449. Timpson, C. G., "On a supposed conceptual inadequacy of the Shannon information in quantum mechanics," Stud. Hist. Phil. Mod. Phys. **33**, 441 (2003).
450. Timpson, C. G., "Quantum information theory and the foundations of quantum mechanics," Ph.D. Thesis, Queen's College, The University of Oxford, 2004; also http://xxx.lanl.gov quant-ph/0412063.
451. Timpson, C. G., "Nonlocality and information flow: The approach of Deutsch and Hayden," Found. Phys. **35**, 313 (2005).
452. Timpson, C. G., "The grammar of teleportation," Brit. J. Phil. Sci. **57**, 587 (2006).
453. Tipler, F. J., "The many-worlds interpretation of quantum mechanics in quantum cosmology," in R. Penrose and C. J. Isham (Eds.), *Quantum concepts in space and time* (Clarendon Press; Oxford, 1986), p. 204.
454. Toner, B. F., and D. Bacon, "Communication cost of simulating Bell correlations," Phys. Rev. Lett. **91**, 187904 (2003).
455. Tsirel'son, B. S., "Quantum generalizations of Bell's inequalities," Lett. Math. Phys. **4**, 93 (1980).
456. Tsirel'son, B. S., "Quantum analogues of the Bell inequalities," J. of Sov. Math. **36**, 557 (1987).
457. Turing, A. M., "On computable numbers, with an application to the Entscheidungsproblem," Proc. London Math. Soc. **42**, 230 (1936); *ibid.*, **43**, 544 (1937).
458. Uffink, J., "Measures of uncertainty and the uncertainty principle," Ph.D. Dissertation, University of Utrecht, 1990.
459. Van Brakel, J., " The possible influence of the discovery of radio-active decay on the concept of physical probability," Arch. Hist. Exact Sci. **31**, 369 (1985).
460. Van Dam, W., "Nonlocality and Communication complexity," D. Phil. thesis, University of Oxford, Dept. of Physics (2000), Ch. 9. http://www.cs.ucsb.edu/˜vandam/publications.html
461. Van Dam, W., "Implausible consequences of superstrong nonlocality," http://xxx.lanl.gov quant-ph/0501159 (27 Jan., 2005).
462. Van der Waerden, B. L., *Sources of quantum mechanics* (North-Holland Publishing Company; Amsterdam, 1967).
463. Van Fraassen, B., *The scientific image* (Clarendon Press; Oxford, 1980).
464. Van Fraassen, B., "The Charybdis of realism: Epistemological implications of Bell's inequality," Synthese **52**, 25 (1982).

465. Van Fraassen, B., *Quantum mechanics* (Clarendon; Oxford, 1991).
466. Vedral, V., "The role of relative entropy in quantum information processing," Rev. Mod. Phys. **74**, 197 (2002).
467. Vedral, V., and E. Kashefi, "Uniqueness of the entanglement measure for bipartite pure states and thermodynamics," Phys. Rev. Lett. **89**, 037903 (2002).
468. Vedral, V., Plenio, M. B., Rippin, and Knight, P. L., "Quantifying entanglement," Phys. Rev. Lett. **78**, 2275 (1997).
469. Vidal, G., "Entanglement monotones," J. Mod. Opt. **47**, 355 (2000).
470. Vidal, G., and J. I. Cirac, "Irreversibility in asymptotic manipulations of entanglement," Phys. Rev. Lett. **86**, 5803 (2001).
471. Vidal, G., D. Jonathan, and M. A. Nielsen, "Approximate transformations and robust manipulation of bipartite pure-state entanglement," Phys. Rev. A **62**, 012304 (2000).
472. Vollbrecht, K.-G. H., and M. M. Wolf, "Conditional entropies and their relation to entanglement criteria," J. Math. Phys. **43**, 4299 (2002).
473. Von Neumann, J., "Mathematische Begründung der Quantenmechanik," Gött. Nach., Session of May 20, 1 (1927).
474. Von Neumann, J., "Allgemeine Eigenwerttheorie Hermitescher Funktionaloperatoren," Math. Ann. **102**, 49 (1929).
475. Von Neumann, J., "Zur Algebra der Funktionaloperatoren und Theorie der normalen Operatoren," Math. Ann. **102**, 370 (1929).
476. Von Neumann, J., "Zur Theorie der unbeschränkten Matrizen," J. Reine Angew Math. **161**, 208 (1929).
477. Von Neumann, J., *Mathematische Grundlagen der Quantenmechanik* (Julius Springer; Berlin, 1932) [English translation: *Mathematical foundations of quantum mechanics* (Princeton University Press; Princeton, NJ, 1955)].
478. Von Neumann, J., and O. Morgenstern, *Theory of games and economic behavior* (Princeton University Press; Princeton, 1947).
479. Von Plato, J., *Creating modern probability* (Cambridge University Press; Cambridge, 1994).
480. Von Weizsäcker, C. F., "Die Quantentheorie der einfachen Alternative," Zeitschrift für Naturforschung **13a**, 245 (1958).
481. Von Weizsäcker, C. F., "The Copenhagen interpretation," in T. Bastin (Ed.), *Quantum theory and beyond* (Cambridge University Press; Cambridge, 1971), p. 25.
482. Wallace, D., "Everettian rationality: defending Deutsch's approach to probability in the Everett interpretation," Stud. Hist. Phil. Mod. Phys. **34**, 415 (2003).
483. Wallace, D., "Epistemology quantized: Circumstances in which we should come to believe in the Everett interpretation," British J. Phil. Sci. **57**, 655 (2006).
484. Wehrl, A., "General properties of entropy," Rev. Mod. Phys. **50**, 221 (1978).
485. Wei, T.-C., J. B. Altepeter, P. M. Goldbart, and W. J. Munro, "Measures of entanglement in multipartite bound entangled states," Phys. Rev. A **70**, 022322 (2004).
486. Wei, T.-C., and P. M. Goldbart, "Geometric measure of entanglement and applications to bipartite and multipartite States," Phys. Rev. A **68**, 042307 (2003).

487. Weihs, G., T. Jennewein, C. Simon, H. Weinfurter, and A. Zeilinger, "Violation of Bell's inequality under strict Einstein locality conditions," Phys. Rev. Lett. **81**, 5039 (1998).

488. Wentzel, L., "Types and tokens," The Stanford Encyclopedia of Philosophy (Summer 2006 edition), Edward N. Zalta (Ed.). http://plato.stanford.edu/entries/types-tokens/ .

489. Werner, R. F., "Quantum states with Einstein–Podolsky–Rosen correlations admitting a hidden-variable model," Phys. Rev. A **40**, 4277 (1989).

490. Werner, R. F., and M. M. Wolf, "Bell inequalities and entanglement," http://xxx.lanl.gov quant-ph/0107093.

491. Weyl, H., *Gruppentheorie und Quantenmechanik* (Hirzel; Leipzig, 1928) [English translation: *The theory of groups and quantum mechanics* (Methuen; London, 1931).]

492. Wheeler, J. A., "Law without law," in [520] (1990), p. 182.

493. Wheeler, J. A., "Information, physics, quantum: The search for links," in W. H. Żurek (Ed.), *Complexity, entropy, and the physics of information*, SFI studies in the sciences of complexity, Vol. VIII (Addison-Wesley; Reading MA, 1990), p. 3.

494. Wheeler, J. A., "Sakharov revisited; It from bit," in L. V. Keldysh and V. Yu. Fainberg (Eds.), *Proceedings of the first international Sakharov conference on physics*, Vol. 2 (Nova Science Publishers; New York, 1991), p. 751.

495. Wheeler, J. A., and W. H. Żurek (Eds.), *Quantum theory and measurement* (Princeton University Press; Princeton, NJ, 1983).

496. Whitaker, A., *Einstein, Bohr and the quantum dilemma* (Cambridge University Press; Cambridge, 1996).

497. Whiteman, J. H. M., "The phenomenology of observation and explanation in quantum theory," in T. Bastin (Ed.), *Quantum theory and beyond* (Cambridge University Press; Cambridge, 1971), p. 71.

498. Wick, D., *The infamous boundary* (Birkhauser; Boston, 1995).

499. Wigner, E., "Remarks on the mind–body question," in I. J. Good (Ed.), *The scientist speculates* (Heinemann; London, 1961), p. 284.

500. Wigner, E., *Symmetries and reflections* (Indiana University Press; Bloomington, 1967).

501. Wigner, E., "On hidden variables and quantum mechanical probabilities," Am. J. Phys. **38**, 1005 (1970).

502. Wigner, E., "The subject of our discussions," in B. d'Espagnat (Ed.), *Foundations of quantum mechanics. International school of physics "Enrico Fermi"* 1970 (Academic Press; New York, 1971).

503. Wittgenstein, L., (C. K. Ogden, translator), *Tractatus logico-philosophicus* (Dover Publications; Mineola, NY, 1998), p. 29.

504. Wootters, W. K., "Statistical distance and Hilbert space," Phys. Rev. D **23**, 357 (1981).

505. Wootters, W. K., "Entanglement of formation of an arbitrary state of two qubits," Phys. Rev. Lett. **80**, 2245 (1998).

506. Wootters, W. K., "Quantum entanglement as a resource for communication," in J. Evans and A. S. Thorndike (Eds.), *Quantum mechanics at a crossroads* (Springer; Heidelberg, 2007), p. 213.

507. Wootters, W. K., and W. H. Żurek, "Complementarity in the double-slit experiment: Quantum nonseparability and a quantitative statement of Bohr's principle," Phys. Rev. D **19**, 473 (1979).

508. Wootters, W. K., and W. H. Żurek, "A single quantum cannot be cloned," Nature **299**, 802 (1982).

509. Wu, C. S., and I. Shaknov, "The angular correlation of scattered annihilation radiation," Phys. Rev. **77**, 136 (1950).

510. Yao, A. C., "Quantum circuit complexity," in *Proceedings of 34th annual IEEE symposium on foundations of computer science* (IEEE Press; New York, 1993), p. 352.

511. Yu, T., and J. H. Eberly, "Qubit disentanglement and decoherence via dephasing," Phys. Rev. B **68**, 165322 (2003).

512. Yuen, H. P. "Amplification of quantum states and noiseless photon amplifiers," Phys. Lett. A **113**, 405 (1986).

513. Zeh, H. D., "On the interpretation of measurement in quantum theory," Found. Phys. **1**, 69 (1970).

514. Zeilinger, A., "Quantum entanglement: A fundamental concept finding its applications," Physica Scripta **T-76**, 203 (1998).

515. Zeilinger, A., "A foundational principle for quantum mechanics," Found. Phys. **29**, 631 (1999).

516. Zeilinger, A., "Experiment and the foundations of quantum physics," Rev. Mod. Phys. **71**, S228 (1999).

517. Zeilinger, A., R. Gähler, C. G. Shull, W. Treimer, and W. Mampe, "Single- and double-slit diffraction of neutrons," Rev. Mod. Phys. **60**, 1067 (1988).

518. Żukowski, M., A. Zeilinger, M. A. Horne, and A. K. Ekert, " 'Event-ready-detectors' Bell experiment via entanglement swapping," Phys. Rev. Lett. **71**, 4287 (1993).

519. Żurek, W. H., "Maxwell's demon, Szilard's engine, and quantum measurements," Los Alamos preprint LAUR 84-2751 (1984); arXiv:quant-ph/0301076v1 (2003).

520. Żurek, W. H. (Ed.), *Complexity, entropy and the physics of information* (Addison-Wesley, Redwood City, CA, 1990).

521. Żurek, W. H., in J. J. Halliwell *et al.* (Eds.), *Physical origins of time asymmetry* (Cambridge Univ. Press; Cambridge, 1994), p. 175.

522. Żurek, W., "Decoherence, einselection, and the origins of the classical," Rev. Mod. Phys. **75**, 715 (2003).

Index

abstraktes Ich, 84, 91
acausality, 59
accuracy
  measurement, 11
admissibility criterion, 65
Aerts, D, 127
agent, 16, 21, 47
Alarm-clock paradox, 34
algebra
  σ, 267
  Boolean, 267, 268
    partial (pba), 65
  Jordan, 269
anti-realism, 100–102, 105, 108, 110,
    177
Aspect, A., 256

B_n, 267
B92 protocol, 88
Ballentine, L., 109, 117, 165–167, 169
Baron, J., 244
bases
  conjugate, 202
basis
  Bell, 225
  computational, 19, 35, 64, 200, 202,
    219
  orthonormal, 260
  pointer, 159
  preferred, 153, 155–157, 159, 161, 162
beable, 39, 173, 178, 289
Bekenstein–Hawking area, 234
Bell inequality, 41, 43, 44, 47, 52, 119,
    168, 251

empirical test, 43
Bell locality, 41, 42, 45, 201, 251, 252,
    256
Bell state, 22, 24, 26, 49, 51, 52, 59,
    201, 225–228, 249, 251, 254, 255
Bell test, 43
Bell's theorem, 17, 24, 37, 40, 41, 104,
    108, 110, 115
  'inequality-free', 45
Bell, J. S., 20, 33, 39–41, 45, 46, 60, 62,
    72, 74, 75, 104, 105, 107, 112, 115,
    118, 138, 145, 148, 156, 160, 162,
    171, 173, 232, 250, 256, 282
Beller, M., 129, 134
Benioff, P., 216
Bennett, C. H., 244
Bethe, H., 8
Birkhoff, G., 63, 64, 66, 67, 123, 217,
    236
bit, 194, 242, 243
black hole, 234
block
  classical, 199
Bohm, A., 262
Bohm, D., 35, 38, 41, 55, 114, 121, 122,
    124, 176, 232
Bohmian mechanics, 114, 123
Bohr, N., viii, ix, 7, 8, 10, 11, 20, 34, 36,
    58, 59, 79, 80, 84, 90, 97–99, 102,
    104, 105, 111, 114, 117, 124–131,
    137, 184, 186, 232, 234, 237, 238
Bohr–Einstein debate, 128, 135, 168
Boltzmann, L., 3

Boolean algebra, 39, 65
  partial, 64
Boolean logic, 61, 64, 221
Borel set, 87
Borel space, 87
Borel, É., 57
Born rule, vii, 14, 38, 61, 62, 68, 69, 73,
    117, 119, 141, 150, 162, 164, 166,
    167, 178, 262, 274
Born, M., vii, 19, 33, 38, 68, 73, 95, 105,
    111, 118, 119, 255
box, 251
  local, 252
  non-local, 251
  non-signaling, 252
  PR, 251
  signaling, 252
bra, 263
Braithwaite, R. B., 176
braket, 263
branching problem, 145
Bruckner, Č, 235, 236
Bub, J., 77, 86, 97, 117, 133, 137, 139,
    140, 154, 157, 158, 178, 185–188,
    217, 220–222, 256
Bunge, M., 128
Busch, P., 87, 89

calibration postulate, 76
Carnot cycle, 52, 245
Carnot, N., 245
category mistake, 70
causality, 1, 34, 35, 55–59, 66, 77,
    120–122, 129, 176, 192, 229, 231,
    250, 253–255
  local, 26, 32, 33, 43, 46, 47, 175, 187,
    249
  strict, 55
causation
  mathematical, 66
CBH, 185
CBH theorem, 185, 187
chance
  objective, 118
channel
  classical, 47
    additive noise, 198
    binary symmetric, 198
    capacity of, 198, 199

  Pauli, 213
  quantum, 211
    amplitude-damping, 214
    bit + phase-flip, 213
    bit-flip, 213
    classically assisted quantum
      capacity of, 212
    entanglement-assisted classical
      capacity of, 213
    noiseless, 211, 214
    noisy, 211
    phase-flip, 213
characteristic equation, 261
CHSH, 43, 256
CHSH inequality, 24, 43, 44, 250, 251
Chuang, I., 229, 230
classical information, 196, 198
classical mechanics, 1, 9, 13, 17, 25, 55,
    62, 72, 76, 77, 132, 163, 183, 191,
    240, 250
classical physics, 2, 11, 18, 55, 58, 63,
    64, 98, 125, 126, 131, 135, 136,
    138, 142, 191, 194, 240
Clauser, J. F., 42, 43
Clauser–Horne inequality, 42
Clifton, R., 185, 247
closed system, 47, 57, 77, 92, 230, 242
coding
  superdense, 225
coefficients
  Schmidt, 18
coherence, 17
coherences, 17
coin toss, 195, 196
collapse
  complexity class, 254
  von Neumann–London–Bauer, 85
  wave-function, 79, 143, 256
communication, 1, 2, 21, 47, 56, 187,
    189, 193–196
  classical (CC), 48
    error-free, 199
  quantum, 191
communication resource, 198
commutator, 9
complementarity, 7–10, 59, 125, 127,
    128, 130, 234, 237, 238, 273
  principle of, 10, 127, 130, 238
completeness

POVM, 87
  quantum state description, 35, 37, 72, 176
completeness criterion, 35, 36
completeness relation, 48
complexity
  communication
    quantum, 250
complexity classification
  collapse of, 254
computation
  classical
    probabilistic, 200
  quantum
    probabilistic, 219
computational basis, 19, 35, 64, 200, 202
concurrence, 25, 26
context
  measurement, 39
contextual hidden-variables, 39
convex roof, 51
convex sum, 21, 22, 38, 49, 261
correlation, 201
  classical, 21
  quantum, 50
CP map, 48
CPTP map, 47, 48, 209
cryptography
  quantum, 206
cut
  Heisenberg, 84, 118, 130

d'Espagnat, B., 35, 71, 105, 107, 138
de Broglie, L., 38, 114
de Finetti theorem, 179
decision theory, 164
decoherence, 2, 5, 17, 29, 52, 80, 83, 121, 154, 156, 157, 159, 162–164, 213, 214
decoherence functional, 163
decoherence-free subspace, 157
decomposition
  bipartite, 84
  mixed state, 15, 31, 32, 51
  operator, 48
  polar, 18
  Schmidt, 18, 19, 27, 50
    generalized, 286

spectral, 23, 87
  tri-, 158
decomposition operator, 48
demon, 67
density
  probability, 22
density matrix, 264
determinate values, 5
determinateness, 90
determinism, 56–59, 102, 103
Deutsch, D., 144, 146, 160, 164, 216, 220
DeWitt, B. S., 139, 141, 142, 144, 145, 148–151, 153, 154, 160
Dieks, D., 12
Dirac notation, 18, 259, 263
Dirac, P. A. M., 2, 4, 37, 61, 79, 95–97, 117, 119–122, 179, 181
Dirac–Jordan transformation theory, 10, 181
discrimination, 197
dispersion, 40
distance
  Hilbert–Schmidt, 261
  Kullback–Leibler, 197, 283
distillation
  entanglement, 25, 52, 246, 247
distinguishability, 2, 30, 198, 203
distribution
  random (vs. non-random), 150
disturbance, 12, 120, 178
disturbance theory, 9
duality
  wave–particle, 6, 7, 12, 125, 127
Dutch book, 71
Duwell, A., 197
dynamical variables, 38, 55, 62, 68, 120, 130, 148, 169

e-bit, 51, 52, 245, 247, 251
eavesdropping, 206
effect, 87
  regular, 89
effects
  coexistent, 89
eigenvalue, 13, 260
  degenerate, 13
  non-degenerate, 13
eigenvalue spectrum, 13

eigenvalue–eigenstate link, 76, 119
eigenvector, 13, 260
Einstein, A., viii, 11, 19, 20, 33–36, 58,
    62, 76, 82, 90, 92, 97, 99, 105, 106,
    110–114, 117, 118, 125, 127, 136,
    137, 154, 162, 165, 166, 173–175,
    179, 183, 245, 255
Einstein, Podolsky and Rosen (see
    EPR), 33
Einstein–Bohr box experiment, 34
Elby, A., 157, 158
element
  identity, 194, 265, 268
  zero, 259, 268
empirical adequacy, 103, 109, 112, 144
empirical content, 146
empirical evidence, 115
empirical facts, 124
empirical predictions, 115
empirical success, 109
empirical test, 168
empirical world, 109
empiricism, 108–110, 131, 142, 144
ensemble, 16, 23, 25, 32, 37, 78, 80, 81,
    83, 206, 211, 212, 214
  Gibbs, 3, 37, 169
  interpretation
  Gibbs, 37
  Maxwell–Boltzmann, 37
  virtual, 181
entanglement, 19, 21, 44, 49–53, 164,
    201, 211, 215, 232, 245–247, 249,
    265
  "fundamental postulate" of, 50, 245
  "thermodynamics" of, 246, 247
  bipartite
    properties of, 50
  bound, 44, 53
  distillation of, 52
  equivalence classes, 265
  free, 53
  multipartite, 27, 49
  negativity measure of, 24
  purification of, 52
  reduction criterion for, 25
entanglement measure, 50
  additivity conjecture, 51
  partial additivity condition for, 51
entanglement monotone, 50

entanglement of distillation, 52, 247
entanglement of formation, 26, 51, 247,
    292
entropy
  classical
    binary, 196
    conditional, 197
    joint, 197
    relative, 197
  quantum
    relative, 209
    von Neumann, 203
  Shannon, 196, 197
    additivity of, 196
    permutation invariance of, 196
  von Neumann, 26, 51, 200, 201
EPR, 19, 20, 33, 44, 104, 184, 232, 256
EPR program, 33
error
  symbol, 198
error correction
  classical, 198
event, 22
  elementary, 22
  signal, 195
Everett, H. III, 38, 80, 85, 139–145, 147,
    148, 150, 153–156, 159, 160, 162
evolution
  deterministic, 57
  irreversible, 211
  joint-state, 143
  non-unitary, 80, 147
  Schrödinger, 47, 77–80, 82, 209, 263
  Schrödinger-like, 171
  state, 80, 84, 210, 219
  stochastic, 22
  temporal, 15, 22
  unitary, 78, 80, 120
existent, 39
expansion
  eigenvalue, 23
expectation catalog, 60
expectation value, 22
experimental metaphysics, 33
explanation
  common cause, 21, 41

factoring algorithm
  Shor, 217

faithful measurement principle, 68, 90, 166
Feyerabend, P., 117
Feynman, R. P., 1, 6, 83, 156, 180, 182, 183, 185, 235
fidelity
  perfect, 207, 247
fidelity of transmission, 211
field, 194
  Galois, 194, 202
Fine, A., 58, 111, 112
finite scheme, 196
Fock, V., 10, 85, 129
form
  Hermitian, 259
    positive, 260
    positive-definite, 260
Fubini–Study metric, 26
Fuchs, C., 100, 172–176, 231
fundamental paradox of quantum theory, 132
fundamental theories, 33

Galois field, 194, 202
gate
  quantum, 216, 218
    $\sqrt{\text{NOT}}$, 218
    Hadamard, 218
    NOT, 218
Gell-Mann, M., 140, 145, 146, 148, 154, 155, 157, 162, 163
Gershenfeld, N., 239
GHJW theorem, 32
GHSZ, 36, 44, 45
GHZ, 45
GHZ state, 19, 45
Gibbins, P., 108
Gibbs, W., 3
Gibbs–Helmholtz equation, 246
Gisin, N., 22, 32
Gleason's theorem, 69, 72, 86, 137, 207, 256, 275
Graham, R. Neill, 139, 148, 154, 160
Greenberger, D., 36, 44, 45
Griffiths, R., 140, 163
Groenwald, H. J., 144
group, 265
Gudder, S., 39

Halvorson, H., 185
Hamilton, W., 55
Hamiltonian, 15, 55, 180
Hardy, L., xi, 115
Hartle, J., 140, 145, 154, 163
Hartley, R. V. L., 194
Healey, R., 68, 103, 130
Heisenberg effect, 9
Heisenberg *Schnitt*, 82, 118, 130, 131, 139
Heisenberg's microscope, 9, 10
Heisenberg, W., ix, 6–11, 59, 62, 79, 80, 95, 97, 105, 111, 124, 125, 128, 129, 131–134, 138, 170, 171, 183
Henderson, L., 248
Heywood, P., 45
hidden variables, 37–41, 73, 74, 100, 168, 169, 188, 230, 255, 272, 276, 283, 292
  contextual, 39
  local, 73, 197
  non-contextual, 39
hidden-variables theory, 20–22, 24, 37–41, 72–74, 104, 162, 167, 168, 204, 237, 252
  non-local, 39
  stochastic, 39
Hilbert space, 13, 259, 260
  projective, 13
  subspace of, 260
histories
  decoherent, 163
  disjoint, 163
  mutually exclusive, 163
Holevo bound, 214
Holevo information, 212, 214
Holevo's theorem, 214, 223
Holevo, A. S., 214
holism, 113
Holt, R., 43
homomorphism
  group, 265
Honner, J., 133, 134
Hooker, C., 109
Horne, M., 36, 42–45
Hughes, R. I. G., 116, 166
Hughston, L. P., 32, 171

IGUS, 155

imprecision, 8
improbability, 194
inaccuracy, 8
incompatibility, 6
incompleteness
    classical state specification, 196
    quantum state description, 20, 33, 35,
        166
independence
    probabilistic, 69
indeterminacy, 59
indeterminism, 35, 56, 59, 67, 75, 118
indistinguishability, 2, 18, 83
individuality, 184
inequality
    Bell, 20, 41, 44, 52, 271
    Bell-type
        violation of, 43
    CHSH, 276
    Clauser–Horne (CH), 42
information, 107
    accessible, 288
    average, 196
    classical, 50, 189, 194–196, 198, 214
        mutual, 197
        unit of, 194
    Holevo, 212, 214
    mutual, 214
    quantum, 193, 196–198, 200, 201,
        203, 233, 248
        mutual, 209
        unit of, 191, 201
    Shannon, 194, 195
    storage of, 192
information content, 195, 196
inner product
    scalar, 259
instrumentalism, 63, 102, 108, 110, 161
inter-subjective agreement, 138
interaction
    non-quantum mechanical, 39
interferogram, 6
interferometer
    double-slit, 2, 3, 5–8, 11, 17, 18,
        27–30, 117, 135, 136, 168, 203, 204
    Hanbury-Brown–Twiss, 6
    Mach–Zehnder, 204, 218
    Young, 2, 3, 5–8, 11, 17, 18, 27–30,
        117, 135, 136, 168, 203, 204

interpretation, 37, 77, 78, 96, 98
    'orthodox', 98
    'unorthodox', 98
    Basic, 72, 76, 92, 117, 142, 143
    Born, 68
    Collapse-Free, 78, 80, 139–141, 145
    Complementarity, 124
    Consistent histories, 154
    Copenhagen, 58, 79, 90, 97, 98, 117,
        124, 132, 135, 138, 141, 142, 173
    Everett, 147
    existential, 159
    Many-worlds, 139
    modal, 98, 158
    Naive, 169
    naive realist, 166
    non-statistical, 33, 38
    Princeton, 117
    quantum logical, 222
    Radical Bayesian, 100, 231
    underdetermination of, 116
isometry, 27
it-from-bit thesis, 171, 233, 234

James, W., 170
Jammer, M., 8, 10, 105
Jarrett, J., 42, 256
Jauch, J. M., 101, 244
Joos, E., 156
Jordan algebra, 269
Jordan, P., 7, 95, 97, 124, 243
Jozsa, R., 32
jump
    quantum, 79, 130

Kant, I., 133
Kent, A., 146
ket, 263
ketbra, 263
Kraus operator, 48
Krips, H., 38, 75, 111
Kuhn, T. S., 175

Lüders rule, 164
Lüders, G., 83
laboratory, 47
Lagrange, J.-L., 55
Lagrangian, 183
Lahti, P., 87, 89

Landé, A., 6
Landauer's principle, 242
Landauer, R., 215, 239
Laplace, P.-S., 57
lattice, 268
    Boolean, 268
    complemented, 268
    non-distributive, 66
    orthomodular, 268
law
    orthomodular, 268
Lloyd, S., 244
local box, 252
local operations (LO), 48
local operations and classical communi-
        cation (LOCC), 48, 50
local realism, 20, 34, 104, 111, 154
locality, 26, 292
    Bell, 41
localization, 34
logic
    Boolean, 64, 217, 221
    common sense, 113
    quantum, 64, 114, 217
London, F., 85, 122
loophole
    locality, 44
Ludwig, G., 85

magnitude
    existent, 39
    informational, 234
    physical, 1, 9, 14
map
    CPTP, 47
    identity, 87
    linear
        completely-positive (CP), 48
        trace-preserving (TP), 48
Margenau, H., 79, 104, 122
matrix
    density, 24, 264
        partially-transposed, 24
    Markov, 211
matter, 239
Maudlin, T., 110, 115, 229
Maxwell's demon, vii, 67, 242
Maxwell, J. C., vii, 242
measure

positive-operator-valued (POVM), 87
projection-operator-valued (PV)
    multiplicativity of, 88
projection-valued (PVM), 88
measurement
    Bell-state, 226
    generalized, 87
    joint, 18
    quantum, 74
        complete, 82
        maximal, 82
        non-maximal, 82, 83
        precise, 83
    selective, 81
    von Neumann, 81
measurement problem, 27, 77, 78, 85,
        86, 92, 122, 123, 130, 137, 140,
        141, 155, 156, 159, 168, 188, 237,
        255
Mittelstaedt, P., 68, 96, 98, 99
mixing, 4, 15, 16
mixture
    quantum, 21, 51, 81
modality, 109
monotone
    entanglement, 50
Multi-verse, 139, 146, 148, 150, 156,
        159, 222
Murdoch, D, 125
mutual information
    classical, 209
    quantum, 208, 209

naive realism, 106, 109, 117, 165, 166
naturalism, 100
    scientific, 58
negativity, 24, 50
Newton, I., 55
Nielsen, M. A., 229, 230
nl-bit, 254
no-cloning theorem, 12
no-go theorem, 284
noise
    classical, 198
    quantum, 211
non-local box, 253, 254
non-locality, 41, 43, 49, 249
non-signaling box, 252
norm

spectral, 261
noumenal world, 177

objective chance, 60
objective indefiniteness, 59, 68, 85, 116,
    119, 194
objective probability, 60
objectivism, 171
observable, 9, 14, 61, 119, 264
    pointer, 76, 77, 79, 98, 148
observer, 16
Ockham's principle, 151
Omnès, R., 102, 140
ontological vagueness, 145
ontology, 97
open system, 209
operation, 47
    collective, 49
    local
        collective and classical communica-
            tion (CLOCC), 49
        invertible (ILO), 266
        stochastic (SLO), 49
        local (LO), 48
        separable, 49
operation element, 48
operationalism, 175, 176, 178
operations
    local
        and classical communication
            (LOCC), 48
        collective and classical communi-
            cation (CLOCC), 52, 53, 246,
            247
        stochastic and classical communica-
            tion (SLOCC), 49, 266
        local (LO), 48
operator
    adjoint, 261
    anti-linear, 260
    anti-unitary, 27
    Hermitian, 261, 264
    inverse, 260
    linear, 260
        self-adjoint, 119
        trace-class, 261
    maximal, 13
    projection, 65, 261
    self-adjoint, 261

unit, 260
unitary, 261
zero, 260
oracle, 219, 224
    quantum, 224
orbit, 265
ordering
    partial, 267
orthocomplementation, 268
orthoembedding, 268
orthorepresentation, 268
Osiander, 110
outcome independence (OI), 42, 251,
    256
outer product, 263

paradox, 60, 67, 69, 85, 93, 100, 101,
    103, 166, 167, 176
    Alarm-clock, 34
    EPR, 20, 35
    Kochen–Specker, 73
    qubit, 230
paradox of the quantum, 132
parallelism
    quantum computational, 215
parameter independence (PI), 42, 251,
    256
partial trace, 25, 212
participatory universe, 171, 234
Pauli 'errors', 214
Pauli group, 265
Pauli matrix, 202, 218, 265
Pauli operator, 213
Pauli, W., vii, 8, 12, 38, 73, 79, 80, 97,
    113, 119, 124, 129, 131, 171, 178,
    238, 241, 255
Pearle, P., 123
Peierls, R., 95, 97, 136, 171
Penrose, R., 91, 121, 123, 229, 231
Peres, A., 24, 100, 158, 172, 175
Peres–Horodečki criterion, 24
phase
    global, 219
phase space, 1, 72
    classical, 61, 64, 163
    quantum, 66
phenomenon (Bohr), 131
physicalism, 92, 123, 147, 190, 240
physics

foundations of, 232
pilot wave theory, 148
Pitowsky, I., 137
Podolsky, B., 19
Poincaré–Bloch sphere, 14, 202, 218
pointer basis, 159
pointer function, 76
pointer observable, 76, 77, 79, 98, 148
pointer reading, 158, 159
polytope, 252
    convex, 253
    correlation, 44, 252
Popescu, S., 22, 245, 250, 253, 254, 256
Popper, K., 70, 107, 113, 124, 128, 134,
        178
poset, 267
    orthomodular, 268
positivism, 99, 103, 111, 128, 162, 175
postmodernism, 235
potentialities, 60
    actualization of, 77, 177
POVM, 87
    additivity of, 87
    completeness of, 87
    positivity of, 87
PPT property, 24
PR non-local box, 251
pre-measurement, 78, 149, 159, 210
prediction, 1, 58
preferred-basis problem, 153, 155–157,
        159, 161, 162
preparation
    quantum state, 2, 3, 28, 32, 65, 68,
        76, 77, 82, 83, 170, 171, 204
    unknown, 179
preparation device, 170
Preskill, J., 239
Price, H., 164
Primas, H., 129
Princeton interpretation, 117
principle of complementarity, 10, 238
principle of indifference, 70
probabilities
    problem of, 151
probability, 69
    classical conception of, 69
    conditional, 5
    de Finetti approach, 71, 171, 176
    frequency conception of, 70

Kolmogorovian, 69
    objective, 118
    propensity conception of, 70
    subjective conception of, 71
probability amplitude, 6
probability mass function, 196
probability space, 69
problem
    coherence, 162
    preferred basis, 153
    ratio, 247
    separability, 21
problem of probabilities, 151
process
    quantum, 181
        type 1, 81
projection
    Lüders, 83, 284
    quantum state, 79–81, 83
projection postulate, 79, 83, 84, 262
projector, 13, 64, 72, 73, 76, 81, 261
property, 61, 89
    sharp, 88, 89
proposition, 61, 62, 68
protocol
    dense coding, 213, 225
    LOCC, 49
    quantum, 224
    state teleportation, 226
psycho-physical parallelism, 83, 84, 93,
        96, 118, 122, 140, 142, 228
purification, 31, 32
    entanglement, 52
purity, 14, 15, 78, 82, 153, 157, 204
Putnam state, 67
Putnam, H., 63, 64, 67, 68, 92, 114, 175,
        184, 221
PV measure, 88

quanglement, 227–229, 231
quanta, 119
quantum cloning, 73
quantum computing, 17, 146, 175,
        215–218, 223, 224, 229
quantum cryptography, 175
quantum dephasing, 213
quantum erasure, 28, 32
quantum information, 193, 196, 198
quantum information science, 244

quantum jump, 79, 130
quantum logic, 63–65, 67, 68, 114, 123, 217, 267
quantum mechanics
  standard postulates of, 259, 262
quantum physics, 1
quantum postulate
  the, 133, 178
quantum protocols, 224
quantum state
  bipartite, 21
  bound, 53
  broadcasting of, 212
  distillable, 44
  entangled, 18
  mixed, 13, 15, 78
  reduced, 25
  separable, 21
qubit, 14, 36, 201, 202, 227, 242, 266
  ancillary, 233
  representation of, 14
  spatial, 204
qutrit, 25, 82

Radical Bayesianism, 114, 115, 176
radius
  spectral, 261
random processes, 22
random variable, 22
randomness, 12, 57
  irreducible, 58
ratio problem, 247
ray, 260
reading
  pointer, 158, 159
readout, 219
realism, 19, 32, 33, 58, 67, 68, 90, 91, 104–108, 111, 114, 115, 117, 122, 128, 154, 175, 256
  dynamic structural, 184
  Einstein's, 112
  Fig-leaf, 177
  local, 34, 115
  motivational, 112
  Naive, 109, 117, 165, 166
  naive, 68, 106, 109, 141, 165
  Peircean, 108
  physical, 90
  Scientific, 33, 100, 138

Redhead, M., 45, 63, 115
registration, 181
  measurement, 76
Reichenbach, H., 56, 97
relative state, 141, 149, 160, 161, 163, 164
relativism, 162
remote steering, 27, 32, 173, 186
repeatability hypothesis, 79, 81
representation
  equivalent, 265
  irreducible, 265
  projective, 265
resource
  communication, 199, 201, 251
  quantum, 53
retrodiction, 58
Rohrlich, D., 22, 245, 250, 253, 254, 256
Rosen, N. (see also EPR), 19
Rosenfeld, L., 98, 99, 102, 176
Rothstein, J., 58, 175, 176, 178
rule
  Born, vii, 14
  Lüders, 83
Russell, B., 236

sample space, 69
Saunders, S., 161, 162
schema
  communication complexity, 200
scheme
  finite, 196
Schmidt measure, 19, 50
Schmidt number, 19
Schnitt
  Heisenberg, 84, 118, 131, 139
Schrödinger equation, 15, 47, 77, 263
Schrödinger picture, 263
Schrödinger's cat, 78, 84, 89–93, 167
Schrödinger, E., 1, 6, 17, 20, 26, 32, 38, 58, 60, 66, 76, 82, 90–92, 95, 106, 109, 113, 124, 169, 171, 201, 208, 232
Schumacher, B., 190, 191
self-excited circuit, 235
self-interpretation, 60, 98, 100, 103, 116, 141, 171
sequence
  Bernoulli, 69

Markov, 69
Shaknov, I., 43
Shannon, C., 190, 194, 195
Shannon–Hartley theorem, 199
Shapere, D., 58
Shimony, A., ix, 26, 35, 36, 39, 41–44,
    59, 60, 85, 123, 132, 153, 155, 157,
    168, 177, 250, 256
Shor, P., 216
sigma algebra, 87, 267
sign, 192
signal, 192
signaling box, 252
singlet state, 35
singular values, 25
Sinha, S., 184
Slater, J. C., 165
solipsism, 85, 94, 160
    radical, 115, 162
Solvay Congress
    Fifth, 11, 135
Sorkin, R., 184
space-time, 150, 152, 228
space-time description, 59, 107, 129
space-time propagation, 228
spectral norm, 261
spectrum
    eigenvalue, 23, 87, 261
    energy, 95
speedup
    quantum, 215–218, 220–224
stabilizer, 265
Stachel, J., 58, 114, 180, 181, 183, 184,
    221
Stairs, A., 63
Stapp, H., 98, 129
state
    Bell-correlated, 22
    complete, 35, 38
    dispersion-free, 40, 73
    EPR-correlated, 22
    GHZ, 19, 45
    Putnam, 67
state space, 55
state steering, 173
statistical operator, 13, 119
    separability of, 24
statistical state, 13
statistics, 3

Steane, A., 217
string
    bit, 194
subalgebra
    Boolean, 268
subensemble, 51, 81, 83, 246
subjectivism, 170, 171, 173, 177, 233
suborthoposet, 268
subspace
    decoherence-free, 157
    vector, 260
subspaces
    logic of, 64
subsystem
    dismissal of, 25
supercorrelation, 209
superoperator, 48
superposition principle, 1, 3, 12, 77, 90,
    147, 152, 187, 191, 215, 232, 233,
    255, 262
superselection, 12
supervenience, 241
surprisal, 195
Svozil, K., 233
symmetry, 27
Szilard's limit, 242, 243
Szilard, L., vii, 194, 239, 242–244

tangle, 26
Taylor, G. I., 4
Taylor, J., 165, 167
teleportation
    quantum state, 53, 224
tensor product, 264
theorem
    Bell's, 17, 20, 24, 37, 40, 41, 104, 108,
        110, 115, 251
    de Finetti's, 179
    EWG meta-, 149
    GHJW, 32
    Gleason's, 71, 72, 86, 207
    Holevo's, 214
    Kochen–Specker, 68, 73, 168
    no-cloning, 12
    no-deleting, 233
    no-go, 37, 40, 121, 168
    noisy channel coding, 199
    nuclear spectral, 23
    Schmidt's, 158

Shannon–Hartley, 199
spectral, 23
tridecompositional, 158
thermodynamic limit, 244
thermodynamics, 10, 51, 52, 243–246,
    248
  laws of, 246
  second law of, 210, 242, 245, 246
  third law of, 247
thermodynamics law of, 247
thought experiment, 75
Timpson, C., 197, 228
Tipler, F., 151–153, 156, 158
trace, 261
  partial, 25
trade-off
  accuracy–disturbance, 12
transformation theory
  Dirac–Jordan, 10, 181
transition matrix, 264
Tsirel'son bound, 250, 252
Tsirel'son, B. S., 250, 251
Tukey, J. W., 194
Turing, A., 194
two-level system, 14, 191, 202, 215, 218,
    223

uncertainty, 8, 59, 162, 164
uncertainty principle, 6, 8
uncertainty relation, 1, 59, 117
  Heisenberg–Robertson, 6, 9
underdetermination, 187
unitarity, 15, 261
universal state, 163
  evolution of, 163

value definiteness, 90
value function, 73
value space, 87
value state, 62
value-definiteness, 72
value-definiteness thesis, 68, 85
van Fraassen, B., 108, 109
vector
  norm of, 259
  unit, 259, 260

vector space, 259
  basis for, 260
  complete, 260
  separable, 260
  subspace of, 260
vectors
  orthogonal, 259
verificationism, 105
visibility, 5
  interference, 5
  two-particle interference, 18
vitalism, 92
von Neumann chain, 84
von Neumann entropy, 26, 51, 200, 201
von Neumann equation, 263
von Neumann, J., 37, 38, 40, 63, 64, 66,
    67, 79, 81, 82, 118, 120, 123, 137,
    217, 255
von Smoluchowski, M., 56
von Weizsäcker, C. F., 134, 135

Wallace, D., 159, 161, 162, 164
wave-function, 14, 231
wave-function collapse, 79, 85, 256
wave-packet
  problem of reduction of, 77
Weaver, W., 190, 192, 195, 240
Werner state, 22
Weyl, H., 8
Wheeler, J., 80, 127, 128, 132, 145, 146,
    176, 233, 234
Whitaker, A., 117
Wigner's friend, 85, 89–91, 93, 94, 140,
    143
Wigner's theorem, 27
Wigner, E., 74, 85, 93, 122, 143
Wittgenstein, L., 134, 236
Wootters, W. K., 11, 12, 32, 248
work, 247
Wu, C. S., 43

Young, T., 2, 3

Zeilinger, A., 36, 44, 45, 92, 235–238
Zeisler, E., 58
Zurek, W. H., 11, 12, 157, 159, 176

THE FRONTIERS COLLECTION

Series Editors:
A.C. Elitzur M. Schlosshauer M.P. Silverman J. Tuszynski R. Vaas H.D. Zeh

Information and Its Role in Nature
By J. G. Roederer

Relativity and the Nature of Spacetime
By V. Petkov

Quo Vadis Quantum Mechanics?
Edited by A. C. Elitzur, S. Dolev,
N. Kolenda

Life – As a Matter of Fat
The Emerging Science of Lipidomics
By O. G. Mouritsen

Quantum–Classical Analogies
By D. Dragoman and M. Dragoman

Knowledge and the World
Edited by M. Carrier, J. Roggenhofer,
G. Küppers, P. Blanchard

Quantum–Classical Correspondence
By A. O. Bolivar

Mind, Matter and Quantum Mechanics
By H. Stapp

Quantum Mechanics and Gravity
By M. Sachs

Extreme Events in Nature and Society
Edited by S. Albeverio, V. Jentsch,
H. Kantz

**The Thermodynamic
Machinery of Life**
By M. Kurzynski

**The Emerging Physics
of Consciousness**
Edited by J. A. Tuszynski

Weak Links
Stabilizers of Complex Systems
from Proteins to Social Networks
By P. Csermely

Mind, Matter and the Implicate Order
By P.T.I. Pylkkänen

Quantum Mechanics at the Crossroads
New Perspectives from History,
Philosophy and Physics
Edited by J. Evans, A.S. Thorndike

Particle Metaphysics
A Critical Account of Subatomic Reality
By B. Falkenburg

**The Physical Basis of the Direction
of Time**
By H.D. Zeh

**Asymmetry: The Foundation
of Information**
By S.J. Muller

Mindful Universe
Quantum Mechanics
and the Participating Observer
By H. Stapp

**Decoherence and the
Quantum-To-Classical Transition**
By M. Schlosshauer

Quantum Superposition
Counterintuitive Consequences of
Coherence, Entanglement, and Interference
By Mark P. Silverman

The Nonlinear Universe
Chaos, Emergence, Life
By A. Scott

Symmetry Rules
How Science and Nature Are Founded
on Symmetry
By J. Rosen

**The Biological Evolution of
Religious Mind and Behaviour**
Edited by E. Voland and W. Schiefenhövel

**Entanglement, Information, and
the Interpretation of Quantum Mechanics**
By G. Jaeger